PHASE-LOCK BASICS

PHASE-LOCK BASICS

WILLIAM F. EGAN, Ph.D.
Lecturer in Electrical Engineering
Santa Clara University

A Wiley-Interscience Publication
JOHN WILEY & SONS, INC.
New York / Chichester / Weinheim / Brisbane / Singapore / Toronto

MATLAB is a registered trademark of The Math Works, Inc., 24 Prime Park Way, Natick, MA 01760; Tel: 978-647-7000, Fax: 978-647-7001; WWW: http://www.mathworks.com; email:info@mathworks.com

Library of Congress Cataloging in Publication Data:

Egan, William F.
    Phase-lock basics / William F. Egan,
        p.   cm.
    "A Wiley-Interscience publication."
    Includes bibliographical references and index.
    ISBN 0-471-24261-6 (cloth : alk. paper)
    1. Phase-locked loops   I. Title.
  TK7872.P38E33   1998
  621.3815′364—dc21   97-38471
                               CIP

Printed in the United States of America

10 9 8 7 6 5 4

To Henry P. Nettesheim
Emeritus Professor of Electrical Engineering
Santa Clara University

CONTENTS

PART 2   PHASE LOCK IN NOISE

11   PHASE MODULATION BY NOISE   295

12   RESPONSE TO NOISE MODULATION   313

15 PHASE-LOCKED LOOP AS A DEMODULATOR   377

16 PARAMETER VARIATION DUE TO NOISE   391

PREFACE

The phase-locked loop (PLL) has become so important in electronics over the past three decades that a working knowledge of its principles is an essential asset for many electrical engineers. This book is here to help you gain that knowledge. It concentrates on achieving an understanding of basic principles that are applicable to a wide range of PLL circuits—old, new, and yet-to-be-invented.

In 1977 I was given the opportunity to teach a graduate course, in the Santa Clara University School of Engineering, on the design of phase-locked frequency synthesizers, an area I had been working in for about fourteen years. The text for that course was my notes, which I later expanded into *Frequency Synthesis by Phase Lock*, published by John Wiley and Sons, Inc. in 1981.

In 1985, Professor Timothy Healy asked if I would teach a two-quarter PLL course, one that he had developed and taught for several years. I have done so since then. I used Blanchard's *Phase-Locked Loops* for a text, but I had to develop problems as well as course notes and I engaged in an ongoing study of original references. By 1991 I had introduced enough new and modified material that I decided to supplant Blanchard with my own text, but I thought I would begin gradually by providing the text for the first quarter only. However, before the end of the quarter, I was surprised (shocked?) to find that Wiley had stopped publishing Blanchard and it would not be available. This resulted in a rather intensive effort that produced the rest of my text in time for the second quarter. I have been using my own text since that time, modifying it each time (every other year) based on student comments and my own re-readings and to introduce new material.

In 1995 I submitted the latest version to Wiley as part of a proposal to publish. As was true with my previous book, the editor had the manuscript reviewed by a knowledgeable professional. In this instance, the reviewer was Professor William Tranter, currently of Virginia Tech, who suggested that I introduce a software component to the book, especially since my biography showed experience in developing computer programs related to PLLs. He recommended MATLAB® for consideration. Since I was then prepared to immediately publish a text that had been carefully refined, I was reluctant to modify it. Nevertheless, I began to experiment with MATLAB and, as I did my enthusiasm grew. Here was an opportunity, using the student edition of MATLAB* on a personal computer, to significantly enhance the learning of PLL fundamentals. It was almost as though the reader would now have an opportunity for hands-on experimentation, being able to modify parameters and observe the effect on loop performance. In addition, MATLAB could provide useful design aids and even open some important areas where analysis is very difficult without a computer.

As a result, I have included MATLAB programs (scripts) and other software in this book in a way that does not detract from the original text, placing them in appendices following the chapters where the corresponding material is discussed. Many of the programs are written explicitly in the book, and they, and others, are available for downloading from a Wiley Internet site. That site also provides other aids in the form of Excel spreadsheets. The MATLAB and Excel files can be read by Macintosh® and IBM®-compatible computers, at least. One other very important benefit of the Internet site: the inevitable errata can be provided and updated.

The software is relatively easy to understand and customize to better suit an engineer's particular analysis or design problem. It adds an additional dimension of instruction by illustrating the use and design of computer-aided engineering (CAE) for PLLs; the simulation programs should be especially instructive in this regard. I have found that new ways to use the software to improve my understanding of the design process continue to occur to me. One must limit the volume of such material in a book, but, when one becomes familiar with the methods by which the software can be used, one is able to apply it as the situation demands.

Phase Lock Basics can be used in many ways by both the professor and the self-directed student. One can concentrate on the development of the theory or on the examples or work the problems, all of which have answers in the back, or can experiment with the MATLAB programs. Of course, a combination will give the most complete learning experience. That experience will also depend on the reader's background. For the graduate engineer, this

---

*The Professional Version of MATLAB, with appropriate toolboxes, may also be used. Those who cannot gain access to either, but can use a program with similar capabilities, may be guided by the text form of the scripts. However, it will be considerably easier to use some version of MATLAB directly.

material not only adds important specific knowledge but it can serve to integrate and solidify previously learned fundamentals, showing how the theory is used in the design of a particular kind of circuit, the PLL. For those lacking some of the desirable background, results, in the forms of equations, graphs, and descriptions of techniques, are still there. And, for you fortunate designers who already possess an understanding of PLLs, the many graphs and the software can help make your work more productive and, I hope, enjoyable.

The text concentrates on the second-order loop. Loops of any greater complexity have too many parameters to permit general discussion. The standard parameters for second-order systems are natural frequency ω_n and damping factor ζ. Commonly, graphs are provided for various kinds of loop responses, in which the x axis is normalized to ω_n and multiple curves are given for various common values of ζ, enabling the reader to denormalize and interpolate to find results appropriate to a particular problem. There is one other necessary parameter, however, which I call α. For practical reasons (too many parameters), when plots of loop responses are given, they must be restricted to a few values of α. Other texts have generally given response curves for type 2 loops, those employing an integrator, and sometimes for loops with low-pass filters. These loops correspond to $\alpha = 1$ and $\alpha = 0$, respectively, but a continuum of useful loop filters exists between those extremes, and I show how transient response curves representing the extreme values of α can be combined to give the response for any value of α. MATLAB can generate responses, in both the time and frequency domain, for any value of α.

Many practical loops are second-order but some are more complex. Often the more complex loops can be approximated as second order so the second-order curves still give valuable information. I show how some of these curves are affected by the introduction of an additional pole, as is common, for example, in frequency synthesis. To illustrate how all the curves are affected by all the possible additional pole frequencies would, again, require too many dimensions to be practical. However, there is a common third-order configuration that can be analyzed with relative ease, and I have provided this analysis and corresponding curves in an appendix to Chapter 10. (I have placed this in an appendix to not detract from the concentration on loops of second order, on which the main development of the theory is based.) One of the great values of the state-space method, and the MATLAB programs that are based on it, is that it can handle higher order loops so, while one learns the general characteristics of PLLs using the more easily handled second-order theory, one can also expand to more complex problems with the help of MATLAB.

Chapter 10 introduces various applications of PLLs and shows how the theory developed in the previous chapters can be applied to them. This is not a substitute for detailed study in particular areas—there are many other things required for a good knowledge of synthesizer design, for example—but does show how the fundamentals apply. Part 2 of the text introduces the effects of noise, often large amounts of noise, on the performance discussed in Part 1. I have found that Part 1 is a good prerequisite for my synthesizer course, reducing the amount of material that has to be absorbed there. The whole book is an even better preparation, since noise concepts are also very important in synthesizer design. However, the noise levels are relatively low in synthesizer design, its elimination being an important goal, whereas communications systems may be designed to operate in the presence of large amounts of noise, such as are treated in Part 2. Thus this book would also be good preparation for a more advanced study of communication systems that depend on PLLs (coherent communications) and must operate in the presence of noise.

It is not uncommon for students, and designers presumably, to make 2π errors in computing loop bandwidth and other parameters, due to the prevalence of both radians and cycles as units of phase (and in units of frequency). By handling units in a certain way, we are able to employ whichever is more convenient and to freely mix them without succumbing to this error. Proper handling of units is also a necessity in development of the equations. This easy movement between units decreases the importance of differentiating between f and ω symbols (although we will attempt to use the more appropriate symbol), especially since we can equate a frequency in one unit to the equivalent in the other.

While the answers provided for all the problems are of obvious help to the reader studying on his or her own, instructors may prefer problems that do not have answers. I suggest that these can be developed by modification of parameters given in the problems. The devoted but uncertain student can then verify his or her problem solving using the parameters in the book before changing to the parameters specified for the homework. The answers in the back of the book should be highly accurate since most have been proved by use. I have taught electrical engineering graduate students using nine two-hour classes plus two periods of exams for each half of the text. However, the software has not been employed in these courses nor has some other new material.

I would like to thank Professor Healy and Santa Clara for the opportunity to teach the Phase-Locked Loops course over the years and the students who have taken it for the contribution they have made, by interaction in the classroom and through their homework as well as by their suggestions and comments, in the development of the material in this book. I hope it will help you to better understand and appreciate the vast and fascinating world of phase-locked loops.

WILLIAM F. EGAN

Cupertino, California
June 1998

SYMBOLS LIST AND GLOSSARY

$\Rightarrow$	imply, become, go to
$\approx$	approximately equal to
$\equiv$	identically equal to, rather than being equal only under some particular condition
$\triangleq$	is defined as
AGC	automatic gain control
AM	amplitude modulation
BM	balanced mixer
B_n	noise bandwidth
C	controlled signal in standard control system terminology; fed-back input to the summing junction
c	cycle
closed-loop transfer function	C/R with the loop operating
damping factor	ζ, a standard parameter of the second-order system, Section 4.2
dB	decibel, $10 \log_{10}(\text{power ratio})$
dBc	decibels relative to carrier power, i.e., $10 \log_{10}(\text{power}/\text{carrier power})$
dBc/Hz	ratio of sideband power density, per hertz bandwidth, to total signal power. Decibels relative to "carrier" per hertz noise bandwidth
dBm	decibels relative to 1 mW

DBM	doubly-balanced mixer
dBr	decibels relative to 1 rad^2
dBr/Hz	phase noise power spectral density in decibels relative to 1 rad^2 per hertz bandwidth
dBV	decibels relative to 1 V
DDS	direct digital synthesizer
DPLL	digital phase-locked loop
E	error signal in standard control system terminology; output from the summer
error	output from summer in standard control system terminology
ExOR	exclusive-OR function, equal to one if and only if the two inputs differ
$E[y]$	expected value of y
$E[y^2(\ )]$	expected value of y^2
f_x	see ω_x for any x
F_x	see Ω_x for any x
$F(s)$	frequency-sensitive part of filter transfer function, which is $K_{LF} F(s)$
frequency power spectral density	$S_f(\omega_m)$, mean square frequency σ_f^2 in a narrow frequency band divided by the bandwidth. The band is narrow enough that the value of $S_f(\omega_m)$ does not change appreciably when the bandwidth decreases
G	transfer function
$|G|$	gain (absolute value of G)
gain margin	reciprocal of open-loop gain at frequency at which open-loop phase has 180° excess phase. The gain increase necessary for oscillation of the loop
G_F	forward G of the loop
G_R	reverse G of the loop (1 except in Chapter 10)
H	closed-loop transfer function, C/R in Fig. 2.5
Hold-in	range $\pm\Omega_H$ of mistuning $\pm\Omega$ over which lock can be maintained as the loop is tuned (Fig. 8.2)
IC	integrated circuit
ICO	current-controlled oscillator
Im	imaginary part of
integrator-and-lead	filter of form $k(\omega_z + s)/s$
K	frequency-independent part of the loop gain
$K' = K'_p K_F K_v$	frequency-independent part of the loop gain when the phase error is such that K_p is maximum

K_{LF}	frequency independent part of loop-filter gain
K_p	phase detector gain, change in voltage divided by change in phase
K_p'	maximum K_p (maximum as phase changes)
K_{pd}	K_p for coherent detector
K_v	VCO gain, change in VCO frequency divided by change in tuning voltage
L	single-sideband power spectral density (noise) relative to the carrier
$\mathscr{L}$	single-sideband power spectral density relative to the carrier under small modulation index. When due to phase modulation, always equals $S_\varphi/2$
$\lim_{a \to b}$	limit as a approaches b
L_φ	single-sideband power spectral density relative to the carrier due to phase noise
low-pass filter	filter with transfer function of the form $K_{LF}/(1 + s/\omega_p)$
m	modulation index, peak phase deviation
mistuning	$\pm \Omega$: difference between the input (reference) frequency and the VCO center frequency (Fig. 8.2)
natural frequency	ω_n: a standard parameter of the second-order system
NCO	numerically controlled oscillator
N_i	input noise power to a limiter
N_0	one-sided noise power spectral density
OA	output accumulator
one-sided spectrum	representation of power spectral density in which all power is shown at positive frequencies
op-amp	operational amplifier
open-loop transfer function	transfer function around the loop as if it were broken to allow the response of the opened loop to be measured
P	signal power
P_c	carrier power; main signal power
PD	phase detector
phase margin	additional phase lag to cause the loop to oscillate, to give $-180°$ at the frequency at which open-loop gain is unity
phase power spectral density	$S_\varphi(\omega_m)$, mean square phase σ_φ^2 in a narrow frequency band divided by the bandwidth. The band is narrow enough that the value of $S_\varphi(\omega_m)$ does not change when the bandwidth decreases

PLL	phase-locked loop
PM	phase modulation
P_o	total output power from a limiter
power spectral density	power per unit frequency, the limit (at a given frequency) of the ratio of power to bandwidth as bandwidth approaches zero
PPSD	phase power spectral density {which see}
PSD	power spectral density {which see}
pull-in	process of acquiring lock. Range $\pm\Omega_{PI}$ of mistunings $\pm\Omega$ over which lock can be acquired (Fig. 8.2)
$P_y(f')$	power spectral density of the parameter y at the frequency f'
Q	quality factor, (energy stored)/(energy dissipated in a cycle), a measure of the sharpness of a resonance curve
R	normalized sweep rate, $(d\Omega/dt)\ \mathrm{rad}/\omega_n^2$ (see Table 9.1)
R	reference signal in standard control system terminology; independent input to the summer
rad	radian
$\mathrm{rad}^2/\mathrm{Hz}$	measure of PPSD, the density that produces a phase variance of $\delta f\ \mathrm{rad}^2$ in a narrow bandwidth δf Hz
reference	input to standard control system
$\mathbb{Re}$	real part of
r_T	relative threshold, ratio of threshold voltage to maximum signal available from a detector (Section 9.5.2)
S	demodulated signal power
s	Laplace variable, $\sigma + j\omega$
S/N	signal-to-noise ratio, sometimes specifically at the output rather than input
SBM	singly-balanced mixer
sec	second
Seize	range $\pm\Omega_s$ of mistuning $\pm\Omega$ over which the loop will acquire lock without skipping a cycle (Fig. 8.2)
$S_f(\omega_m)$	frequency power spectral density {which see}
S_i	input signal power to a limiter
spectrum analyzer	an instrument that displays power vs. frequency by measuring power passing through a filter whose center frequency is swept in synchronism with the abscissa of the display

S_φ	phase power spectral density {which see}
$S_{\varphi n}(\omega_m)$	phase power spectral density of noise at frequency ω_m
$S_{\varphi s}(\omega_m)$	phase power spectral density of signal at frequency φ_m
S_ω	frequency power spectral density {which see}
$S_y(f')$	power spectral density of y at frequency f'
T_{PI}	Pull-in time
two-sided spectrum	representation of power spectral density in which half of the power is shown at positive frequencies and half is shown at negative frequencies
T_m	mean time between cycle skipping
u_1	phase detector output voltage
u_2	VCO tuning voltage
v	subscripts indicate responses to noise at VCO input
$\tilde{v}$	root mean square are (rms) value of v
V_T	threshold voltage
V_v	VCO signal amplitude
α	0 for lag filter; 1 for integrator and lead, Eq. (6.5) otherwise
α	efficiency of the phase detector as a multiplier
$\Delta\varphi$	$\varphi_{\text{in}} - \varphi_{\text{out}}$
$\Delta\omega$	$\omega_{\text{out}} - \omega_{\text{in}} = -\omega_e$
$\phi_i(t)$	modulation-related input phase
$\phi_o(t)$	modulation-related output phase
φ_e	phase error, change in $\Delta\varphi$
φ_{in}	loop input phase (at phase detector)
φ_0	output phase response to additive input noise, linear response assumed (loop parameters are not changed by the noise)
φ_n	output phase response to additive input noise, nonlinear response assumed (loop parameters are changed by the noise)
φ_{out}	loop output phase (at VCO output)
ρ	signal-to-noise ratio, sometimes specifically at the output rather than input
ρ_{L0}	signal-to-noise ratio in a bandwidth equal to the loop noise bandwidth under linear conditions, $1/\sigma_{\varphi 0}^2$
σ_φ^2	mean square phase deviation, variance of phase
σ_ω^2	mean square frequency deviation, variance of frequency
$\sigma_{\omega s}^2$	variance of the signal frequency

τ_1	$(1/\omega_p)$ time constant of the loop pole
τ_2	$(1/\omega_z)$ time constant of the loop zero
ω_c	VCO center frequency; spectral center or carrier frequency
ω_e	frequency error, frequency at output of loop summing junction $(\omega_{in} - \omega_{out})$
ω_{in}	loop input frequency (at phase detector)
ω_L	unity-gain bandwidth of the loop; frequency were open-loop gain is one
ω_m	modulation frequency
ω_n	natural frequency; a standard parameter of the second-order system
ω_{out}	loop output frequency (at VCO output)
ω_p	loop-filter pole frequency
ω_{po}	pull-out frequency, the largest frequency step that can be tolerated without cycle skipping
ω_z	loop-filter zero frequency
Ω	$\pm$ mistuning: difference between the input (reference) frequency and the VCO center frequency (Fig. 8.2)
Ω	normalized modulation frequency, ω_m/ω_n (Section 7.6)
Ω_H	$\pm$ hold-in range: maximum $\pm$ change in ω that still permits lock to be maintained
Ω_{PI}	$\pm$ pull-in range: frequency range over which eventual lock is certain
Ω_S	$\pm$ seize frequency range: frequency range over which loop locks without skipping a cycle
ζ	damping factor, a standard parameter of the second-order system

Subscript (shown with x):

x_0	x under linear conditions
x_n	effective value of x, changed from x_0 due to the effects of noise
x_v	response to noise at VCO input

GETTING FILES FROM THE WILEY ftp AND INTERNET SITES

To download the files listed in this book and other material associated with it, use an ftp program or a Web browser.

FTP ACCESS

If you are using an ftp program, type the following at your ftp prompt:

```
ftp://ftp.wiley.com
```

Some programs may provide the first "ftp" for you, in which case you can type:

```
ftp.wiley.com
```

Log in as anonymous (e.g., User ID: anonymous). Leave password blank
   After you have connected to the Wiley ftp site, navigate through the directory path of:

```
/public/sci_tech_med/phase_lock
```

WEB ACCESS

If you are using a standard Web browser, type URL address of:

```
ftp://ftp.wiley.com
```

Follow the directions in the ftp section to navigate through the directory path of:

`/public / sci_tech_med / phase_lock`

FILE ORGANIZATION

Under the phase_lock directory are subdirectories that include MATLAB files for PC, Macintosh, and UNIX systems and Microsoft® Excel files. Important information is included in README files.

If you need further information about downloading the files, you can call Wiley's technical support at 212-850-6753.

PART 1

PHASE LOCK WITHOUT NOISE

CHAPTER 1

INTRODUCTION

1.1 WHAT IS A PHASE-LOCKED LOOP (PLL)?

A phase-locked loop is a circuit that synchronizes the signal from an oscillator with a second input signal, called the reference, so that they operate at the same frequency. The synchronized oscillator is commonly a voltage-controlled oscillator (VCO), so we will usually use these terms interchangeably. The loop synchronizes the VCO to the reference by comparing their phases and controlling the VCO in a manner that tends to maintain a constant phase relationship between the two. In some types of phase-locked loops (PLLs) this phase relationship is held constant. In other types it is allowed to vary somewhat. But the frequency is always synchronized—otherwise the loop is said to be "out of lock."

1.2 WHY USE A PHASE-LOCKED LOOP?

Phase-locked loops are often used because they provide filtering to the phase or frequency of a signal that is similar to what is provided to voltage or current waveforms by ordinary electronic filters. The designer has some control over the manner in which the phase (or frequency) of the VCO follows a changing reference phase (or a changing reference frequency—one cannot occur without the other). The loop can be made to follow quickly or to follow sluggishly. This capability is particularly valuable in removing the effects of noise on the reference or on the synchronized oscillator or, with astute design, on both.

If the reference signal is supposed to be a constant-amplitude phase-modulated signal [e.g., a frequency-modulated (FM) radio signal], the spectrum of the VCO will be a cleaned-up version of the reference spectrum. The PLL, while reproducing the reference signal, rejects all amplitude modulation noise and all other noise that is separated sufficiently in frequency from the signal. It acts like a filter that tracks the signal frequency. In fact, it can provide filtering that ordinary filters cannot because it can follow a signal whose frequency varies slowly by an amount that is greater than the filter bandwidth. For example, a PLL could be designed to filter out noise that is more than 1 kHz from a signal as the signal drifts by say 10 kHz.

Another use of the PLL is in phase or frequency modulating and demodulating signals. This is possible because there exist, within the PLL, a voltage that is proportional to the frequency of the reference and another that follows its phase. Not only can these be extracted for demodulation, but the loop can be forced to produce phase or frequency changes that are proportional to voltages that are injected into the loop, and thus to provide modulation.

Phase-locked loops are also used in frequency synthesis where the VCO oscillates at a selectable multiple of the reference. In this case the oscillator is synchronized with the reference without having the same frequency. However, the combination of the controlled oscillator and the subsequent frequency divider, which outputs the reference frequency, can be looked upon as a synchronized oscillator whose frequency is equal to the reference frequency.

Other PLLs, which are seemingly very different, involve things as diverse as rotating machinery and computer programs.

1.3 SCOPE OF THIS BOOK

This text describes the fundamentals of PLLs, principles that apply to all types of PLLs. It emphasizes the details that apply to the processing of analog electronic signals, as discussed in the first three paragraphs of the previous section, leaving details that are more peculiar to frequency synthesis to other works [Egan, 1981; Manassewitsch, 1976]. However, at the end of Part 1, we will illustrate how the theory developed here can be used to analyze many applications, including synthesizers and other circuits usable in wireless and telecommunications.

The text concentrates on second-order loops because higher order loops have too many parameters to make general discussions practical and because higher order loops can often be approximated by second-order loops. The effect of such approximations will be discussed, too. A special class of third-order loops is analyzed and described in Appendix 10.A.

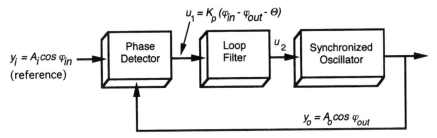

$$u_1 = K_p (\varphi_{in} - \varphi_{out} - \Theta)$$

$y_i = A_i cos \, \varphi_{in}$ (reference) → Phase Detector → Loop Filter → u_2 → Synchronized Oscillator →

$y_o = A_o cos \, \varphi_{out}$

Fig. 1.1   Basic phase-locked loop; Θ is an offset that depends on the type of phase detector.

While the main body of text stands by itself, appendices show how to use MATLAB or Excel as aids in understanding and designing PLLs. MATLAB programs are based on the student edition of MATLAB, which is available to students at a much lower price than the standard version. The standard version with appropriate toolboxes can also be used. Refer to the instruction manual that comes with your edition to learn how to use the program and to expand and enhance the material presented here. Information on the functions and processes used by MATLAB are available in the manual, in the help software, and by reading code for various functions.[1] These too are important learning resources. Excel is a very adaptable program that can be of great help to the engineer. The graphs of time-domain and frequency-domain loop responses in this book were generated using Excel. Many of the programs described in the appendices can be entered into MATLAB or Excel based on the information given there, but these and additional programs are available for downloading from the Internet. Refer to Appendix 20.B for instructions on downloading this software. Appendix 20.A brings together various representations of signals, modulations, and noise that are discussed throughout the book.

Answers to all the problems are given at the end of the text.

1.4   BASIC LOOP

A block diagram of the basic PLL is shown in Fig. 1.1. While the synchronized oscillator is usually a VCO, one whose frequency changes in response to a control voltage, u_2, it is also possible that u_2 could be a current and the synchronized oscillator would thus be a current-controlled oscillator (ICO).

---

[1]It should become apparent that some programs have user-interface features that could be used in others. There is a tradeoff between user friendliness and simplicity, but these programs can be customized and enhanced with relative ease.

However, we will consider it a VCO—the transition between the two is easily understood.

The phase detector compares the reference to the VCO's output and produces a signal u_1, which changes in proportion to the difference in their phases. This is processed by the loop filter to provide the oscillator control signal u_2. The loop filter can be as simple as a conductor $(u_2 = u_1)$ or a flat amplifier $(u_2 = K_{LF}u_1)$, but it is usually designed to provide some advantageous response characteristic.

If the output frequency, $\omega_{out} = d\varphi_{out}/dt$, should be greater than the reference frequency, $\omega_{in} = d\varphi_{in}/dt$, then $u_1 \sim (\varphi_{in} - \varphi_{out} - \Theta)$ would necessarily decrease with time, causing u_2 to decrease, which, in turn, would cause ω_{out} to decrease, bringing ω_{out} down toward ω_{in}. Thus the PLL provides negative feedback to keep the output frequency ω_{out} equal to the reference frequency ω_{in}. The output amplitude A_o is constant and independent of the input amplitude A_i.

1.5 PHASE DEFINITIONS

Figure 1.2 shows one cycle of a sinusoid and various ways of measuring the abscissa value. The first measure is time. The other measures are of phase, which indicates the time relative to the waveform's period. The figure illustrates how phase can be measured in any of three common units, cycles

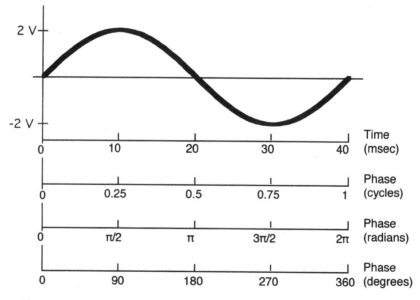

Fig. 1.2 A 2-V 25-Hz sine wave showing various measures of the abscissa.

(c), radians (rad), or degrees (deg). Because of this we should be careful to carry the units in our calculations. Otherwise errors, typically of 2π, begin to appear in important computed parameters. For example, if φ equals $10t^2$ rad/sec^2 (a linear frequency sweep), then the frequency is

$$\omega = \frac{d\varphi}{dt} = 20t \text{ rad/sec}^2. \tag{1.1}$$

The frequency can also be expressed as[2]

$$f = (20t \text{ rad/sec}^2)(1 \text{ c}/2\pi \text{ rad}) = 3.18t \text{ c/sec}^2 = 3.18t \text{ Hz/sec.} \tag{1.2}$$

In this example the change of symbol between ω and f is a clue to the units, but it is more efficient to carry units than to memorize a different set of formulas for each set of units. The task of maintaining units is made more difficult, however, because of the widespread practice of not considering the radian a unit (i.e., considering angles measured in radians to be without units) and the common practice of not incorporating units in equations involving Laplace transforms (see Section 1.9) or derivatives and integrals of some functions. An alternative to carrying units might be to employ always a set of formulas that use one type of measure, say radians, and each time to convert to those units. This can be awkward in a field that deals so much with angle units and in which various angle units are more natural to use in various situations. In this text the usual procedure will be to carry units, with one exception that we will discuss at the end of this chapter.

Example 1.1   Preservation of Units   The time derivative of 1 V sin kt is commonly said to be

$$\frac{d}{dt} 1 \text{ V} \sin kt = k \text{ V} \cos kt \tag{1.3a}$$

but the units are wrong. For example, if $k = 5$ Hz, Eq. (1.3a) gives a time derivative at $t = 0.01$ sec of [5 c-V/sec cos 0.05 c]. The cosine could conceivably be obtained from a table written in cycle units or any table could be used with conversion of 0.05 c to the correct units (e.g., 18°), but cycle-volts/second (c-V/sec) are not proper units for the slope of a voltage waveform. We are able to use Eq. (1.3a) correctly only because we know that it requires that k be in radians/second and because we know that radian units may not be carried. We would therefore begin by converting 5 Hz to 10π rad/sec. Then we would drop the radian unit to obtain the correct answer without explicitly converting units.

---

[2]Note that these expressions produce correct units for phase and frequency when time units are included with t. For example, at $t = 3$ sec, $\varphi = 10(3 \text{ sec})^2 \text{rad/sec}^2 = 90$ rad, $\omega = 20(3 \text{ sec})\text{rad/sec}^2 = 60$ rad/sec, and $f = 3.18(3 \text{ sec})\text{Hz/sec} = 9.54$ Hz.

The formula that correctly carries units gives the derivative of $\cos kt$ as $(k/\mathrm{radian})\sin kt$ so that Eq. (1.3a) becomes

$$\frac{d}{dt} 1\ \mathrm{V} \sin kt = \frac{k}{\mathrm{rad}} \mathrm{V} \cos kt \Rightarrow \frac{5c\text{-V}}{\mathrm{rad\text{-}sec}} \cos 0.05\ c = 10\pi\ \mathrm{V/sec} \cos 0.05c.$$

(1.3b)

Equation (1.3b) permits the use of any units.

Figure 1.3 shows how the phase difference between two sine waves is defined. At time t_x the phase difference is

$$\Delta\varphi(t_x) = \varphi_2(t_x) - \varphi_1(t_x) = \sin^{-1} v_2/A_2 - \sin^{-1} v_1/A_1.$$ (1.4)

This is true even if the frequencies of the two sinusoids are not the same. However, some phase detectors are sensitive to the zero crossings of the waveforms. For these the quantity measured is really

$$\Delta\varphi = (\Delta t/T)\ \text{cycles}.$$ (1.5)

If the frequencies differ, the phase of one of the waveforms is used as a reference and T is taken from that waveform. In that case what we are actually measuring is the phase of the reference waveform at the zero crossing of the second waveform.

Equation (1.5) not only illustrates the carrying of units but also the efficiency of using whatever units are most convenient.

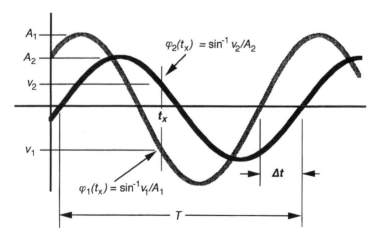

Fig. 1.3 Defining relative phase.

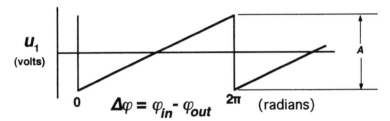

Fig. 1.4 A "Linear" phase detector characteristic. The characteristic is linear but only over one cycle of phase range.

1.6   PHASE DETECTOR

Figure 1.4 shows the type of response that we would like to get from a phase detector (PD). It produces a voltage proportional to the difference in phases of the reference φ_{in} and the VCO output, which is also the loop output, φ_{out}. The constant of proportionality, K_p, is the gain of the phase detector, relating its output voltage change to its input phase change. For this case

$$K_p = A/(2\pi \text{ rad}) = A/(1 \text{ c}). \tag{1.6}$$

For example, if $A = 3$ V, then $K_p = 3$ V/c $= 0.48$ V/rad.

The PD response will always involve a nonlinearity. Of course, all representations of the responses of physical things involve a nonlinearity for large enough values, but, in the case of the PD, this nonlinearity can easily occur in a region of operation, or at least of consideration. The PD in Fig. 1.4 has a linear range of 1 cycle. Often the linear range is smaller than a cycle, but some types have a linear range of 2 cycles.

A common type of PD, the balanced mixer, may produce a sinusoidal response like that shown in Fig. 1.5. The gain is the slope of the characteristic,

$$u_1 = B \cos \Delta\varphi, \tag{1.7}$$

$$K_p = \frac{du_1}{d\,\Delta\varphi} = \frac{d(B \cos \Delta\varphi)}{d\,\Delta\varphi} = \frac{-B \sin \Delta\varphi}{\text{rad}}, \tag{1.8}$$

where

$$\Delta\varphi = \varphi_{in} - \varphi_{out}. \tag{1.9}$$

The most desirable operating point is at $\Delta\varphi = -\pi/2$, where the slope is maximum and equals

$$K'_p \equiv K_p(\Delta\varphi = -\pi/2) = B/\text{rad} \tag{1.10}$$

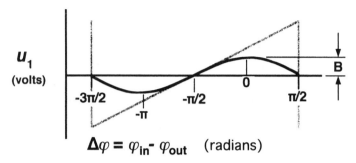

Fig. 1.5 Sinusoidal phase-detector characteristic, superimposed on the linear response. Both of these have the same gain K_p' at $\pi/2$ radians.

so Eq. (1.7) can be rewritten

$$u_1 = K_p' \text{ rad} \cos \Delta\varphi. \tag{1.11}$$

Thus K_p', the maximum value of K_p, for a sinusoid equals its amplitude per radian. Often K_p is used to mean this maximum value. Obviously, this characteristic has no truly linear range, but $K_p' \geq K_p \geq 0.7K_p'$ for a range of $\pm 45°$ about $\Delta\varphi = -\pi/2 = -90°$.

1.7 COMBINED GAIN

In its simplest form, the loop filter in Fig. 1.1 consists merely of an amplifier, in which case

$$\frac{du_2}{du_1} \equiv K_{\text{LF}}. \tag{1.12}$$

Just as the PD can have a nonlinear response characteristic, so can the VCO. But it too can be characterized at some operating point by the gain there:

$$\frac{d\omega}{du_2} \equiv K_v, \tag{1.13}$$

where K_v has units of radians per second per volt, or

$$\frac{df}{du_2} \equiv K_v, \tag{1.14}$$

where the units are hertz per volt.

If we break the loop's feedback path in order to determine the gain around the loop (the "open-loop gain"), we see that

$$\frac{d\omega_{out}}{d\Delta\varphi} = K_p K_{LF} K_v \triangleq K. \tag{1.15}$$

Checking the units of K_p, K_{LF}, and K_v, we can see that the units of K should be reciprocal seconds, sec^{-1}. This is not quite the entire open-loop transfer function, which is the ratio of $d\omega_{out}$ at the output of the opened loop to $d\omega_{out}$ at its input, and which has no units.

Example 1.2   Combined Gain   A phase detector, in the region of operation, has a characteristic with a slope of 0.1 V per radian so

$$K_p = 0.1 \text{ V/rad}.$$

The loop filter is an amplifier with a voltage gain of 5 so

$$K_{LF} = 5.$$

The slope of the VCO tuning curve at the operating frequency is 2 MHz/V so

$$K_v = 2 \times 10^6 \text{ Hz/V}.$$

The combined gain is, from (1.15)

$$K = 10^6 \frac{\text{Hz}}{\text{rad}} = 10^6 \frac{c}{\text{sec-rad}} = 10^6 \frac{c}{\text{sec-rad}} \frac{2\pi \text{ rad}}{c} = 6.28 \times 10^6 \text{ sec}^{-1}.$$

If the input phase φ_{in} suddenly changes by 0.01 radian, the immediate result will be a change of voltage at the phase detector output of

$$0.01 \text{ rad}(K_p) = 10^{-3} \text{ V}.$$

This will cause a change K_{LF} larger at the loop filter output, that is 5×10^{-3} V. This will produce a frequency change at the VCO output of

$$5 \times 10^{-3} \text{ V}(K_V) = 10^4 \text{ Hz}.$$

Or we can also simply multiply the phase change by K to obtain the frequency change

$$0.01 \text{ rad}(6.28 \times 10^6 \text{ sec}^{-1}) = 6.28 \times 10^4 \text{ rad/sec} = 1 \times 10^4 \text{ Hz}.$$

In a locked loop, this frequency change causes the phase at the VCO output to begin to advance more rapidly than the input phase and thus to begin the process of restoring the phase difference that existed before the step in input phase.

1.8 OPERATING RANGE

Assuming a PLL that is initially locked, what limitation is there on the range of ω_{in} over which lock can be maintained? As the input frequency ω_{in} is slowly lowered, the operating point will move down the ramp in Fig. 1.4 to lower the VCO's frequency ω_{out} and keep it equal to ω_{in}. At some point ω_{in} may become so low that lock will be lost because it is not within the capability of the PD to generate a voltage that is low enough to cause ω_{out} to equal ω_{in}. Likewise there is a limit on how high ω_{in} can go before the operating point runs off the upper extremity of the PD characteristic. The difference between these two extremes is called the "hold-in" or "synchronization" range.

From Fig. 1.4 we can see that the output of a PD with the sawtooth characteristic can vary over a total range of A. By Eq. (1.6), this equals $2\pi K_p$ rad. We can also write this as $\pm \pi K_p$ rad about the midpoint. We obtain the corresponding change in ω_{out} by multiplying the change in PD output by $K_{LF}K_v$, giving a total hold-in or synchronization range of

$$\pm \Omega_{H, \text{ sawtooth}} = \pm \pi K_p K_{LF} K_v \text{ rad} = \pm \pi K \text{ rad}. \qquad (1.16)$$

Note that, since the units of K are reciprocal seconds, the units of Ω_H are radians/second, as they should be. If there is significant curvature in the VCO tuning characteristic over Ω_H, an average value of K_v should be used. In practice, however, the actual values of u_2 that correspond to the extremes of the PD range must be found before the region of operation of the VCO can be determined so it is as easy to find Ω_H from the tuning curve as it is to determine the average value of K_v. This is illustrated in Fig. 1.6.

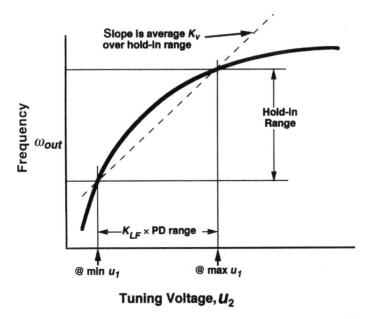

Fig. 1.6 Typical VCO tuning curve with the average slope over the hold-in range shown.

From Fig. 1.5 we can see that the output of a PD with a sinusoidal characteristic can vary its output over a total range of $2B$. By Eq. (1.11), this equals $2K'_p$ rad. The hold-in range for this PD is therefore

$$\pm \Omega_{H,\,\text{sinusoidal}} = \pm K' \text{ rad}, \qquad (1.17)$$

where $K' = K'_p K_{\text{LF}} K_v$, the maximum value of K. Thus, for the same maximum gain, the hold-in range of the sawtooth PD is π larger than that of the sinusoidal PD. Assuming other components are linear, we can write the forward transfer function for a loop with this phase detector as

$$\Delta f = K' \cos \Delta \varphi. \qquad (1.18)$$

Here the Δf and $\Delta \varphi$ are deviations from midrange. In practice, the range of other components, such as the VCO or an operational amplifier, can be more limiting than that of the phase detector. However, these limitations are not fundamental—they can normally be increased, whereas the range of any phase detector has an inherent limit. For this reason, the phase detector is assumed to be the range-limiting element in most PLL analyses.

While we have assumed a simple loop filter, Eqs. (1.16)–(1.18) apply also for more complex loops whose low-frequency filter gains are K_{LF}.

Example 1.3 Hold-in Range The parameters in Example 1.2 describe a loop when the phase is in the center of the phase detector's range (output midway between maximum and minimum). Assume the other components have linear responses.

If the phase detector is sinusoidal, (1.17) indicates a hold-in range of

$$\pm K' \text{ rad} = \pm (6.28 \times 10^6 \text{ sec}^{-1}) \text{ rad} = \pm 6.28 \times 10^6 \text{ rad/sec} = \pm 1 \text{ MHz}.$$

Thus the VCO can be tuned ± 1 MHz without loss of lock.

If the phase detector has a sawtooth characteristic, (1.16) gives the hold-in range as

$$\pm \pi K \text{ rad} = \pm 3.14 (6.28 \times 10^6 \text{ sec}^{-1}) \text{ rad}$$

$$= \pm 1.97 \times 10^7 \text{ rad/sec} = \pm 3.14 \text{ MHz}.$$

1.9   UNITS AND THE LAPLACE VARIABLE s

The Laplace variable s represents d/dt, so it should have units of reciprocal seconds. When $1/s$ represents $\int dt$, again, s should have units of reciprocal seconds. But then, for consistency of units, $s = j\omega + \sigma$ should be written $s = j\omega/\text{rad} + \sigma/\text{neper}$. The open-loop transfer function should be written $-G = -K/(j\omega/\text{rad})$. How can we prevent the confusion and wrong answers that can be caused by not carrying units and yet not be overburdened by nontraditional rules involving radian units? Perhaps the best solution is to *drop or add radians, and only radians, as required.* For example, if I want the hold-in range in hertz and I use Eq. (1.16) but drop the "rad," the hold-in range is $2\pi K$. The unit of K is reciprocal seconds, but this is not a proper unit for frequency. To obtain proper units, we add "rad," obtaining $2\pi K$ rad with units of radians/second for hold-in range. We cannot mistake this for the hold-in range in hertz because we do not treat cycles as arbitrarily as we do radians. To obtain hold-in range in hertz, we must convert Eq. (1.17) as follows:

$$\Omega_{H, \text{ sawtooth}} = 2\pi K \text{ rad} \frac{c}{2\pi \text{ rad}} = K \text{ c} \tag{1.19}$$

This has units of cycles/second = hertz (c/sec = Hz), as desired. In fact, even though the equalities above are valid, Ω, which we might assume from the symbol to be in radians/second, has here been equated to hold-in range in hertz. This gives a hint of the danger in relying on the symbol to differentiate between radians/second and hertz.

PROBLEMS

1.1 Refer to Fig. 1.1. For the phase detector, $\Theta = \pi$ radians. The VCO has $\varphi_{out}(t) = (\omega_c + K_v u_2)t$. The loop is locked. Fill in the remainder of the following table. Row (a) is done as an example, with the answers shown in *italic* script.

	K_p	$\varphi_{in} - \varphi_{out}$	$u_1 = u_2$ volts	ω_c or f_c	K_v	$d\varphi_{in}/dt$
(a)	0.02 V/deg	220°	0.8	10^3 rad/sec	100 rad/sec V	*1080 rad/sec*
(b)	0.16 V/rad	3.14 rad		10^4 Hz	100 Hz/V	
(c)	10 V/rad		−12		2 Hz/V	180 Hz
(d)	20 V/c			100 Hz	2 Hz/V	104 Hz
(e)	2 V/c	0.75 c		2 GHz		2020 MHz
(f)	2 V/c			17 MHz	1 MHz/V	18.5 MHz

1.2 Over what range of frequency (lowest and highest possible values of $d\varphi_{in}/dt$) can the loop of row (f) in Problem 1.1 maintain lock if the phase detector has a sawtooth characteristic as in Fig. 1.4 (with $\Theta = \pi$ radians still)? Is the state described by row (f) possible?

1.3 Over what range of frequency (lowest and highest possible values of $d\varphi_{in}/dt$) can the loop of row (f) in Problem 1.1 maintain lock if the phase detector had a sinusoidal characteristic as in Fig. 1.5? The midvoltage corresponds to f_c.

CHAPTER 2

THE BASIC LOOP

We begin by studying the basic phase-locked loop (PLL), whose loop filter is represented by a frequency-independent gain. We will look at its transient response and its frequency response in the same way that we will later look at those for the more complicated cases, but it is important that we first gain a clear understanding by studying this simpler case.

2.1 STEADY-STATE CONDITIONS

The center frequency of the voltage-controlled oscillator (VCO) f_c is the frequency that occurs when the phase detector is in the center of its output range or the frequency midway between the frequencies that occur at the edges of the phase detector range. If the transfer function from the phase detector to the VCO output frequency is linear, these are the same thing. The output from the phase detector is not necessarily zero at that point; it depends on the particular phase detector realization. Even if it is zero, a voltage shift may be implemented between the phase detector and the VCO. In fact, most VCOs are operated with a tuning voltage that is always positive or always negative. While these details are of considerable practical concern, they are easily understood, and our studies will concentrate on changes from some initial or average condition.

Figure 2.1 is similar to Fig. 1.6 except that the independent variable has been changed from the tuning voltage u_2 to the phase difference $\Delta\varphi$. It illustrates the one-to-one correspondence between the VCO's frequency and its phase (relative to the reference phase) in steady state. In this simple

17

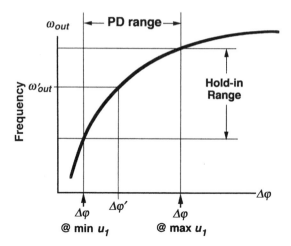

Fig. 2.1 Loop output frequency vs. phase difference at the phase detector with a typical VCO. The hold-in range, over which the frequency will vary as the phase detector covers its output range, is indicated.

first-order, loop, the one-to-one relationship holds not only in steady state but always. If we know one variable, we know the other because there is no storage, no memory, in the filter. The loops that we will study later will be different in this regard; the filter will permit the relationship between the VCO's frequency and phase to change with time. That will make those loops more versatile, and more challenging.

2.2 CLASSICAL ANALYSIS

Before using Laplace transforms to analyze the loops' dynamics, let us first determine the response by more directly solving Eq. (1.15). We will find that the PLL responds to changes in phase or frequency in a manner similar to that in which other simple circuits respond to changes in voltage or current. Thus we find that, whereas, in other circuits voltage and current are the variables that describe the state of the circuit and phase and frequency are parameters of those variables, in the PLL frequency and phase are state variables. They too, nevertheless, have frequencies and phases, which are their modulation parameters. For example, Eq. (2.1),

$$\omega_{\text{out}} = \omega_c + A \sin(\omega_m t + \theta_0), \qquad (2.1)$$

describes the state variable ω_{out}, which represents the output frequency of the loop, but ω_{out} can be seen to possess a frequency (its modulation frequency) ω_m and a modulation phase, θ_0. These parameters must also be parameters of u_1 and u_2, other state variables in the loop, since u_1 and u_2 are proportional to ω_{out}.

2.2.1 Transient Response

Equation (1.15) can be rewritten as

$$d\omega_{\text{out}} = K\, d\,\Delta\varphi. \tag{2.2}$$

The time derivative of this equation is

$$\frac{d\omega_{\text{out}}}{dt} = K\frac{d\,\Delta\varphi}{dt} = K(\omega_{\text{in}} - \omega_{\text{out}}), \tag{2.3}$$

where Eq. (1.9) was used. We rewrite (2.3) as

$$d\omega_{\text{out}} = K(\omega_{\text{in}} - \omega_{\text{out}})\, dt. \tag{2.4}$$

If ω_{in} is constant, it may be subtracted from ω_{out} on the left side, since the differential will not be affected by inclusion of a constant. This gives

$$d(\omega_{\text{out}} - \omega_{\text{in}}) = -K(\omega_{\text{out}} - \omega_{\text{in}})\, dt \tag{2.5}$$

or

$$d\,\Delta\omega = -K\,\Delta\omega\, dt, \tag{2.6}$$

where

$$\Delta\omega \equiv \omega_{\text{out}} - \omega_{\text{in}}. \tag{2.7}$$

We now write Eq. (2.6) as

$$\frac{d\,\Delta\omega}{\Delta\omega} = -K\, dt \tag{2.8}$$

and integrate both sides to obtain

$$\int \frac{1}{\Delta\omega}\, d\,\Delta\omega = -\int K\, dt \tag{2.9}$$

with the solution

$$\ln\,\Delta\omega = -Kt + C. \tag{2.10}$$

Taking the exponential of both sides, we obtain

$$\Delta\omega = e^{-Kt+C} = e^{C}e^{-Kt}. \tag{2.11}$$

We can determine the value of e^C by evaluating this expression at $t = 0$. When we do so and substitute the results in Eq. (2.11), we obtain

$$\Delta\omega(t) = \Delta\omega(0)e^{-Kt}. \tag{2.12a}$$

The output frequency is the input frequency plus $\Delta\omega$,

$$\omega_{out} = \omega_{in} + \Delta\omega(t) = \omega_{in} + \Delta\omega(0)e^{-Kt}. \tag{2.12b}$$

Thus we see that this simple loop responds to an initial frequency error (difference between the output and reference frequencies) $\Delta\omega(0)$ by exponentially decreasing the error with a time constant equal to $1/K$. As a result, the step response shown in Fig. 2.2a is obtained. This is the same response to a frequency step that a single-pole low-pass circuit has to a voltage step (Fig. 2.2b).

We might intuit this exponential response through the following considerations. After the input frequency steps, there is a frequency error $\Delta\omega$ that causes $\Delta\varphi$ to change. This change in $\Delta\varphi$ (multiplied by K) causes the VCO frequency to change in such a direction as to decrease the frequency error (Fig. 2.3). This, in turn, decreases the rate of phase change and thus the rate at which the VCO frequency changes. As always, when the rate of change of a variable is proportional to the value of the variable, it changes exponentially. Of course, the exponential shape will be perturbed to the degree that there is curvature in the gain characteristics over the operating region.

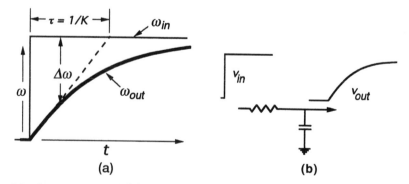

Fig. 2.2   Step response of (a) a simple loop compared to the step response of (b) a low-pass filter. The filter in this loop is a simple amplifier. The frequency of the loop responds like the voltage of the RC circuit.

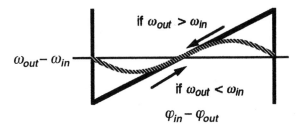

Fig. 2.3 Output frequency vs. phase for two phase detector characteristics. Since frequency is the derivative of phase, the direction of change is determined by the sign of the frequency difference.

Example 2.1 Transient Response When the input frequency to a particular loop steps to a new frequency, we require that the output achieve that frequency, to within an accuracy of 1%, in 1 msec. What are the range of K values that allow a simple loop to comply with this requirement?

Restating the problem, we require

$$\Delta\omega(1 \text{ msec})/\Delta\omega(0) \leq 0.01.$$

From (2.12a) this requires

$$(e^{-K\ 10^{-3}\ \text{sec}}) \leq 0.01 \Rightarrow K \geq -\ln(0.01)/10^{-3}\ \text{sec} = 4.6 \times 10^{3}\ \text{sec}^{-1}.$$

2.2.2 *Modulation Response*

Rearranging Eq. (2.3) we can write

$$\frac{1}{K}\frac{d\omega_{\text{out}}}{dt} + \omega_{\text{out}} = \omega_{\text{in}}. \tag{2.13}$$

If the input is frequency modulated, then we expect the response to be frequency modulated also, so we assume a response of the form

$$\omega_{\text{out}}(t) = \Omega_0 + A\sin(\Omega_m t), \tag{2.14}$$

where Ω_0 is the center, or carrier, frequency, A is the peak frequency deviation, and Ω_m is the modulation frequency. Putting Eq. (2.14) into (2.13), we obtain

$$A\frac{\Omega_m}{K}\cos(\Omega_m t) + \Omega_0 + A\sin(\Omega_m t) = \omega_{\text{in}}. \tag{2.15}$$

We rewrite (2.14) and (2.15) in terms of deviations from the mean:

$$\Delta\omega_{\text{out}} \equiv \omega_{\text{out}}(t) - \Omega_0 = A\sin(\Omega_m t), \tag{2.16}$$

$$\Delta\omega_{\text{in}} \equiv \omega_{\text{in}} - \Omega_0 = A\frac{\Omega_m}{K}\cos(\Omega_m t) + A\sin(\Omega_m t). \tag{2.17}$$

The amplitude of $\Delta\omega_{\text{out}}$ is A and that of $\Delta\omega_{\text{in}}$ is

$$|\Delta\omega_{\text{in}}| = A\sqrt{\left(\frac{\Omega_m}{K}\right)^2 + 1}, \tag{2.18}$$

so the magnitude of the frequency response is

$$|H(\Omega_m)| = 1/\sqrt{\left(\frac{\Omega_m}{K}\right)^2 + 1}. \tag{2.19}$$

Since $\Delta\omega_{\text{in}}$ contains a component equal to $\Delta\omega_{\text{out}}$ plus one that leads it by 90° and is larger by Ω_m/K, the phase angle of $\Delta\omega_{\text{in}}$ relative to $\Delta\omega_{\text{out}}$ is $\tan^{-1}[\Omega_m/K]$ and the transfer phase is

$$\angle H(\Omega_m) \equiv \varphi_{\text{out}}(\Omega_m) - \varphi_{\text{in}}(\Omega_m) = -\tan^{-1}[\Omega_m/K]. \tag{2.20}$$

As with the time response, the response of this loop to a sinusoidally modulated frequency is the same as the response of a low-pass filter to a sinusoidal voltage. In each case the corner (-3 dB) frequency is $\omega_C = K$.

Example 2.2 Modulation Response A simple loop has a sinusoidal phase detector characteristic (Fig. 1.5) and $K = 1000$ sec^{-1}. The input (reference) frequency is frequency modulated at an unknown rate with a peak deviation of 50 Hz. What is the phase shift between the VCO modulation and the input modulation when the deviation at the output is 10 Hz?

   We will use the gain to determine the modulation frequency and then use the modulation frequency to determine the phase shift. The closed-loop gain is

$$|H(\Omega_m)| = 10 \text{ Hz}/50 \text{ Hz} = 0.2.$$

From (2.19), this occurs at

$$\sqrt{\left(\frac{\Omega_m}{K}\right)^2 + 1} = 5,$$

from which we obtain,

$$\frac{\Omega_m}{K} = \sqrt{24},$$

giving the modulation frequency as $\Omega_m = 4899$ rad/sec. Inserting Ω_m/K into (2.20), we obtain the phase shift as $\angle H(\Omega_m) = -\tan^{-1}(4.9) = -78.5°$. Thus the output modulation lags the input modulation by 78.5°.

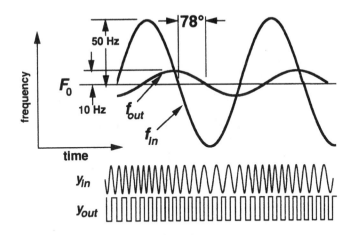

The usual input and output state variables (y for voltage or current) are shown below the plot of instantaneous frequency. The input happens to be a sinusoid while the output is a square wave. These are intended to illustrate the meaning of the frequency deviations plotted above. Note that the magnitude of the output frequency deviation is smaller than that of the input deviation and how the times of occurrence of maximum frequency (minimum period) correspond to the peaks in the frequency plots.

2.3   MATHEMATICAL BLOCK DIAGRAM

The mathematical relationships in the loop can be shown by means of a block diagram such as Fig. 2.4. Here the input variable is the reference, or input, instantaneous phase φ_{in}, which is, in general, a function of time, perhaps a sinusoidal function such as

$$\varphi_{in}(t) = \varphi_i + A_i \sin(\Omega_m t + \theta_i) \qquad (2.21)$$

with average value φ_i, peak deviation A_i, modulation frequency Ω_m, and phase θ_i. Generally, as in any control system, responses are given in terms of the deviation from the average or initial or steady-state value, and φ_i is only of interest in establishing the steady-state operating point. A block is shown preceding φ_{in} in order to establish the integral relationship ($1/s$) between $\omega_{in}(s)$ and $\varphi_{in}(s)$. The output from the summer (a subtracter in this case) is

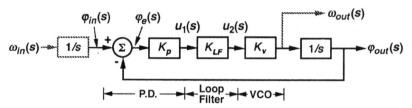

Fig. 2.4 Mathematical block diagram of the simple loop with phase variables as input and output. The frequency state variables are shown to indicate how they relate to the phase variables.

the phase error $\varphi_e(s)$,[1] the difference between the input phase and the VCO phase, φ_{out}. The phase error is converted to voltage in the phase detector, which is represented by the gain $K_p(s) = u_1(s)/\varphi_e(s)$. For this simple case the loop filter is merely an amplifier with gain K_{LF}; K_v represents the tuning sensitivity $\omega_{out}(s)/u_2(s)$ of the VCO. The output from this block is the output frequency ω_{out}. However, φ_{out} is needed to complete the loop so $\omega_{out}(s)$ is integrated (multiplied by $1/s$) to produce it. Then the loop is completed by subtracting $\varphi_{out}(s)$ from $\varphi_{in}(s)$ in the summer. Note that the minus sign at the summer represents $-180°$ of phase shift around the loop that does not appear in the transfer functions of the individual blocks.

The generic control system block diagram is shown in Fig. 2.5. The well-known equations describing its transfer function are the response of the controlled variable C to the reference R,

$$\frac{C}{R} = \frac{G_F}{1 + G_F G_R}, \tag{2.22}$$

where G_F and G_R are forward and reverse transfer functions, respectively, and the response of the error E to the reference[2]

$$\frac{E}{R} = \frac{1}{1 + G_F G_R}. \tag{2.23}$$

The correspondence between the usual designations for loop signals, C, R, and E, and state variables of the phase-locked loop depends on the input and output of interest. Most often, in this text, we will be interested in the configuration shown in Fig. 2.4 so we identify C as $\varphi_{out}(s)$ and R as $\varphi_{in}(s)$. Because the feedback path has unity gain, we have also $G_R = 1$ and $-G_F$ is

---

[1] With reference to Eq. (1.9), $\varphi_e(s) = \Delta\varphi(s)$ except that $\varphi_e(s)$ is used to refer to a change without reference to the steady-state phase difference.
[2] Note: $-G_F G_R$ is the open-loop gain and C/R and E/R are closed-loop gains.

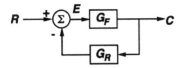

Fig. 2.5   Generic control system diagram.

the entire open-loop transfer function $-G(s)$,

$$G(s) = K_p K_{LF} K_v/s \equiv K/s. \tag{2.24}$$

Thus from Eq. (2.22) we obtain

$$\frac{\varphi_{out}(s)}{\varphi_{in}(s)} = \frac{G(s)}{1 + G(s)} \tag{2.25a}$$

$$= \frac{K/s}{1 + K/s} = \frac{1}{1 + s/K}, \tag{2.25b}$$

while from Eq. (2.23) we obtain

$$\frac{\varphi_e(s)}{\varphi_{in}(s)} = \frac{1}{1 + G(s)} \tag{2.26a}$$

$$= \frac{1}{1 + K/s} = \frac{s}{s + K}. \tag{2.26b}$$

When we are interested in inputs or outputs at points other than those used above, we can rewrite the equations or, what is often simpler, relate the desired input or output to that shown in Fig. 2.4 by the transfer function of the segment that connects them.

Equation (2.25) represents a low-pass characteristic with a cutoff (-3 dB) frequency of $\omega_c = K$. This is analogous to the low-pass filter of Fig. 2.6a in which the voltages have been given names corresponding to the phases. Equation (2.26) says that the error phase has a high-pass characteristic, analogous to Fig. 2.6b. It is generally true, even in more complex loops, that the output has a low-pass relationship to the input while the error has a high-pass relationship. This reflects the fact that the error responds immediately to a change in the input while, and because, the output response is delayed.

Figure 2.7 is essentially the same as Fig. 2.4 except that the two $1/s$ blocks have been moved forward through the summer; since both integration and summing are linear operations, either can be done first. The input and output are now frequencies rather than phases, illustrating that the response of output frequency to input frequency is the same as the response of output

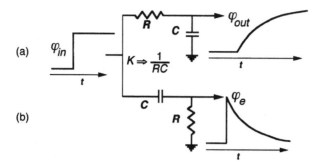

Fig. 2.6 Electrical analogs to the PLL. Voltages have names of analogous loop variables.

phase to input phase. We can show the same results as follows:

$$\frac{\varphi_{out}(s)}{\varphi_{in}(s)} = \frac{s\varphi_{out}(s)}{s\varphi_{in}(s)} = \frac{\omega_{out}(s)}{\omega_{in}(s)}. \tag{2.27}$$

A similar relationship holds for error responses.

It is apparent from Figs. 2.4 and 2.7 that

$$\varphi_{in} = \varphi_{out} + \varphi_e \tag{2.28a}$$

and

$$\omega_{in} = \omega_{out} + \omega_e \tag{2.28b}$$

and thus

$$1 = \frac{\omega_{out}}{\omega_{in}} + \frac{\omega_e}{\omega_{in}} = \frac{\varphi_{out}}{\varphi_{in}} + \frac{\varphi_e}{\varphi_{in}} \tag{2.28c}$$

so that the output can easily be obtained from error and vice versa in most cases.

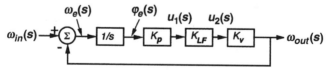

Fig. 2.7 Mathematical block diagram of the simple loop with frequency variables as input and output.

Example 2.3   Modulation Response   Use (2.25b) to obtain the results of
Example 2.2. At $s = j4899$, we have

$$\frac{\varphi_{out}}{\varphi_{in}} = \frac{1}{1 + j4899/1000} = 0.2\angle-78.5°$$

as before.

_____

2.4   BODE PLOT

One of the most important tools with which we will work is a logarithmic plot
of open-loop gain $|G_F G_R|$ and phase shift $\angle G_F G_R$ versus modulation fre-
quency, $s \rightarrow \omega_m$. This plot, called the Bode plot, is shown for this simple loop
in Fig. 2.8. Sometimes we will forgo the phase plot since the phase shift is
implied by the slope of the gain. A linear increase in gain with ω_m originates
in a term such as $s = j\omega$ and therefore implies both a gain that is increasing
proportionally with frequency and a 90° phase shift. In general the segments
of the Bode plot correspond to approximations of the form

$$F(\omega) = C(j\omega)^n = C\omega^n \angle n(90°). \tag{2.29}$$

Thus, for example, when $n = -2$, the gain drops as the square of the
frequency and the phase is $-180°$. The slope is often expressed logarithmi-
cally. The magnitude is described by

$$|F(\omega)| \Rightarrow 20 \text{ dB} \log_{10}(C\omega^n) = 20n \text{ dB} \log_{10}(\omega) + C'. \tag{2.30}$$

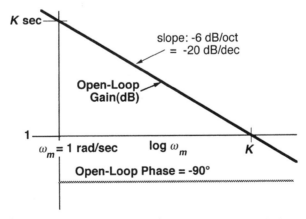

Fig. 2.8   Bode plot for the simple loop. Note that the abscissa axis is marked with the
values of ω_m (e.g., K), rather than the actual distance, $\log(\omega_m \text{ sec/rad})$. Similarly, the
ordinate axis is marked with gain rather than its logarithm or value in dB.

Note that we precede $\log_{10}$ by 20 dB rather than 10 dB because $G_F G_R$ is a voltage gain. We can break the loop at a point where the variable is voltage and $G_F G_R$ will describe the open-loop gain there.

In each decade of frequency change the function changes by

$$|F(10\omega)/F(\omega)| \Rightarrow 20n \text{ dB } \log_{10}(10) = 20n \text{ dB} \qquad (2.31)$$

so the slope is $20n$ dB per decade. Similarly, a two-to-one change in frequency corresponds to

$$|F(2\omega)/F(\omega)| \Rightarrow 20n \text{ dB } \log_{10}(2) = 6n \text{ dB} \qquad (2.32)$$

or $6n$ dB per octave (an octave is a two-to-one change).

Figure 2.8 also illustrates that K is both the gain at $\omega = 1$ and the radian frequency at which the open-loop gain is unity. As has been shown, this is the frequency at which the closed-loop gain is -3 dB. At frequencies well below $\omega_c = K$, Eq. (2.25) can be approximated as

$$\frac{\varphi_{\text{out}}(s)}{\varphi_{\text{in}}(s)} = 1, \qquad (2.33)$$

whereas at frequencies well above $\omega_c = K$ the equation is approximately

$$\frac{\varphi_{\text{out}}(s)}{\varphi_{\text{in}}(s)} = \frac{K}{s} = -j\frac{K}{\omega}. \qquad (2.34)$$

This equals the open-loop transfer function, Eq. (2.24). Thus, at low frequencies, where the gain is high, the output follows the input faithfully, whereas, at high frequencies, where the gain becomes low, the loop is essentially open and the response is as if there were no loop (except that the low-frequency gain keeps it locked—otherwise none of these equations would be valid). At the loop corner frequency, $\omega_c = K$, it is easy to show from Eq. (2.25b), that the closed-loop gain is $1/\sqrt{2}$, or -3 dB, and the phase shift is $-45°$, just as in the case of the low-pass filter.

Similarly, the high-pass characteristic of Eq. (2.26) approaches unity at high frequencies and zero at low frequencies and, at $\omega_c = K$, also has a gain of -3 dB, but the phase shift is $+45°$.

The more complex loops that we will study also have these general tendencies, but we will be able to shape their characteristics more exactly to our needs because of the increased number of parameters at our disposal.

2.5 NOTE ON PHASE REVERSALS

It is not uncommon that the transfer functions of the various blocks in Fig. 2.4 have negative signs. It is only necessary that the total phase shift around the loop not be altered. The negation of an even number of transfer functions has no important effect. In fact, when a balanced mixer is used, it will automatically operate on the proper slope to produce a correct number of phase reversals.

2.6 SUMMARY OF TRANSIENT RESPONSES OF THE FIRST-ORDER LOOP

Here we will summarize the transient responses of the first-order loop based on previous material and simple extensions of that material.

The unit step response is given in Eq. (2.12b) and shown in Fig. 2.2. Since a unit ramp t is the time integral of a unit step, we can obtain the time response to t by integrating the time response to the unit step. Similarly, the integral of t is a parabola $t^2/2$, and we can obtain the response to this by integrating the ramp response. Alternately, we can take the inverse Laplace transform of the transform obtained by multiplying the transfer function of Eq. (2.25b) by $1/s$, $1/s^2$, and $1/s^3$ to obtain step, ramp, and parabola responses. (The parabolic input is important because it represents the phase corresponding to a frequency ramp.)

The error response can be obtained by following the above process but starting with equations for the error response, Eq. (2.12a) or the Laplace transform of the error response, Eq. (2.26b).

Either the output response or the error response can be obtained from the other by subtraction according to Eq. (2.28c).

Table 2.1 summarizes these responses. If an input is a 5-Hz step, since that is 5 Hz times the given unit input, the output or error will be 5 Hz times

TABLE 2.1 Summary of First-Order Loop Transient Responses

	Input		Response	
	$f(t)$	$F(s)$	Error	Output
Step	1	$\dfrac{1}{s}$	e^{-Kt}	$1 - e^{-Kt}$
Ramp	$\dfrac{t}{\sec}$	$\dfrac{1}{s^2}$	$\dfrac{1}{K}(1 - e^{-Kt})$	$t - \dfrac{1}{K}(1 - e^{-Kt})$
Parabola	$\dfrac{t^2}{2\sec^2}$	$\dfrac{1}{s^3}$	$\dfrac{1}{K}[t + \dfrac{1}{K}(e^{-Kt} - 1)]$	$\dfrac{t^2}{2} - \dfrac{1}{K}[t + \dfrac{1}{K}(e^{-Kt} - 1)]$

the given output or error. If the input is a 2-rad or 2° step, then the output will be 2 rad or 2° times the given response. A 10-Hz step is also a 10-cycle/sec ramp (10 cycles × t/sec), 10 cycles times the given ramp input, so the output phase will be 10 cycles times the given ramp response. If the input is a frequency ramp of 100 rad/sec/sec [(100 rad/sec) × t/sec], the output frequency will be obtained by multiplying the ramp output response by 100 rad/sec. However, we can also obtain the output phase by considering the input as a parabolic phase [100 rad × t^2/(2 sec^2)] and obtaining the phase of the response by multiplying the given response by 100 radians.

PROBLEMS

2.1 A first-order PLL has an open-loop gain of 50 at $f_m = 100$ Hz. Show that its open-loop gain $|G(1 \text{ Hz})|$ at 1 Hz is 5000. At what frequency is the gain one? What is the phase shift $\angle G(1 \text{ Hz})$?

2.2 In Fig. 2.4, $K_p = 0.1$ V/c, $K_{LF} = 3$, $K_v = 1$ MHz/V.

   (a) Show that unity open-loop gain occurs at 47.75 kHz.

   (b) If K_p is changed to 1 V/rad, what is the unity-gain frequency (include units—always)?

2.3 In Fig. 2.7, $K = K_p K_{LF} K_v = 1000$ sec^{-1}, and ω_{in} has a step increase of 100 rad/sec.

   (a) Sketch the transient at ω_{out} and indicate its amplitude.

   (b) How long will it take ω_{out} to come within 1 rad/s of its final steady-state value?

   (c) Sketch the transient frequency error at ω_e, label its amplitude, and show its final value.

   (d) How long will it take for the frequency error to come within 10 Hz of its final steady-state value?

   (e) Sketch the change in the phase error φ_e, label its amplitude and show its final value.

CHAPTER 3

LOOP COMPONENTS

In this chapter we consider the characteristics of the major components of a phase-locked loop (PLL).

3.1 PHASE DETECTOR

We will consider several of the many types of phase detectors (PDs) that are in common use [Egan, 1981, Chapter 5]. We begin with three types that employ logic circuits because they are both useful and relatively easy to understand. The most important type for the processing of analog signals, however, is the balanced mixer, so we will then study its operation in some detail.

3.1.1 Flip-Flop Phase Detector

Figure 3.1 illustrates the operation of the flip-flop PD. The symbol for an RS flip flop is shown in Fig. 3.1*a*. This type changes state in response to state change at either input, going to the $Q = 1$ state with a 1 input at A and to the $Q = 0$ state with a 1 input at B. When both inputs are 1, the state is undefined,[1] so the input waveforms are shown as very narrow pulses to avoid overlap of 1 states at the two inputs. The duration of the $Q = 1$ state depends on the time from the A input to the B input, as can be seen in

---

[1]The state is undefined in general but can be discerned for a particular realization of the circuit.

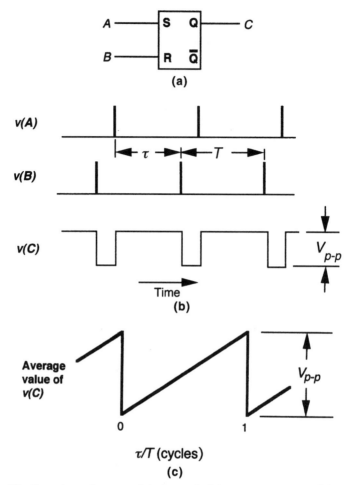

Fig. 3.1 Flip-flop phase detector: (*a*) the symbol for an RS flip-flop, (*b*) waveforms, and (*c*) phase detector characteristic.

Fig. 3.1*b*. Therefore, the duration of the 1 state, and thus the average output voltage, is proportional to the time difference between inputs and thus to their phase difference. The average value is the useful phase detector output. Higher frequency components must be filtered out. The average voltage has a sawtooth-shaped characteristic versus phase, as shown in Fig. 3.1*c*. The linear phase range is 2π radians.

One useful realization employs a flip-flop with an edge triggered clock for one input. That input need not be narrow because the flip-flop changes states on one of the transitions of the input, but the other input must still be kept narrow.

3.1.2 Exclusive-OR Gate Phase Detector

Figure 3.2 illustrates the operation of the Exclusive-OR (ExOR) phase detector. The ExOR gate outputs 1 when the two input states differ and outputs 0 when they are the same. Since the part of the time when they are the same (or different) indicates their relative phase, the ExOR is useful as a phase detector. However, deviation from a perfect in-phase condition (or a perfect out-of-phase condition) in either an increasing or a decreasing direction of phase change, produces the same change in duty factor, and thus in average voltage. Therefore, the characteristic is triangular and each linear slope extends for only half a cycle, so the linear phase range is π radians.

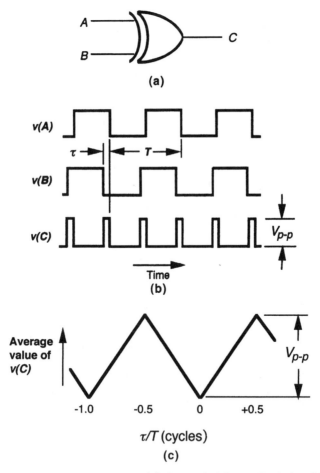

Fig. 3.2 Exclusive-OR phase detector: (*a*) the symbol for an Exclusive-OR gate, (*b*) waveforms, and (*c*) the phase detector characteristic.

3.1.3 Charge-Pump Phase Detector

One widely used type of PD has a characteristic like the flip-flop PD, as shown in Fig. 3.1c, except for one very important addition [Gardner, 1980; Egan, 1981, pp. 115–123]. It generates two different pulse outputs, one when the transition of $v(B)$ lags that of $v(A)$, as shown in Fig. 3.1b, and another when it leads. (The initial definition of whether a particular transition is leading or lagging is arbitrary, but the loop will eventually establish a phase relationship such that one or the other of these outputs is being produced during lock.) The normal operating point is near-zero phase difference ($\tau = 0$). As τ increases, the characteristics represented by Figs. 3.1b and c occur. But, if τ decreases from 0, $v(C)$ stays low, and another output generates pulses that extend from the $v(B)$ transition to the $v(A)$ transition, thus beginning very narrow for small $-\tau$ and widening as $v(B)$ occurs earlier and earlier. Effectively we see the Q output of the flip-flop when $v(B)$ lags and the $\overline{Q}$ output when it leads. If we just added these two outputs, we would get something like the characteristic of the Exclusive-OR PD in the vicinity of zero phase difference—not a useful operating point. But, by converting one of the outputs to a positive analog signal and the other to a negative signal and adding them, we obtain a characteristic that is linear, being positive or negative depending on the sign of the phase and passing through zero at zero phase difference.

The characteristic is further illustrated by Fig. 3.3. Here B initially leads A but has a lower frequency so eventually achieves a lagging position, as in Fig. 3.1. Here we see that $v(U)$ is similar to Fig. 3.1 when A lags, but, when A leads, its pulses disappear and are replaced by pulses at $v(D)$. Voltage v_p is produced by combining U and D *using* the appropriate sign $[V(U) - V(D)]$. Note how the average value of this waveform produces a ramp as the phase difference increases, illustrating the linear relationship between average voltage or current and phase. Moreover, this linear relationship has a range of $\pm 2\pi$ (Fig. 3.4).

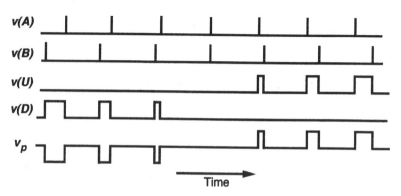

Fig. 3.3 Charge-pump waveforms.

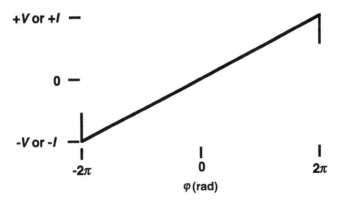

Fig. 3.4 Charge-pump phase detector characteristic.

These pulses are used in various ways to "pump charge" into the loop filter. A phase detector that provides this kind of output is often called a charge-pump phase detector and is commonly part of a phase-frequency detector (see Section 9.4). Sometimes an actual switched current source is used, as illustrated in Fig. 3.5a. Sometimes voltages are switched and converted to current by a resistor, as illustrated in Fig. 3.5b. In the first case the PD gain is $K_p = I/\text{cycle}$, since the average current would be $\pm I$ if the phase error were ± 1 cycle and would be 0 for 0 phase error, and the characteristic is linear between these points. Similarly, in the second case, if $V_x = 0$, the PD gain is $K_p = V/\text{cycle}$. This circuit should not ordinarily be used where V_x does not stay midway between the two voltage sources (near ground for $\pm V$ sources as shown); otherwise the gain will differ for positive and negative phase errors and will vary with V_x. Gain K_p could also be written in current units as $(V/R_x)/\text{cycle}$. Resistor R_x can be considered part of the PD or part of the loop filter (but not both). Gains K_p and K_{LF} will both change if R_x is moved conceptually from one block to the other.

Unless otherwise stated, assume a voltage-source phase detector throughout this book (in most places it will make no difference).

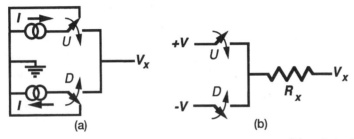

Fig. 3.5 Charge pumps: (*a*) switched current sources and (*b*) switched voltage sources.

3.1.4 Sinusoidal Phase Detector

The PD with a sinusoidal response of voltage to phase was introduced in Chapter 1. This response can be obtained from any mixer.[2] The object of a mixer is to provide signals at the sum or difference of the two frequencies that are injected into it,[3] both the sum and difference being provided simultaneously. In the case of a phase detector, for which the two input frequencies are the same, the difference frequency goes to zero and the output at this "low" frequency becomes the desired voltage that is proportional to phase. Note, in Fig. 1.5, that if $\varphi_{in} = \omega_{in}t$ and $\varphi_{out} = \omega_{out}t$, then $\Delta\varphi = -\Delta\omega t$, and the output voltage will be a sinusoid at frequency $\Delta\omega$. Thus the same characteristic that describes the PD output also describes the output at the difference frequency.

Other signals will accompany the desired low- or zero-frequency output and must be suppressed by a low-pass filter. For a balanced mixer (BM) the strongest of these will normally be at the sum frequency. Of next greatest concern will be leakage at the input frequency. There will also be various harmonics of the input signals. In many applications the harmonics can be easily filtered. When undesired components are a problem, which is often the case in synthesizer applications, other types of phase detectors that have better rejection of these components are used.

The undesired output of most concern in a PD is that which is due to rectification of the input signals because it is a direct current (DC) signal and cannot be rejected by a filter without the filter also rejecting the desired signal. Of course, compensation is possible—the rectification term can be canceled—but variations with temperature and with input amplitude sometimes make this difficult. Balanced mixers greatly reduce this rectification component as well as some of the other undesired signals and therefore are generally used for PDs. However, simpler mixers also have phase detection capability, and they form the components of which the BMs are composed, so we begin by discussing the simpler types and work our way to the BM.

3.1.4.1 *Phase Detection in a Simple Mixer.* Figure 3.6 shows a simple mixer composed of a resistor and a diode. The nonlinearity of the diode produces the desired difference-frequency signal when two sinusoids are injected through the resistor. We will assume that the driving voltage v_i consists of two sinusoids, $A\cos\varphi_1(t)$ and $B\cos\varphi_2(t)$, where, in each case, $\varphi_i(t) = \omega_i t + \theta_i$. It is also likely that there would be a significant load, used to extract the desired energy from the diode, but the combination of the diode and a shunt load resistor can still be represented as a nonlinearity.

---

[2]A mixer is a nonlinear device used for frequency conversion in radios, for example.
[3]Occasionally, the desired output involves a harmonic of one of the input signals, but the results, for our present purposes, are fundamentally the same as would be obtained by multiplying the frequency of that signal and then mixing to obtain the sum or difference. The PLL theory is the same.

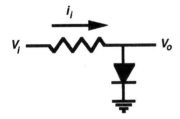

Fig. 3.6 Simple mixer.

The important feature of the diode is that it is nonlinear. We use the general MacLaurin expansion to represent the nonlinearity of the circuit:

$$v_o = a + bv_i + cv_i^2 + dv_i^3 + \cdots \tag{3.1}$$

with

$$v_i = A \cos \varphi_1(t) + B \cos \varphi_2(t). \tag{3.2}$$

The output component of greatest interest to us is that generated by the square-law term:

$$v_{oc} = c \left[A \cos \varphi_1 + B \cos \varphi_2 \right]^2 \tag{3.3}$$

$$= c \left[\frac{A^2}{2} (1 + \cos 2\varphi_1(t)) + 2AB \cos \varphi_1(t) \cos \varphi_2(t) \right.$$

$$\left. + \frac{B^2}{2} (1 + \cos 2\varphi_2(t)) \right] \tag{3.4}$$

$$= c \left\{ \frac{A^2 + B^2}{2} + AB[\cos(\varphi_1(t) + \varphi_2(t)) + \cos(\varphi_1(t) - \varphi_2(t))] + \cdots \right\}. \tag{3.5}$$

The second-harmonic terms in Eq. (3.4) are to be filtered and have not been shown in Eq. (3.5). The first term in Eq. (3.5) is the rectification term, the highly undesirable DC component that can be confused with the desired output. The next term comes from the middle term in Eq. (3.4) by a trigonometric identity. It contains the sum-frequency term, which, at phase lock, is at the same frequency as the second-harmonic terms and will be filtered with them,[4] and the desired difference-frequency term that contains the phase information.[5]

---

[4] Before filtering, the sum frequency term is often the largest of the undesired terms, having an amplitude equal to the desired difference frequency term.
[5] Direct current and sum and difference frequency outputs are also available from higher even-ordered terms in Eq. (3.1). Their strengths, relative to those produced by the second-order nonlinearity, are dependent on the details of the nonlinearity and the strengths of the signals, but their effects are essentially the same as described here.

3.1.4.2 Balanced Mixers. Since the rectification term cannot be filtered, it must be reduced by balancing. The process is illustrated in Fig. 3.7, which shows a single-balanced mixer, (SBM). The difference is taken between the outputs of two identical single-diode mixers. Each has the same two input signals except that the phase of one of them has been inverted at one diode relative to the other. Since the rectification term in Eq. (3.5) is not affected by the signs of A or B, it will be the same for both diodes and thus will cancel in the subtraction. The desired term is affected, however, being proportional to the product of A and B and therefore changing signs with either of them. Thus, the desired difference-frequency term from each diode will have an opposite sign, and subtraction of the two outputs will cause the desired terms to add.

A more practical realization of the SBM is shown in Fig. 3.8. The results are similar, with e_s in Fig. 3.8 injected into the two diodes in phase (positive polarity to the anodes), acting as v_1 in Fig. 3.7, and e_r being injected out of phase (to the anode of one and the cathode of the other), acting as v_2. Resistor R_2 takes the difference of the diode voltages (averaging the anode voltage of one with the cathode voltage of the other). A still more practical realization is shown in Fig. 3.9 where radio frequency (RF) transformers are use to accomplish the same summations.

It is common, in frequency mixing applications, to operate with a large, local oscillator (LO) input and a weak "signal" input. In that case the LO is often sufficiently strong that the diodes may be considered to be switches that change between conduction and nonconduction as the LO signal reverses polarity. Thus, in the circuit of Fig. 3.9, if e_s were the strong input, the diodes could be considered to connect and disconnect the output load to e_r under control of e_s. With e_s driving current through the diodes, assuming a well-balanced circuit, the center tap of the transformer would be at the same voltage as the center point between diodes, thus passing e_r to the output. When the polarity of e_s would reverse, the back-biased diodes would disconnect the load. Thus e_r would be passed to the output only half of the time, with the exact part of its waveform that is transmitted depending on the phase relationship between e_r and e_s. This is illustrated in Fig. 3.10, where e_r

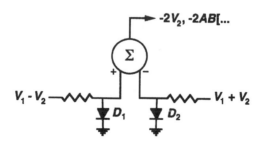

Fig. 3.7 Balanced mixer concept.

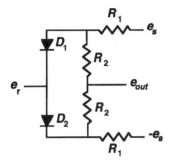

Fig. 3.8 Possible balanced mixer realization.

is multiplied by 1, starting at a phase α and ending one-half cycle later; α would here be the phase difference between the two signals. After low passing to eliminate the high-frequency outputs from the PD, the DC or low-frequency component would remain. This would be proportional to the time average over a cycle:

$$V_{\text{avg.}} = \frac{1}{2\pi} \int_{\alpha}^{\alpha+\pi} \sin \theta \, d\theta = \frac{1}{2\pi} [\cos \alpha - \cos(\alpha + \pi)] = \frac{1}{\pi} \cos \alpha. \quad (3.6)$$

Again we see the cosinusoidal output as a function of the phase difference between the two signals. Similar arguments could be made if e_r were stronger.

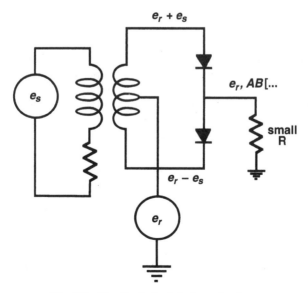

Fig. 3.9 Practical singly balanced mixer.

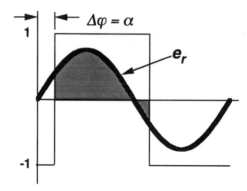

Fig. 3.10 Averaging of mixer output. The shaded area is used for a singly balanced mixer. In a doubly balanced mixer, the area under the rest of the sinusoid is subtracted from this.

While a large difference between signal strengths is common for frequency mixers, this is not the best operating condition for PDs. This is because the desired output is proportional to AB while the highly undesired rectification component is proportional to $(A^2 + B^2)$. As one input drops much below the other, the desired output will fall proportionally to this dropping input, but the undesired component will become approximately proportional to the square of the unchanging stronger input and will not drop appreciably. Therefore, it is optimum for a PD to have the two inputs approximately of equal strength ($A \approx B$).

While the SBM balances out the e_s term, the e_r term is not balanced out and appears quite strongly in the output. To balance it also, doubly balanced mixers (DBMs) are common. A DBM is shown conceptually in Fig. 3.11 and schematically in Fig. 3.12. It is necessary that the BM selected for a PD application have a DC output, like that shown in Fig. 3.12. Usually one port will be DC connected and, while interchanging ports may be acceptable in some frequency mixing applications, only the DC-connected port can be chosen as the output port for a PD.

If one signal is much stronger than the other, it can be thought of as switching the polarity of the weaker signal as it is seen at the output. This is apparent from Fig. 3.12; the strong signal will cause current to flow through one half of the diode bridge at a time, depending on its polarity. Since each half of the bridge is connected to a different polarity of the weaker signal, the voltage delivered at the output will be proportional to the weaker signal multiplied by the polarity of the stronger. We could easily repeat Eq. (3.6) for this case and come up with a similar result. If the weaker signal is a square wave, rather than a sinusoid, as assumed until now, the DBM can be seen to perform an ExOR function; when the two signals have the same polarity

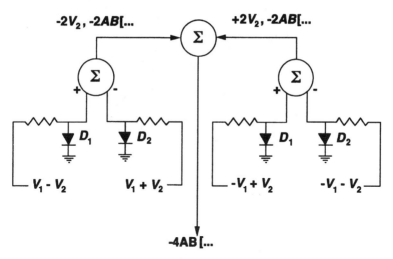

Fig. 3.11   Doubly balanced mixer.

(state), one output occurs, and when they have different polarities, the other occurs. When the two input signals are sinusoidal but have similar amplitudes, the ExOR's triangular characteristic is also approached, [Blanchard, 1976, p. 14], as illustrated in Fig. 3.13.

The sinusoidal shape is applicable under the usual frequency mixing conditions where one signal (the LO) is at some relatively high level appropriate to the particular mixer design and the other signal is relatively weak. Under such conditions K_p is relatively independent of the strength of the stronger signal but is directly proportional to the amplitude of the weaker signal. The relative independence of K_p from the stronger signal is related to the higher order terms in Eq. (3.1).

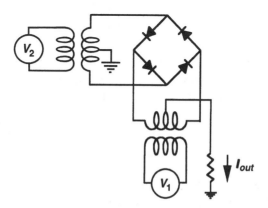

Fig. 3.12   Practical doubly balanced mixer.

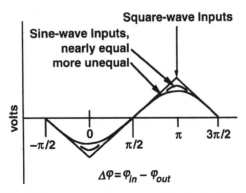

Fig. 3.13   PD responses for various drive levels, DB mixer.

3.1.4.3 Analog Multipliers. The desired term in Eq. (3.5) is the result of multiplication of the two input sinusoids. An analog multiplier can perform this function without producing the additional undesired components that are created when the even-order terms in the expansion of the diode characteristic are used. Such multipliers are available in integrated circuit (IC) form [Gray and Meyer, 1977, pp. 561–575]. Their maximum operating frequency is, however, limited, whereas that of the diode mixer is practically unlimited.

Analog multipliers generally make use of the diode current–voltage characteristic. The diode current can be expressed as

$$I = I_s(e^{v/V_T} - 1). \tag{3.7}$$

For $I \gg I_s$ this implies

$$\ln \frac{I}{I_s} \approx \frac{v}{V_T}. \tag{3.8}$$

As an example of how this equation might be used for multiplication, consider a circuit that sums the voltage across two diodes, each with a different current. Then the total voltage can be written

$$v = v_1 + v_2 = V_T \left[\ln \frac{I_1}{I_{s1}} + \ln \frac{I_2}{I_{s2}} \right] = V_T \left[\ln \frac{I_1 I_2}{I_{s1} I_{s2}} \right]. \tag{3.9}$$

Note that an expression has been obtained that can contain the product of two signals represented by the currents I_1 and I_2, and v can then be linearized by various means. For example, a current might be passed throughout a third diode and controlled in such a manner that the diode voltage would equal v. Then the current could be written, from Eqs. (3.7) and (3.9), as

$$\frac{I}{I_s} = \left(\frac{I_1}{I_{1s}} \frac{I_2}{I_{2s}} - 1 \right),$$

(3.10)

thus providing a signal, in the form of the current I, that contains the product of two other signals, I_1 and I_2.

3.1.4.4 *Integrated Circuit Doubly Balanced Mixer.*

To understand the IC BM we begin with the bipolar transistor differential pair in Fig. 3.14. For small excursions of the base voltage, the collector current I_1 can be expressed as

$$I_1 = \frac{I_0}{2} + \frac{v_1}{r_e} = \frac{I_0}{2} \left(1 + \frac{v_1}{V_T} \right),$$

(3.11)

where r_e is the differential emitter resistance, obtained by differentiating I with respect to v in Eq. (3.7), $r_e = V_T/I$. The objectives will be to make one signal proportional to v_1, which is easily done, and the other proportional to I_0, a little more complicated, and to eliminate all but the product term. The details of how this is accomplished are given in Appendix 3.A. Here we will note that a circuit such as shown in Fig. 3.14 can be used to generate the current I_0 for a second such circuit.

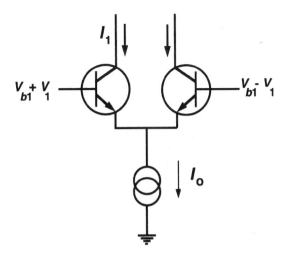

Fig. 3.14   Basic circuit in the IC doubly balanced mixer.

3.2 VOLTAGE-CONTROLLED OSCILLATOR (VCO)

The voltage-controlled oscillator (VCO) is an essential part of every PLL. While we will not study oscillator design [Manassewitsch, 1976, pp. 370–398] in this text, we will here consider the various types of oscillators that are useful in PLLs and discuss their properties.

The VCO may, in fact, be a current-controlled oscillator, an ICO. There is no difference to the extent that the response of the loop is concerned. At low frequencies a current-controlled astable multivibrator may be used to produce square waves at a frequency that is proportional to the control current over a wide range. Such a circuit is illustrated in Fig. 3.15.

When a negative transition occurs at the collector of Q_1, it is coupled to the base of Q_2, which it turns off. The cross-coupling capacitor at the base of Q_1 then recharges through the controlled current source Q_3. When it recharges to the point where the base of Q_1 becomes forward biased, the collector of Q_1 moves negative, causing the base of Q_2 to go negative and its collector to go positive. This reinforces the positive-going waveform at the

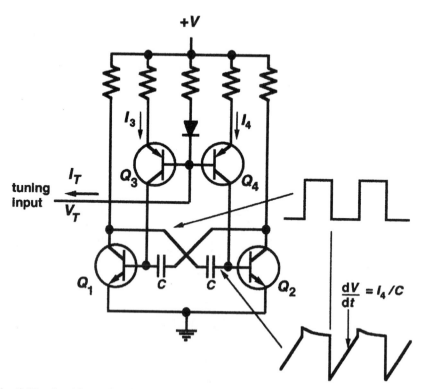

Fig. 3.15 Astable multivibrator. Auxiliary circuitry may be needed to ensure that both transistors do not saturate and cause the oscillator to lock up.

base of Q_1. Thus a regenerative state change occurs. The time between such transitions is the time required for the base voltage to rise to a diode drop above ground. The rate of change is proportional to the charging current, $I_c = I_3 = I_4$, and the time required for a transition is therefore $T = \Delta V/I_c$. Thus the frequency, $f = 2/T$, is directly proportional to I_c, which can be controlled either by a voltage or a current at the tuning control input, but it will be somewhat more linear relative to a tuning current, I_T. Since both I_T and I_c pass through a resistor and a diode with a common voltage across them, they tend to be proportional, but the tuning voltage is less so because of the base-to-emitter voltage of Q_3 and Q_4.

While the circuit is conceptually simple and good linearity of frequency versus control current can be obtained, it has relatively poor spectral purity. That is, its phase (and therefore its frequency) is modulated by noise, producing a broader spectral line. In addition, this simple version has a potential startup problem—if both of the cross-coupled transistors should be saturated at the same time, they will remain so—which may require additional circuitry to solve.

A second form of ICO, used at microwave frequencies, is the YIG-tuned oscillator. Like the astable, this has an inherently linear tuning characteristic and can easily cover more than an octave of frequency range. In addition, the tunable resonator, a YIG (yttrium iron garnet) sphere, has a very high Q. High oscillator Q is necessary for good spectral purity, and this requires high component Q. The resonant frequency of the sphere is proportional to the strength of the magnetic field in which it is immersed, and that is in turn proportional to the control current. Disadvantages are that the magnetic circuitry is relatively bulky and susceptible to stray magnetic fields. Isolation in the form of bulky shielding and physical separation from potential interfering fields may be necessary to prevent frequency modulation of the oscillator.

Solid-state oscillators can be tuned by control of bias (supply) voltages that alter active-device capacitances or by changing bias on a varactor, a diode designed for use of its voltage-variable junction capacitance, incorporated into the tuning circuit. The varactor-tuned oscillator is the most common VCO. It sometimes has a quite nonlinear tuning characteristic, especially when tuned over a wide frequency range. A linearizer circuit can be used in series with the VCO if linearity is important. Hyperabrupt junction varactors can give a much more linear tuning curve than ordinary varactors, but they tend to have lower Q.

A representative VCO package and tuning curve are shown in Figs. 3.16 and 3.17. Figure 3.18 shows a representation of a VCO circuit. The output is fed back through R_{FB} to the shunt resonator "tank" circuit consisting of L_0, C_0, and C_{TUNE} (assuming C_{BLOCK} is large enough to be ignored as a frequency-determining element). A tap on the inductor L_0 (or sometimes a capacitive divider) sends a portion of the resonator voltage to the active element G by which it is amplified to provide the output, thus completing the loop. Oscillation occurs at the resonant frequency of the tank; there the

Fig. 3.16 VCO package. This TO-8 can is 0.27 in. high by 0.45 in. diameter. (Courtesy of VARI-L Company, Inc.)

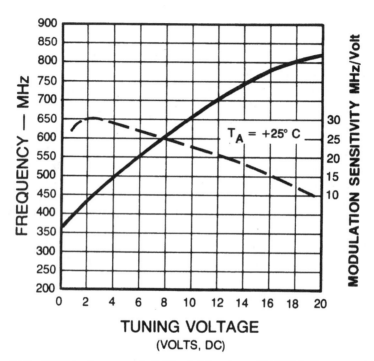

Fig. 3.17 VCO tuning curve (solid line). (Courtesy of VARI-L Company, Inc.)

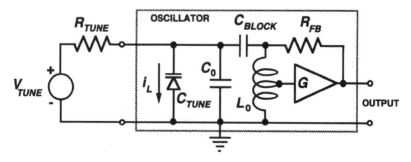

Fig. 3.18 VCO circuit representation.

voltage across the tank, and at the tap, is in phase with the output voltage. The gain G compensates for the loss in the feedback circuit and the inductor tap ratio to give unity gain around the loop with zero phase shift at the resonant frequency. Changing the tuning voltage V_{TUNE} changes the depletion capacitance of the varactor diode C_{TUNE}, thus modifying the resonant frequency of the tank circuit and the frequency of oscillation.

Some impedance is needed at R_{TUNE} to prevent the tank circuit from being shorted, but, if it is too large (e.g., ≥ 1 kΩ; experiment or consult the manufacturer), the noisy leakage current i_L from the varactor diode junction will produce a voltage across R_{TUNE} that will modulate the oscillator frequency and can thereby produce significant additional phase noise.

Other methods of oscillator tuning also exist, particularly in microwave tubes, klystrons, magnetrons, and backward wave oscillators (BWOs). Where the required range is only a few hundred parts per million or less, a voltage-controlled crystal oscillator (VCXO) can be used. This provides many of the advantages of the crystal oscillator, high Q and long-term frequency stability, although to a degree that is lessened by the voltage control. Voltage control is achieved by modifying the resonant frequency, which is primarily determined by the crystal, by varying the capacitance of a varactor that is also part of the resonant circuit.

Other high-Q circuits also exist, dielectric resonator and cavity oscillators at microwave frequencies and surface-accoustic-wave (SAW) resonator oscillators in the region between these and the highest crystal oscillator frequencies (≈ 100 MHz). These oscillators, like the VCXO, are usually tuned by perturbing a high-Q resonator with a varactor, and, like the VCXO, their tuning range is therefore limited; however, not, on a relative basis, to the extent of the VCXO.

3.3 LOOP FILTER

We have considered the operation of the loop with a simple amplifier as the loop filter. Now we will consider the filters that give us more flexibility in controlling the characteristics of the PLL. These will be of two types, passive and active.

3.3.1 Passive Loop Filter

A general form of the passive filter that we will discuss is shown in Fig. 3.19. Its transfer function is

$$F(s) = \frac{u_2(s)}{u_1(s)} = \frac{R_2 + 1/(Cs)}{R_1 + R_2 + 1/(Cs)} = \frac{1 + R_2Cs}{1 + (R_1 + R_2)Cs} \quad (3.12)$$

$$= \frac{1 + \tau_2 s}{1 + \tau_1 s} = \frac{1 + s/\omega_z}{1 + s/\omega_p} \quad (3.13)$$

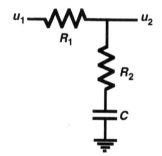

Fig. 3.19   Passive lag-lead loop filter.

where

$$\tau_1 \equiv 1/\omega_p = (R_1 + R_2)C \tag{3.14}$$

and

$$\tau_2 \equiv 1/\omega_z = R_2 C. \tag{3.15}$$

The gain and phase of $F(\omega)$ are shown in Fig. 3.20.

If $R_2 \Rightarrow 0$, this becomes a low-pass filter (Fig. 3.21) with transfer function

$$F(s) = \frac{1}{1 + R_1 Cs} = \frac{1}{1 + s/\omega_p}, \tag{3.16}$$

and response as shown in Fig. 3.22.

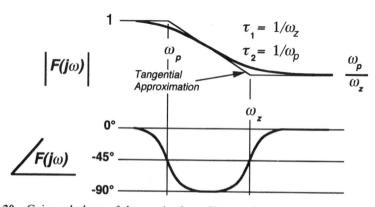

Fig. 3.20   Gain and phase of the passive loop filter vs. frequency, with tangential gain approximation shown.

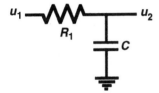

Fig. 3.21   Low-pass loop filter.

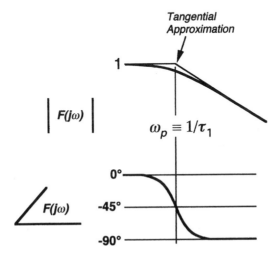

Fig. 3.22   Gain and phase of the low-pass loop filter vs. frequency, with tangential gain approximation shown.

3.3.2   Active Loop Filter

The active loop filter can also implement Eq. (3.13), but, in addition to possibly providing gain, a wider range of values for ω_p and ω_z can be more easily obtained. We will look at the general equations that describe the active loop filter, then study various special cases to illustrate its properties, and finally consider the full implementation of Eq. (3.13). But first we will briefly consider the nature of the typical operational amplifier (op-amp) since the performance of the active filter depends on it.

3.3.2.1   The Op-Amp.   The traditional op-amp is a high-gain, high-input-impedance amplifier intended for use in circuits in which part of the output signal is fed back to the input (as in Fig. 3.23a).

At low frequencies the gain is very high, often more than 100 dB. If it were uncompensated, its transfer gain and phase would be as shown in Fig. 3.24, curves 1 and 2. Unfortunately, it would tend to oscillate when feedback was applied because the gain would be too high when the multiple, unavoidable, poles within its circuitry produced 180° excess phase shift (beyond f_x). To

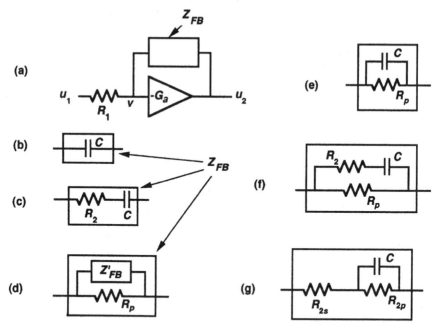

Fig. 3.23 Active loop filter: (*a*) with generic feedback, (*b*) capacitive feedback for integrator, (*c*) series *RC* feedback for integrator-and-lead, (*d*) resistor in parallel with generic feedback for DC gain control, (*e*) shunt *RC* feedback for low pass, (*f*) series-*RC*-in-parallel-with-*R*-feedback for lag-lead, (*g*) shunt-*RC*-in-series-with-*R*-feedback for lag-lead.

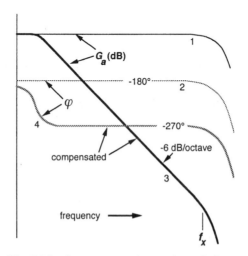

Fig. 3.24 Op-amp open-loop gain and phase.

control this, a single-pole roll-off is incorporated within the op-amp, beginning at perhaps a few hertz and producing the gain and phase shown in curves 3 and 4. Thus the gain can be reduced to a tolerable level while maintaining phase margin relative to 360°.

Besides the inverting input shown at v in Fig. 3.23a, op-amps have a noninverting input. The output depends on the difference between the two inputs. For now we assume zero at the noninverting input.

A current feedback op-amp differs in that it is a transimpedance amplifier, producing an output voltage proportional to the current into the inverting ($-$) input [Franco, 1989; Little, 1990]. (The noninverting input is not shown in Fig. 3.23.) The input impedance into the inverting input is low, but that point is virtual ground in the configurations that we will study anyway. We will begin with the traditional op-amp.

3.3.2.2   General Equations, Voltage Feedback. The active filter is shown in Fig. 3.23a in generic form with the feedback impedance Z_{FB} as yet unspecified. We will begin by assuming only that the amplifier has ideal infinite input impedance and develop the equations for that rather general case. We will then proceed to simplify the equations by making various other assumptions that apply in many practical situations. While the simpler equations will commonly be most useful, we will thus be aware of the modifications that will be necessary under conditions where they do not apply.

The output voltage can be expressed in terms of the op-amp input voltage as

$$u_2 = -G_a v. \tag{3.17}$$

Since the same current flows through all the passive components, the voltage drops are proportional to their impedances:

$$\frac{u_2 - v}{v - u_1} = \frac{Z_{FB}}{R_1}. \tag{3.18}$$

From these last two equations we can eliminate v to obtain

$$\frac{u_2(1 + 1/G_a)}{-u_2/G_a - u_1} \equiv -\frac{u_2(1 + G_a)}{u_2 + G_a u_1} = \frac{Z_{FB}}{R_1}, \tag{3.19}$$

$$\frac{u_2}{u_2 + G_a u_1} = -\frac{Z_{FB}}{(1 + G_a)R_1}. \tag{3.20}$$

From this we now obtain the ratio u_2/u_1:

$$u_2 = -\frac{Z_{FB}}{(1 + G_a)R_1}u_2 - \frac{G_a}{1 + G_a}\frac{Z_{FB}}{R_1}u_1, \quad (3.21)$$

$$\left(1 + G_a + \frac{Z_{FB}}{R_1}\right)u_2 = -G_a\frac{Z_{FB}}{R_1}u_1, \quad (3.22)$$

$$-K_{LF}F(s) \equiv \frac{u_2(s)}{u_1(s)} = \frac{G_aF_d}{1 + G_a - F_d}, \quad (3.23)$$

where F_d is the desired response,

$$F_d \triangleq -\frac{Z_{FB}(s)}{R_1}. \quad (3.24)$$

The minus sign in (3.23) implies that part of the filter transfer function $K_{LF}F(s)$ is an inversion somewhere else in the loop. This detail allows $K_{LF}F(s)$ to represent both active and passive filters.

Equation (3.23) can be rearranged as

$$\frac{G_a}{-K_{LF}F(s)} = \frac{1 + G_a}{F_d} - 1 \quad (3.25)$$

to show that, if the amplifier's gain is much greater than the magnitude of the desired transfer function, then

$$\frac{G_a}{-K_{LF}F(s)} \approx \frac{1 + G_a}{F_d} \quad (3.26)$$

so that

$$-K_{LF}F(s)|_{|1+G_a| \gg |F_d|} \approx \frac{G_a}{1 + G_a}F_d. \quad (3.27)$$

If, also, $|G_a| \gg 1$, then

$$-K_{LF}F(s)|_{|1+G_a| \gg |F_d|, |G_a| \gg 1} \approx F_d. \quad (3.28)$$

When $(1 + G_a)$ drops well below F_d, Eq. (3.25) shows that

$$-K_{LF}F(s)_{|1+G_a| \ll |F_d|} \approx -G_a. \quad (3.29)$$

That is, the filter transfer function becomes the op-amp's open-loop gain. We can see the transition in Fig. 3.25.

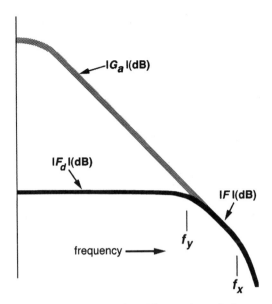

Fig. 3.25   Op-amp closed-loop gain and phase.

3.3.2.3   *General Equations, Current Feedback.* The current feedback op-amp is a more recent version of the traditional op-amp. Refer again to Fig. 3.23*a*. In this case $v = 0$ (it has the same DC value as the noninverting input—the input appears to be shorted). The output voltage equals the current out of the inverting port multiplied by the op-amp's transimpedance, Z_{21}. Thus we can write the current out of the inverting port as

$$\frac{u_2}{Z_{21}} = I_- = -\frac{u_1}{R_1} - \frac{u_2}{Z_{FB}}, \qquad (3.30)$$

which we solve as

$$-K_{LF}F(s) \triangleq \frac{u_2}{u_1} = -\frac{1}{R_1(1/Z_{21} + 1/Z_{FB})} = F_d\frac{1}{1 + 1/G_a}, \qquad (3.31)$$

where $G_a = Z_{21}/Z_{FB}$. This has the same form as Eq. (3.27), but an advantage claimed for these amplifiers is that the gain can be changed by varying R_1 without affecting the bandwidth. In the previous type of op-amp, the input resistor R_1 forms part of a voltage divider that is the feedback circuit [Eq. (3.18)]. The closed-loop bandwidth (Fig. 3.25) depends on the gain, which depends on R_1. In this type, the effective short at the input isolates the feedback from R_1. If $|G_a| \gg 1$, then $-K_{LF}F(s) \approx F_d$ and if $|G_a| \ll 1$, then $K_{LF}F(s) \approx -Z_{21}/R_1$ so, much as with voltage feedback, the desired transfer function F_d is obtained until the frequency increases to the point where the gain drops too much, and then the closed-loop transfer function becomes equal to the open-loop transfer function.

3.3.2.4 *High-Frequency Poles.* The filter will normally be designed using Eq. (3.28), but the transition from (3.28) to (3.29) at f_y in Fig. 3.25 represents an additional pole in the transfer function. Likewise, additional poles in various parts of the op-amp circuitry cause phase shift to accumulate, often quite rapidly once a critical frequency is reached, as shown at f_x. These frequencies must be high enough compared to the bandwidth of the PLL that they do not have a significant detrimental effect.

3.3.2.5 *Integrator.* The first particular implementation will be one that cannot be obtained with a passive filter. The circuit of Fig. 3.23b has

$$Z_{FB} = \frac{1}{Cs} \tag{3.32}$$

so that Eq. (3.23) becomes

$$K_{LF}F(s) = \frac{G_a}{1 + s/\omega_p}, \tag{3.33}$$

where

$$1/\omega_p \equiv \tau_1 = (1 + G_a)R_1C. \tag{3.34}$$

This is a low-pass filter, but, as G_a becomes large, it approaches

$$K_{LF}F(s)|_{|s| \gg \omega_p} \approx \frac{1}{R_1Cs}, \tag{3.35}$$

and the circuit is therefore often called an integrator, although a true integrator has infinite gain at zero frequency ($\omega_p \Rightarrow 0$). Note that the difference between active and passive filter here is really relative. Neither can be a true integrator, but it is easier to come closer with an active filter. While $K_{LF}F(s)$ is defined by Eq. (3.35), the DC gain, K_{LF}, approaches infinity:

$$K_{LF|integrator} \rightarrow \infty. \tag{3.36}$$

Equation (3.35) can also be obtained directly from Eq. (3.28).

3.3.2.6 *Reducing the Size of C.* One benefit that an active filter can provide is to reduce the size of the capacitor in very narrow bandwidth applications where a very low frequency low-pass filter may be required. By using an active low-pass filter, represented by Eqs. (3.33) and (3.34), in place of a passive low-pass filter, represented by Eq. (3.16), the same value of τ_1 can be achieved with a capacitor of value $C' = \tau_1/[(1 + G_a)R_1]$ rather than $C'' = \tau_1/R_1$. Thus, for the same corner frequency ω_p, C' can be smaller than C'' by the very large number $G_a + 1$. If it would be necessary to reestablish the same gain, the active filter would be followed by an attenuator with gain $1/|G_a|$. A phase reversal is also added. In many applications these changes are easily accommodated.

3.3.2.7 *Integrator-and-Lead Filter.*

The integrator will revert to an amplifier at high frequencies if a resistor R_2 is placed in series with C. This type of filter (Fig. 3.23c) has

$$Z_{\text{FB}} = R_2 + \frac{1}{Cs} = \frac{1 + R_2 Cs}{Cs} \tag{3.37}$$

so that Eq. (3.28) becomes

$$K_{\text{LF}}F(s) = \frac{1}{R_1 C}\frac{1 + s/\omega_z}{s} \tag{3.38}$$

where Eq. (3.36) applies and

$$1/\omega_z \equiv \tau_2 = R_2 C. \tag{3.39}$$

This is an integrator-and-lead filter, the lead being produced by the zero at $\omega_z = 1/\tau_2$, the same as given for the passive filter by Eq. (3.15). The response is shown in Fig. 3.26. Of course, as with the integrator, there is really a pole, given approximately by Eq. (3.34), so Fig. 3.20 actually applies with $\omega_p \Rightarrow 0$. The high-frequency gain is $K_{\text{LF}}F(\omega \gg \omega_z) = 1/(R_1 C\omega_z)$ and the low-frequency gain is $K_{\text{LF}}F(\omega \ll \omega_z) = -j/(R_1 C\omega)$.

3.3.2.8 *Control of Zero-Frequency Gain.*

It sometimes is important to control the zero-frequency (DC) gain of the op-amp, G_a, which is not ordinarily a controllable parameter. To do this, a resistor can be added across the op-amp (Fig. 3.23d), giving

$$Z_{\text{FB}} = \frac{1}{1/R_p + 1/Z'_{\text{FB}}} = \frac{R_p Z'_{\text{FB}}}{R_p + Z'_{\text{FB}}}. \tag{3.40}$$

Equation (3.28) gives

$$K_{\text{LF}}F(s)|_{|G_a| \gg |F_d|, 1} \approx \frac{(R_p/R_1)Z'_{\text{FB}}}{(R_p/R_1)R_1 + Z'_{\text{FB}}}, \tag{3.41}$$

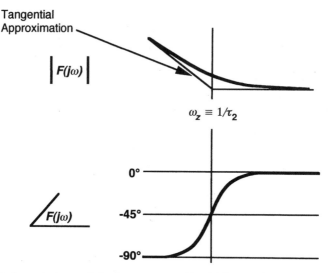

Fig. 3.26 Gain and phase of the integrator-and-lead filter vs. frequency with tangential gain approximation.

which, for $|G_a| \gg 1$, is the same as Eq. (3.23) but with

$$Z_{FB} \Rightarrow Z'_{FB} \tag{3.42}$$

and

$$G_a \Rightarrow R_p/R_1. \tag{3.43}$$

In other words, the presence of R_p just causes the effective op-amp gain to be reduced from $|G_a| \gg |F_d|$ to R_p/R_1.

3.3.2.9 Lag (Low-Pass) Filter.
We can obtain the transfer function for an active low-pass filter, whose feedback network is shown in Fig. 3.23e, by a simple extension of an equation that we have already developed. Comparing Fig. 3.23e to Fig. 3.23d, we see that $Z'_{FB} = 1/(Cs)$, so Eq. (3.41) gives

$$K_{LF} F(s)|_{|G_a| \gg |F_d|, 1} \approx K_{LF} \frac{1}{1 + s/\omega_p}, \tag{3.44}$$

where

$$K_{LF} = R_p/R_1 \tag{3.45}$$

and

$$1/\omega_p = \tau_1 = R_p C. \tag{3.46}$$

The resulting low-pass response is shown in Fig. 3.22.

3.3.2.10 Lag-Lead Filter. We can obtain the lag-lead filter by inserting resistance in series with the capacitor of Fig. 3.23e, using either Fig. 3.23f or 3.23g. The former is equivalent to the substitution of Fig. 3.23c for Z'_{FB} in Fig. 3.23d and can be analyzed as in the last section but with $Z'_{FB}(s) = R_2 + 1/(Cs)$ in Eq. (3.41). The result is

$$K_{LF}F(s)|_{|G_a| \gg |F_d|,1} \approx K_{LF}\frac{1 + s/\omega_z}{1 + s/\omega_p}, \tag{3.47}$$

where K_{LF} is given by Eq. (3.45) and the corners are at

$$1/\omega_p \equiv \tau_1 = (R_2 + R_p)C \tag{3.48}$$

and

$$1/\omega_z \equiv \tau_2 = R_2C. \tag{3.49}$$

For Fig. 3.23g, substitution of

$$Z = R_{2s} + \frac{1}{Cs + 1/R_{2p}} \tag{3.50}$$

into Eq. (3.28) gives Eq. (3.47) again but with

$$R_p = R_{2s} + R_{2p} \tag{3.51}$$

in Eq. (3.45),

$$1/\omega_p \equiv \tau_1 = R_{2p}C, \tag{3.52}$$

$$\frac{1}{\omega_z} \equiv \tau_2 = \frac{R_{2s}R_{2p}}{R_{2s} + R_{2p}}C. \tag{3.53}$$

Note that if $R_2 = 0$ in Eqs. (3.48) and (3.49) or $R_{2s} = 0$ in Eq. (3.53), then Eq. (3.47) becomes the same as Eq. (3.44), representing a low-pass filter. Also note that if $R_p \Rightarrow \infty$ in Eq. (3.48) or $R_{2p} \Rightarrow \infty$ in Eqs. (3.51)–(3.53), then Eq. (3.47) becomes the same as Eq. (3.38), representing an integrator-and-lead.

3.3.2.11 *Note on the Form of the Filter Equations.* Note that Eq. (3.47) can be written in another form:

$$K_{\mathrm{LF}} F(s) = K_{\mathrm{LF}} \frac{1 + s/\omega_z}{1 + s/\omega_p} = \left(K_{\mathrm{LF}} \frac{\omega_p}{\omega_z} \right) \frac{s + \omega_z}{s + \omega_p}. \tag{3.54}$$

It is important to recognize that the constant term on the right side above is not K_{LF}. If $F(s)$ is written in the form on the right, the constant there is likely to be mistaken for K_{LF}. Then equations, which we will derive, that contain K_{LF}, or K, will yield incorrect answers. The appropriate form of $F(s)$, except in the limiting cases of an integrator with or without lead, gives $F(0) = 1$.

3.3.2.12 *Filter Stability.* While the filter characteristics are important to the performance and stability of the PLL, the stability of the active filter itself is also important. Stability considerations for the filter loop are similar to those for the PLL, so they may be more easily understood after loop stability has been studied in Chapter 5. Nevertheless, the material is presented here because it is an essential part of loop filter design.

The open-loop G of the active-filter loop (through the op-amp and back through Z_{FB}) with a voltage feedback op-amp can be obtained from Eqs. (3.17) and (3.18). If a voltage $v = v'$ is applied at the input, by (3.17) it will cause an output

$$u_2 = -G_a v'. \tag{3.55}$$

By (3.18), with $u_1 = 0$, the value of $v = v''$ resulting from this value of u_2 is given by

$$\frac{-G_a v' - v''}{v''} = \frac{Z_{\mathrm{FB}}}{R_1}, \tag{3.56}$$

from which can be obtained the open-loop transfer function,

$$\frac{v''}{v'} = -\frac{G_a}{Z_{\mathrm{FB}}/R_1 + 1} = \frac{G_a}{F_d - 1}. \tag{3.57}$$

Unity open-loop gain occurs when the numerator and denominator have the same magnitude. For a general lag-lead filter and $|F_d| \gg 1$, this corresponds to point x in Fig. 3.27. The phase of $-F_d$ will be approximately $0°$ in this flat-gain region so the open-loop phase will be the phase of $-G_a$, and this should be more than $-360°$ for stability (much more if the filter is to be well behaved). Otherwise the open-loop gain will be greater than 1 when the phase shift reaches $-360°$, which will very likely cause instability.

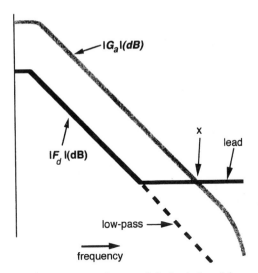

Fig. 3.27 Op-amp open-loop and desired closed-loop gains.

If the closed-loop filter characteristic is low-pass, as shown by the dashed line, then the denominator of Eq. (3.57) will approach -1 as F_d becomes small compared to one. Then the open-loop G becomes equal to $-G_a$. Many op-amps are designed to be stable with unity feedback, so circuits using such op-amps will be stable.

The open-loop G of the current feedback op-amp can be obtained beginning with an input current I_- that produces an output voltage u''_2,

$$u''_2 = Z_{21}I_- = Z_{21}u'_2/Z_{FB}. \tag{3.58}$$

where I_- is derived from the output voltage u'_2 through the feedback impedance Z_{FB}. The open-loop transfer function is thus

$$\frac{u''_2}{u'_2} = -\frac{Z_{21}}{Z_{FB}}. \tag{3.59}$$

This differs from Eq. (3.57) in that the denominator lacks a minimum value of 1. Therefore, if a lag filter (Fig. 3.23e) is used with a current feedback op-amp, there is a danger of instability because F_d and G_a can intersect in the region where G_a has developed considerable excess phase shift. Physically, the impedance of the feedback capacitor continues to decrease as frequency increases, so the output voltage is converted to ever-increasing current, which maintains the open-loop gain as G_a falls. If a lead configuration is used (Fig. 3.23c), if ω_z is much less than the frequency where G_a acquires a phase of $-360°$, and if R_p is in the recommended range of feedback resistances for the op-amp, the circuit should be stable.

Example 3.1 Active Filter Requirement: Integrator-and-lead filter with $f_p < 10$ Hz, $f_z = 5000$ Hz, $K_{LF} F(f \gg f_z) = 20$.

A solution: Use Fig. 3.23c. Choose a convenient value of $R_2 = 10$ kΩ. From Eq. (3.39),

$$C = \frac{1}{R_2 \omega_z} = \frac{1}{(10^4 \text{ V/A})(2\pi 5 \times 10^3/\text{sec})} = 3.18 \times 10^{-9} \text{C/V} \approx 3300 \text{ pF}.$$

High-frequency gain is R_2/R_1 so $R_1 = 500$ Ω to give 20. The op-amp must have more than the desired gain at 10 Hz if the pole is to be lower than 10 Hz. Using Eq. (3.38), the value of $k = 1/(R_1 C)$ can be obtained from the high-frequency gain, and the gain at 10 Hz can then be obtained using k.

$$|K_{LF} F(f \gg 5000 \text{ Hz})| = 20 \approx \frac{k}{\omega_z} = \frac{k}{2\pi \, 5000 \text{ rad/sec}};$$

$$k = 2\pi \times 10^5 \text{ rad/sec}$$

$$|K_{LF} F(f = 10 \text{ Hz})| \approx \left|\frac{k}{s}\right| = \frac{2\pi \times 10^5 \text{ rad/sec}}{2\pi \times 10 \text{ rad/sec}} = 10^4 \Rightarrow 80 \text{ dB}.$$

The DC gain of the op-amp must exceed 80 dB, therefore. For stability of the filter, its open-loop phase shift must be $\gg -360°$ when its open-loop gain is 20 (Fig. 3.27). In addition, the frequency at which that gain is 20 must be high enough that the additional pole at that frequency will not be of importance to the loop, certainly much higher than 5000 Hz.

3.3.2.13 *Noninverting Input.* We can extend our development of active filter performance to the noninverting case. The op-amp responds to the difference between the inverting ($-$) and noninverting ($+$) inputs. We now allow a voltage v_+ to be present on the $+$ input to the op-amp. We can treat this as a change in voltage reference from ground to v_+ and write the input and output voltages relative to this new reference. Equation (3.23) then becomes

$$\frac{u_2 - v_+}{u_1 - v_+} = \frac{G_a F_d}{1 + G_a - F_d} \triangleq -H_a, \tag{3.60}$$

which gives

$$u_2 = v_+(1 + H_a) - u_1 H_a. \tag{3.61}$$

Thus the response to a signal on the op-amp's + input equals unity minus the response from u_1.

How might we obtain a differential response, the same response from two inputs except that one is the negative of the other? From (3.61) we can see that such a response can be obtained from a voltage u_1' if it is related to v_+ by

$$v_+ = \frac{H_a}{1 + H_a} u_1'. \tag{3.62}$$

Thus we want to introduce u_1' to the op-amp by means of a voltage divider that gives Eq. (3.62). However, H_a is a function of frequency. Nevertheless, for large G_a, Eq. (3.60) becomes

$$H_a \approx -F_d \equiv \frac{Z_{FB}(s)}{R_1}, \tag{3.63}$$

which describes the voltage divider that we actually use and with which Eq. (3.62) becomes

$$v_+ = -u_1' \frac{F_d}{1 - F_d}. \tag{3.64}$$

The divider is illustrated in Fig. 3.28. Equation (3.61) now becomes

$$u_2 = -u_1' F_d \frac{1 + H_a}{1 - F_d} - u_1 H_a, \tag{3.65}$$

which gives the desired results as long as G_a is large enough so $H_a \approx -F_d$.

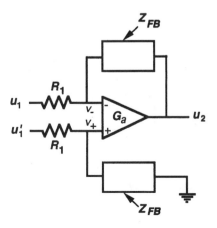

Fig. 3.28 Op-amp using both inputs.

Under the assumed conditions, the common mode response is zero. That is, if $u_1 = u'_1, u_2 = 0$. To the degree that the impedances (R_1 and Z_{FB}) on the two sides of the op-amp are not equal, however, the circuit will have a nonzero common-mode response (Endres and Kirkpatrick, 1992).

3.3.2.14 *Impractical Loop Filters.* The circuits shown in Fig. 3.29 have

serious problems for use as loop filters. While consideration of the pure integrator (Fig. 3.29a) was of academic value, the integrator is not a practical loop filter because there is no frequency at which it does not have a transfer phase of $-90°$. As we will see in the following chapters, this would inevitably put the loop on the verge of instability.

The transfer function of Eq. (3.47) could be implemented by the configuration shown in Fig. 3.29b if the op-amp could be prevented from "hitting the rails" (i.e., for the output to reach an extreme value set by the power supply voltages). However, it should come as no surprise that a DC connection is necessary to maintain steady-state conditions. Assuming that phase lock is somehow established, the op-amp's input bias current or board leakage I_b will charge C_1 while the loop causes the circuit input voltage (PD output) to change in order to maintain zero input voltage at the op-amp. Eventually the circuit input voltage will run out of range and will no longer be able to compensate. At this point the loop will be out of lock. The leakage current will then charge C_2 until the op-amp output hits a rail. The slope of the voltage ramp at the filter input, I_b/C_1, can vary widely depending on components employed, but it would be an unusual application that would cause us to be satisfied with a loop with such a built-in failure mechanism, especially where the alternative discussed in Section 3.3.2.10 is available.

3.3.3 Filters Driven by Current Sources

Ideally, R_1 in Fig. 3.23a transforms u_1 into a current that multiplies Z_{FB} to produce $-u_2$ and the transfer function given by Eq. (3.24). Alternately, this

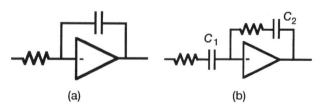

(a)   (b)

Fig. 3.29   Loop filters to be avoided: (a) pure integrator and (b) filter with capacitor in the input circuit.

current can be provided by a current source, or a circuit that acts like a current source, that is part of the phase detector as in the charge pump PDs in Section 3.1.3. These PDs can drive the active loop filter circuits we have considered (R_x in Fig. 3.5b would become R_1 in Fig. 3.23a), but here we consider them as driving Z_{FB} directly. Then K_p has units like amperes/radian and the filter transfer function is Z_{FB}, with units of ohms. In either case, $K_p K_{LF} F(s)$ has units like volts per radian.

The charge pump in Fig. 3.5a can provide a high output impedance, especially in its usual switched-off state. When it drives a capacitor (Fig. 3.23b), a near integrator will be produced. The finite impedance introduced in parallel with the capacitor by finite impedances of the current source and other loading will cause the pole to be nonzero, just as finite gain does when the capacitor is a feedback element with an op-amp (recall Section 3.3.2.5). However, a better approximation to an integrator—a lower pole frequency— may be available with the current source. This can be true also of the PD configuration in Fig. 3.5b. However, that circuit can complicate analysis of the loop filter, since its impedance changes when the switch closes or opens, introducing a nonlinear resistance into the filter. Sometimes the impact of that complexity is reduced by connecting the charge pump immediately to a part of the filter that does not influence the critical poles (e.g., a very high frequency pole). Often a low-pass filter is placed immediately after the charge pump to keep its often short but large pulses from momentarily overdriving a following op-amp.

Charge-pump PDs have been know to have nonlinearities in a small region near zero phase. Since many loops drive the phase error to zero, even a tiny region of zero or excessive gain can cause problems there. Sometimes the phase error is purposely offset by the introduction of leakage current or a narrow pulse for which the PD must compensate by operating at a slight offset from zero phase.

3.3.4 Capacitors in Loop Filters

There are two deviations from the ideal that we should be aware of in choosing capacitors for use in loop filters. The first is leakage current, which can cause a resistor to effectively appear in parallel with the capacitor, especially where electrolytic capacitors (e.g., tantalum) are used in order to obtain large values in a small size and especially at higher temperatures. This could effectively turn the integrator-and-lead filter employing the feedback circuit of Fig. 3.23c into a lag-lead filter with the feedback circuit shown in Fig. 3.23g.

The second is dielectric absorption [Pease, 1982; Buchanan, 1983], which can cause the voltage across a capacitor to continue to change after current has stopped flowing through it. High dielectric constant ceramic capacitors (e.g., CK05, CK06) especially exhibit this phenomenon. It can be a problem if the PLL is required to settle rapidly.

The need for large capacitance values in a small package tends to aggravate both problems. Plastic capacitors should be considered as alternatives where necessary.

3.A APPENDIX: INTEGRATED CIRCUIT DOUBLY BALANCED MIXER—DETAILS

To understand the IC BM we begin with the bipolar transistor differential pair in Fig. 3.14. For small excursions of the base voltage, the collector current I_1 can be expressed as

$$I_1 = \frac{I_0}{2} + \frac{v_1}{r_e} = \frac{I_0}{2}\left(1 + \frac{v_1}{V_T}\right), \tag{3.11}$$

where r_e is the differential emitter resistance, obtained by differentiating I with respect to v in Eq. (3.7), $r_e = V_T/I$. The objectives will be to make one signal proportional to v_1, which is easily done, and the other proportional to I_0, a little more difficult, and to eliminate all but the product term. To this end we generate I_0 for the differential pair in which the multiplication will occur by creating it in another differential pair. The current so generated, I_1 in Fig. 3.14, is proportional to v_1 and becomes the total emitter current for the second pair, as shown in Fig. 3.A.1. The collector current I_{21} can then be written

$$I_{21} = \frac{I_1}{2}\left(1 + \frac{v_2}{V_T}\right) = \frac{I_0}{4}\left(1 + \frac{v_1}{V_T}\right)\left(1 + \frac{v_2}{V_T}\right) \tag{3.A.1}$$

$$= \frac{I_0}{4}\left[1 + \frac{v_1}{V_T} + \frac{v_2}{V_T} + \left(\frac{v_1}{V_T}\right)\left(\frac{v_2}{V_T}\right)\right]. \tag{3.A.2}$$

Note that I_{22} could be represented by a similar equation except that the terms involving v_2 would be negative. To eliminate the undesired terms, we generate I_{22} and two more currents like I_{21} but with terms having different signs, as shown in Fig. 3.A.2. Table 3.A.1 shows the signs of the various terms as they appear in the various currents. From the table can be seen that the differential output voltage contains only the desired product.

While the preceding has been a small-signal analysis, it is not difficult to see how a large signal into the lower stage could alternately turn off the left and right upper stages, thus inverting the polarity of the part of the output current that is proportional to v_2. This is similar to what happens in the diode DBM. Similarly, with both v_1 and v_2 large, the circuit could act like an

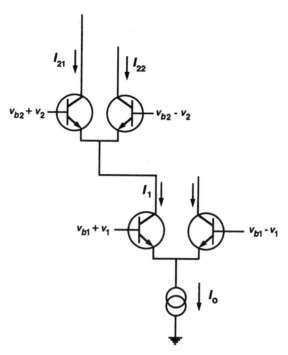

Fig. 3.A.1   Two basic circuits interconnected for the IC doubly balanced mixer.

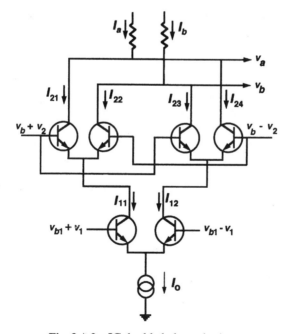

Fig. 3.A.2   IC doubly balanced mixer.

TABLE 3.A.1 Signs of the Currents and Voltages in Fig. 3.A.2 Showing Cancellation of Undesired Components and Reinforcement of Desired Components

	$\dfrac{I_0}{4}\Big[$	1	$\dfrac{v_1}{V_T}$	$\dfrac{v_2}{V_T}$	$\left(\dfrac{v_1}{V_T}\right)\left(\dfrac{v_2}{V_T}\right)\Big]$
I_{21}		$+$	$+$	$+$	$+$
I_{22}		$+$	$+$	$-$	$-$
I_{23}		$+$	$-$	$+$	$-$
I_{24}		$+$	$-$	$-$	$+$
$I_a = I_{21} + I_{24}$		$+2$	0	0	$+2$
$I_b = I_{22} + I_{23}$		$+2$	0	0	-2
$v_b - v_a$		0	0	0	$+4$

ExOR gate, as was true for the diode circuit. Consider the relationship between the collector currents I_{11} and I_{12} and the input voltage v_1 under the large signal condition, $v_1, v_2 \gg V_T$.

$$I_{11} = (I_0/2)(1 + \text{sign } v_1); \qquad I_{12} = (I_0/2)(1 - \text{sign } v_1). \quad (3.A.3)$$

The currents are either equal to I_0 or to zero, depending on v_1. Let us call the condition where they equal I_0 the 1 state and the other condition the 0 state. Then we can represent the state of the currents I_{11} and I_{12} by their "logic" states, $L(I_{11})$ and $L(I_{12})$. Let us also define the logic state of v_1, $L(v_1)$, to be 1 when the sign is positive and 0 when it is negative. Then Eq. (3.A.3) can be written in shorthand notation as

$$L(I_{11}) = L(v_1); \qquad L(I_{12}) = \overline{L(v_1)}. \quad (3.A.4)$$

This says that if v_1 is positive (large being understood), I_{11} is on and I_{12} is off. Continuing in the same manner, we can describe the state of the other current as a function of the large input voltages:

$$L(I_{21}) = L(I_{11}) \cdot L(v_2) = L(v_1) \cdot L(v_2), \quad (3.A.5)$$

$$L(I_{22}) = L(I_{11}) \cdot \overline{L(v_2)} = L(v_1) \cdot \overline{L(v_2)}, \quad (3.A.6)$$

$$L(I_{23}) = \overline{L(I_{11})} \cdot L(v_2) = \overline{L(v_1)} \cdot L(v_2), \quad (3.A.7)$$

$$L(I_{24}) = \overline{L(I_{11})} \cdot \overline{L(v_2)} = \overline{L(v_1)} \cdot \overline{L(v_2)}, \quad (3.A.8)$$

$$L(I_a) = L(I_{21}) + L(I_{24}) = \overline{L(v_1) \oplus L(v_2)}, \quad (3.A.9)$$

$$L(I_a) = L(I_{22}) + L(I_{23}) = L(v_1) \oplus L(v_2), \quad (3.A.10)$$

where $+$ and $\oplus$ represent logical OR and ExOR, respectively. The algebraic sum of currents at I_a and I_b is equivalent to an OR function because, for the assumed large signals, only one of the constituent currents I_{2i} can be on at a time. Thus we see that both the IC- and diode-type DBMs act like ExOR circuits under large signal conditions, although the latter requires square-wave inputs to give a true triangular characteristic.

PROBLEMS

3.1 In Fig. 3.2b, V_{p-p} = 1.5 V. What is K_p (include units—always)?

3.2 In Fig. 3.5a, I = 3 mA.
  (a) What is K_p?
  (b) In Fig. 3.5b, V_x = 0 and V = 2 V. What is R_x to give the same K_p as in 3.5a?
  (c) If V_x changes to 2 V, what change should be made in Fig. 3.5b to maintain the same PD characteristic?

3.3 In Fig. 3.19, R_1 = 1 kΩ, R_2 = 100 Ω, C = 1 μF.
  (a) Write $F(s)$ for this filter.
  (b) What are its gain and phase shift at 500 Hz?

3.4 In Fig. 3.23, R_1 = 1 kΩ.
  (a) What are the values of R_2 and C at Fig. 3.23c to give a high-frequency gain of 5 and a 1-Hz gain of 100?
  (b) What value would R_p have at Fig. 3.23d to limit the gain to 1000?
  (c) What are the values of the components at Fig. 3.23g to give a pole at 20 Hz, a zero at 100 Hz, and a DC gain of 100?

3.5 In Fig. 3.21, R_1 = 10 kΩ and C = 1000 pF.
  (a) What configuration for Z_{FB} in Fig. 3.23 will give the same response (excepting an inversion) and what will be the component values if R_1 = 1 kΩ in Fig. 3.23?
  (b) What passive configuration will give the same response as in Fig. 3.21 when driven by the circuit of Fig. 3.5a? What will be the capacitor value if the resistor is 10 kΩ?
  (c) What will be I in Fig. 3.5a for $K_p K_{LF}$ = 0.2 V/rad?

CHAPTER 4

LOOP RESPONSE

In this chapter we will develop the equations that describe both the frequency and transient responses for loops that use the filters studied in Chapter 3 (Fig. 4.1).

4.1 LOOP ORDER AND TYPE

The order of the loop equals the number of poles in the open-loop transfer function $G(s)$ [Klapper and Frankle, 1972, pp. 85–86]. This is also the highest power of s in its denominator. Its type is the number of such poles that are at the origin ($s = 0$). Thus the simple loop described in Chapter 2 by (2.24) is a first-order, type 1, loop. Because of this pole at the origin and an additional pole contained in the loop filter, the loops that we will study here, and throughout the remainder of the book, are second order. This restriction will permit us to make use of the long and well-developed theory of second-order systems.

The pole due to conversion from frequency to phase is truly the only pole at the origin so the phase-locked loop (PLL) is strictly always a type 1 loop. However, the active filter can have a pole at a very low frequency, which is often approximated as being at the origin, giving the loop then two such poles and thereby making it a type 2 loop. A primary difference between the two types lies in the nature of the phase error. In response to a frequency step, the steady-state phase error of a type 1 loop will change but that of a type 2 loop will not—after the transient settles the phase error will have returned to zero. (If such an error could exist in a type 2 loop, the integrator in the loop filter would continue to change its output and steady state would therefore

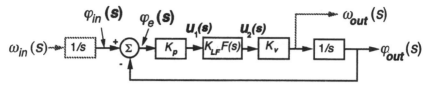

Fig. 4.1   Mathematical block diagram of a loop with a filter. This is the same as Fig. 2.4 except for the filter transfer function $F(s)$.

not exist.) When following a frequency ramp, a type 2 loop has a steady phase error whereas a type 1 loop has an ever increasing phase error.

While PLLs may actually be of an order higher than second, performance estimates can often be made based on similar second-order loops. (See examples in Sections 6.12 and 7.10.) In practical circuits, multiple poles exist at high frequencies where the gains of the various elements begin to fall. Fortunately, these may be of little importance in determining performance in the range of interest. The phase transfer function has been given by

$$\frac{\varphi_{\text{out}}(s)}{\varphi_{\text{in}}(s)} = \frac{G(s)}{1 + G(s)}. \tag{2.25a}$$

At low frequencies, where $G(\omega)$ is very large, this function is approximately equal to one, regardless of the details of $G(\omega)$. At high frequencies, where $G(\omega)$ is small, the function is approximately equal to $G(\omega)$, as if there were no feedback. Thus, frequency response can be easily predicted there, and transient response, which is affected by the various frequencies of which the transient wave is composed, tends to be affected relatively little because of the low amplitude of $G(\omega)$. Only in the transition region, near the frequency where $G(\omega) \approx 1$, is the gross performance in question. Therefore, if $G(\omega)$ is approximately the same in that region as for some second-order loop, we expect the gross performance of the two loops to be the same. The two will differ in some fine details, for example, how long is required to settle to a very small fraction of a step change or the value of a small error in following some low-frequency signal that is being followed very closely, but gross performance will be similar.

Similar statements can be made concerning the error response, Eq. (2.26a), except it is approximately one at high frequencies and approximately equal to $1/G(\omega)$ at low frequencies

4.2   CLOSED-LOOP EQUATIONS

The equations describing the loop will be the same as Eqs. (2.25) and (2.26) except that the gain K is replaced by $KF(s)$ to account for the filter.

Therefore the open-loop transfer function is

$$G(s) = \frac{KF(s)}{s} \tag{4.1}$$

$$= \frac{K}{s}\left(\frac{1 + s/\omega_z}{1 + s/\omega_p}\right), \tag{4.2}$$

and the closed-loop transfer function is

$$H(s) \triangleq \frac{\varphi_{\text{out}}(s)}{\varphi_{\text{in}}(s)} = \frac{G(s)}{1 + G(s)} = \frac{1}{1 + s/[KF(s)]} \tag{4.3}$$

or

$$1 - H(s) = \frac{\varphi_e(s)}{\varphi_{\text{in}}(s)} = \frac{1}{1 + G(s)} = \frac{s}{s + KF(s)}. \tag{4.4}$$

Substituting Eq. (4.2) into (4.3), we obtain

$$H(s) = \left[1 + \frac{s}{K}\left(\frac{1 + s/\omega_p}{1 + s/\omega_z}\right)\right]^{-1} = K\omega_p \frac{1 + s/\omega_z}{s^2 + s\omega_p(1 + K/\omega_z) + K\omega_p} \tag{4.5}$$

The standard second-order equation form for the denominator is

$$s^2 + 2\zeta\omega_n s + \omega_n^2 = s^2 + \omega_p(1 + K/\omega_z)s + K\omega_p \tag{4.6}$$

where ζ is the damping factor and ω_n is the natural frequency.
From this it is apparent that

$$\omega_n^2 = K\omega_p \tag{4.7}$$

and

$$2\zeta\omega_n = \omega_p\left(1 + \frac{K}{\omega_z}\right) = \omega_p + \frac{\omega_n^2}{\omega_z}, \tag{4.8}$$

$$\zeta = \frac{1}{2}\left(\frac{\omega_p}{\omega_n} + \frac{\omega_n}{\omega_z}\right). \tag{4.9}$$

In terms of Eqs. (4.7) and (4.9), Eq. (4.5) can be written

$$H(s) = \omega_n^2 \frac{1 + s/\omega_z}{s^2 + 2\zeta\omega_n s + \omega_n^2}.$$

(4.10)

The loop error response can also be written from Eq. (4.4) as

$$1 - H(s) = \frac{s[s + \omega_n(2\zeta - \omega_n/\omega_z)]}{s^2 + 2\zeta\omega_n s + \omega_n^2} = \frac{s(s + \omega_p)}{s^2 + 2\zeta\omega_n s + \omega_n^2},$$

(4.11)

where Eq. (4.9) was used in the last simplification.

4.3   OPEN-LOOP EQUATIONS—LAG-LEAD FILTER

Now that we have used the closed-loop response to establish the values of the traditional second-order parameters in terms of the parameters of our loop, we will determine how the open-loop transfer function can be interpreted in these terms. Equations (4.1) and (4.2) at $s = j\omega$ become

$$G(\omega) = \frac{K}{j\omega} F(j\omega) = \frac{K}{j\omega} \left(\frac{1 + j\omega/\omega_z}{1 + j\omega/\omega_p} \right).$$

(4.12)

The straight-line approximation of the magnitude of the filter response $F(\omega)$ and of the open-loop transfer function from Eq. (4.12) $[G(\omega)]$ are plotted in Fig. 4.2 on logarithmic scales.[1] The open-loop gain has the same shape as the filter response except for the additional $1/\omega$ term. The filter gain is flat to $\omega = \omega_p$, at which point it begins to fall linearly with ω. After ω reaches ω_z, the two frequency-dependent terms dominate and the filter response becomes approximately ω_p/ω_z. When $\omega = 1$, the filter gain is K_{LF} and the open-loop gain is K. This assumes that ω_p and ω_z are greater than 1; otherwise K and K_{LF} are along the straight-line extension of the plots from the lowest frequencies. Since the extension of the lowest frequency portion of the open-loop gain curve falls proportionally to frequency, the fact that it intersects $\omega = 1$ rad/sec at $|G(\omega)| = K$ sec implies that it also intersects $|G(\omega)| = 1$ at $\omega = K$. This can be verified by setting the part of Eq. (4.12) that dominates at low frequencies, $|K/\omega|$, equal to unity and solving for ω. Note that the steep part of the open-loop gain curve crosses unity gain at ω_n. This is because, by Eq. (4.7), ω_n is the geometric mean of K and ω_p and

[1]For compactness, we will use the notation $Q(\omega)$ to represent $Q(s)$ with $s \Rightarrow j\omega$, although $Q(j\omega)$ would show the relationship more exactly.

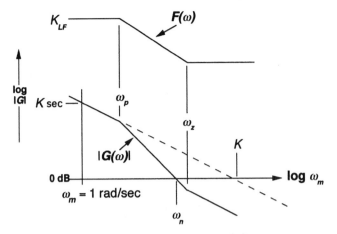

Fig. 4.2 Gain of the open-loop transfer function $G(\omega)$ and of the corresponding lag-lead filter $F(\omega)$. Tangential approximations are shown. Values of $|G|$ and ω are marked on the axes although distances are logarithms of $|G|$ and ω in rad/sec.

therefore is midway between them on a logarithmic plot. In other words, from geometry,

$$\log(\omega_n) - \log(\omega_p) = \log(K) - \log(\omega_n), \qquad (4.13)$$

$$2\log(\omega_n) = \log(K) + \log(\omega_p), \qquad (4.14)$$

$$\omega_n^2 = K\omega_p. \qquad (4.7)$$

If this -12 dB/octave portion of the open-loop gain curve does not cross the unity gain level, the same geometric arguments show that ω_n occurs at the intersection of the extension of this region, as illustrated in Fig. 4.3.

We can also relate the damping factor ζ to the open-loop gain curve. The first term in (4.9) dominates the second if

$$\frac{\omega_p}{\omega_n} \gg \frac{\omega_n}{\omega_z} \Rightarrow \omega_n \ll \sqrt{\omega_p \omega_z} \equiv \omega_{\text{mid}}. \qquad (4.15a)$$

In other words, the natural frequency is significantly to the left of the midpoint ω_{mid} between ω_p and ω_n in the log plot of Fig. 4.3, as in curve c. Conversely, if the natural frequency is significantly to the right of the midpoint, as in curve a, the second term dominates (4.9).

For the former case (curve c) we have

$$\zeta\big|_{\omega_n \ll \omega_{\text{mid}}} \approx \frac{1}{2}\left(\frac{\omega_p}{\omega_n}\right), \qquad (4.15b)$$

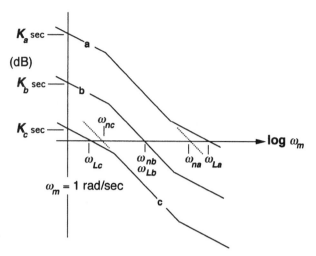

Fig. 4.3 Gain vs. modulation frequency ω_m for several values of K; ω_L is the frequency at which $|G(\omega)|$ is one.

whereas in the latter case (curve a) we have

$$\zeta\,|_{\omega_n \gg \omega_{\text{mid}}} \approx \frac{1}{2}\left(\frac{\omega_n}{\omega_z}\right).$$
(4.16)

Writing these last two equations in logarithmic form, we obtain

$$\log\bigl(2\zeta\,|_{\omega_n \ll \omega_{\text{mid}}}\bigr) = \log(\omega_p) - \log(\omega_n)$$
(4.17)

and

$$\log\bigl(2\zeta\,|_{\omega_n \gg \omega_{\text{mid}}}\bigr) = \log(\omega_n) - \log(\omega_z).$$
(4.18)

This is illustrated in Fig. 4.4. From the graph can be seen that, when a critical frequency, either a pole or a zero, is at unity gain, the damping factor is 0.5 (the logarithm of 2ζ is zero) approximately. The accuracy of the approximation requires significant separation between ω_p and ω_z. Otherwise the inequality in (4.17) or (4.18) will not hold with the critical frequency at unity gain and ζ will be larger than 0.5 there. Assuming these inequalities, as the unity-gain frequency ω_L moves into the steep region between ω_p and ω_z, the logarithm becomes negative (arrow pointing toward the left), corresponding to a fractional value of 2ζ, that is, $\zeta < 0.5$. Along this steep slope, once the frequency is far removed from both ω_p and ω_z, the total phase shift in the open-loop transfer function approaches $-180°$, the value for instability. In this region the loop is underdamped. As the unity-gain frequency ω_L in-

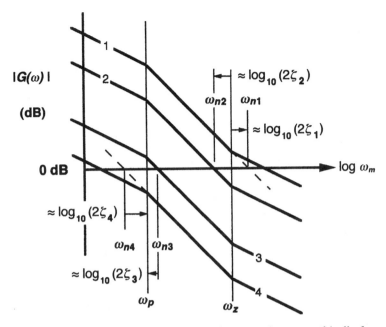

Fig. 4.4 Natural frequency ω_n and damping factor ζ shown graphically for several values of gain factor K. The graphical representation of ζ is accurate when $\omega_z \gg \omega_p$ and unity gain occurs far from ω_{mid}.

creases beyond ω_z, or decreases below ω_p, ζ increases until it equals one. At $\zeta = 1$, the loop is said to be critically damped. At higher or lower values, respectively, of ω_L it is overdamped.

4.4 LOOP WITH A LAG FILTER

If the zero frequency goes to infinity, then the open-loop transfer function, Eq. (4.2), becomes

$$G(s) = \frac{KF(s)}{s} = \frac{K}{s}\left(\frac{1}{1 + s/\omega_p}\right), \tag{4.19}$$

the Bode plot is as shown in Fig. 4.5, the filter is represented by Eq. (3.16), and the closed-loop transfer function, Eq. (4.10), becomes

$$H(s) = \omega_n^2 \frac{1}{s^2 + 2\zeta\omega_n s + \omega_n^2}, \tag{4.20}$$

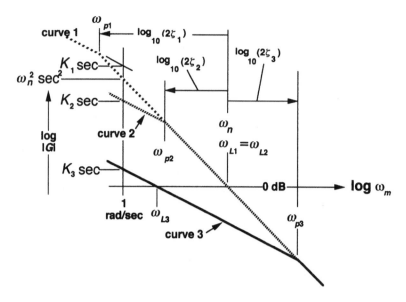

Fig. 4.5 Gain of the open-loop transfer function $G(\omega)$, loop with a lag filter, at three values of K.

where ω_n is still given by Eq. (4.7) and Eq. (4.9) becomes Eq. (4.15b) exactly. The loop error is obtained from Eq. (4.11) as

$$1 - H(s) = \frac{s(s + 2\zeta\omega_n)}{s^2 + 2\zeta\omega_n s + \omega_n^2} = \frac{s(s + \omega_p)}{s^2 + 2\zeta\omega_n s + \omega_n^2}. \tag{4.21}$$

4.5 LOOP WITH AN INTEGRATOR-AND-LEAD FILTER

If the pole moves toward zero frequency, Eq. (4.2) becomes, for frequencies well beyond ω_p,

$$G(s) \approx \frac{K\omega_p}{s^2}\left(1 + \frac{s}{\omega_z}\right). \tag{4.22}$$

This form illustrates that an integrator can only be an approximation—there really is a pole back there somewhere.[2] However, once we have lost track of, or interest in, the position of that pole, we should be able to express $K\omega_p$ as a single constant. Therefore, we use Eq. (4.7) to write G in a form that does

---

[2]Blanchard (1976) uses this form. It has the advantage that the expression for ω_n remains the same for the various filter configurations.

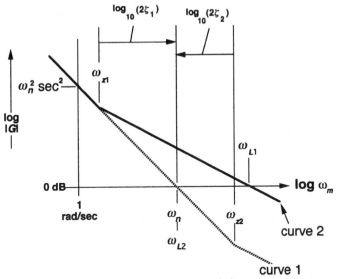

Fig. 4.6 Gain of the open-loop transfer function $G(\omega)$, loop with an integrator-and-lead filter, at two values of ω_z.

not explicitly involve ω_p,

$$G(s) \approx \frac{\omega_n^2}{s^2}\left(1 + \frac{s}{\omega_z}\right). \tag{4.23}$$

This is represented by Fig. 4.6. By substituting $K_p K_v K_{LF}$ for K in Eq. (4.1) and then substituting $(1 + s/\omega_z)/(R_1 Cs)$ for $K_{LF} F(s)$ in that result [according to Eq. (3.38)] and comparing the result to Eq. (4.23), we see that

$$\omega_n^2 = \frac{K_p K_v}{R_1 C}. \tag{4.24}$$

Equation (4.9) becomes Eq. (4.16) exactly. Equation (4.10) remains a valid expression for $H(s)$ and Eq. (4.11) becomes

$$1 - H(s) = \frac{s^2}{s^2 + 2\zeta\omega_n s + \omega_n^2}. \tag{4.25}$$

4.6   SUMMARY OF EQUATIONS

The equations for the second-order loop parameters and closed-loop responses that were developed above are summarized in Table 4.1. Figure 4.7 shows the corresponding Bode gain plots.

TABLE 4.1. Summary of Second-Order Loop Equations

Filter Type	(a) Lag	(b) Lag-Lead	(c) Integrator and Lead
[a]$\alpha =$	0	$0 < \alpha < 1$	1
$F(s) =$	$\dfrac{1}{1 + s/\omega_p}$	$\dfrac{1 + s/\omega_z}{1 + s/\omega_p}$	$\dfrac{1 + s/\omega_z}{R_1 C s}$
$\zeta =$	$\dfrac{1}{2}\left(\dfrac{\omega_p}{\omega_n}\right)$ Eq. (4.15b)	$\dfrac{1}{2}\left(\dfrac{\omega_p}{\omega_n} + \dfrac{\omega_n}{\omega_z}\right)$ Eq. (4.9)	$\dfrac{1}{2}\left(\dfrac{\omega_n}{\omega_z}\right)$ Eq. (4.16)
$\omega_n^2 =$	$K\omega_p$ Eq. (4.7)	$K\omega_p$ Eq. (4.7)	$\dfrac{K_p K_v}{R_1 C}$ Eq. (4.24)
$H(s) =$	$\omega_n^2 \dfrac{1}{s^2 + 2\zeta\omega_n s + \omega_n^2}$ Eq. (4.20)	$\omega_n^2 \dfrac{1 + s/\omega_z}{s^2 + 2\zeta\omega_n s + \omega_n^2}$ Eq. (4.10)	$\omega_n^2 \dfrac{1 + s/\omega_z}{s^2 + 2\zeta\omega_n s + \omega_n^2}$ Eq. (4.10)
$1 - H(s) =$	$\dfrac{s(s + \omega_p)}{s^2 + 2\zeta\omega_n s + \omega_n^2}$ Eq. (4.11)	$\dfrac{s(s + \omega_p)}{s^2 + 2\zeta\omega_n s + \omega_n^2}$ Eq. (4.11)	$\dfrac{s^2}{s^2 + 2\zeta\omega_n s + \omega_n^2}$ Eq. (4.25)

[a]α will be defined by Eq. (6.5)

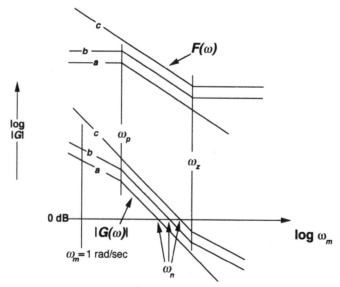

Fig. 4.7 Gain of the open-loop transfer function $G(\omega)$ and of the corresponding loop filter $F(\omega)$ for: (a) lag, (b) lag-lead, and (c) integrator-and-lead filter. Tangential approximations are shown.

PROBLEMS

4.1 (a) What are the damping factor ζ and natural frequency ω_n for a loop having an open-loop gain of 1000 at $\omega = 1$ rad/sec and a lag-lead filter with $\omega_p = 10$ rad/sec and $\omega_z = 200$ rad/sec?

(b) What are ζ and ω_n for such a loop without a zero (lag filter)?

(c) What are ζ and ω_n for the loop in (a) except that it has no nonzero filter pole (integrator-and-lead filter)?

(d) What are ζ and ω_n for the loop in (a) if the same value of $K\omega_p$ is maintained while $\omega_p \to 0$?

4.2 (a) Write the phase transfer function $\varphi_{out}(\omega)/\varphi_{in}(\omega)$ for a closed loop with a lag-lead filter as shown in Fig. 3.19 and with the following parameters: $R_1 = 2$ kΩ; $R_2 = 200$ Ω; $C = 10$ μF; $K_p = 0.5$ V/rad; $K_v = 10^4$ (rad/sec)/V.

(b) If a phase modulation of peak deviation 0.01 rad appears at the input with a modulation frequency of 50 Hz, what will be the peak deviation of the output phase at 50 Hz?

(c) What if the modulation frequency increases to 5 kHz?

(d) If the input is modulated with 10-Hz peak frequency deviation at a 50-Hz rate, what will be the output peak frequency deviation?

CHAPTER 5

LOOP STABILITY

The shapes of the closed-loop frequency and transient responses indicate the stability of the loop, but during design, stability is assessed from the open-loop characteristics.

5.1 OBSERVING THE OPEN-LOOP RESPONSE

In general open-loop response is observed by injecting a signal at some point in a loop and observing the response around the loop to the same point. At least conceptually, the loop is broken so that the observed response can be differentiated from the injected excitation. This presents a problem for loops of type 1 and above because they can have no steady-state "position" error. Any deviation of the integrator input from its steady-state value will cause it to integrate. For example, in the simplest phase-locked loop (PLL) (Fig. 2.7), which is type 1, if the loop is broken at the voltage-controlled oscillator (VCO) input, there will always be some error in the injected direct current (DC) voltage relative to the steady-state value that would be there with the loop closed. As a result, the VCO frequency will be different from the reference frequency by some amount ($\omega_e \neq 0$), causing the phase difference (φ_e) to change with time. This will eventually drive some component out of its linear range, usually so quickly that testing is not practical.

Practically, the open-loop response must be ascertained through closed-loop testing. Figure 5.1 illustrates this. Here a signal is injected at the VCO input, and the response is observed at the output of a low-impedance amplifier, possibly the op-amp used in the loop filter. If R_2/R_1 is large

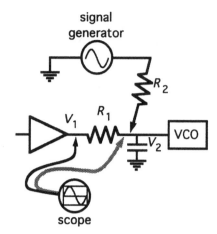

Fig. 5.1 Measuring open-loop response with the loop closed.

enough, R_2 will not appreciably affect the loop response, and if R_1 is large enough compared to the amplifier output impedance, the injected signal will not appreciably affect v_1. The ratio v_1/v_2 is not quite the open-loop response, however. The transfer function of the low-pass must be accounted for. This can be done by multiplying v_1/v_2 by the theoretical response of the low-pass, but often its cutoff frequency will be high enough to have little effect.

Type 2 loops are even more difficult to test. Not even the loop filter alone can be tested open loop because any error at that point will cause the output of the integrator eventually to saturate.

Responses involving the integrating filter can be measured by reducing the op-amp's gain, by placing a resistor in the feedback path (R_p in Fig. 3.23d), to the point where the DC level at its input u_1 can be controlled well enough to keep it from saturating. Of course, the proper DC level must be applied along with the test signal to do this.

5.2 METHODS OF STABILITY ANALYSIS AND MEASURES OF STABILITY

We will consider three commonly used methods—Bode plot, Nyquist diagram, and Evans (root locus) plot—as they apply to the PLL. Throughout the text we will concentrate on the Bode plot, which seems to be the most useful for PLL design, but at this point we will compare it to the other two. This should be helpful to those who are familiar with either of the other methods. See the References for more complete treatments.

The most common measures of stability are phase margin and gain margin. Both unity gain and 180° *excess* phase shift (i.e., more negative than the −180° that is built in with negative feedback) are necessary for oscillation so the margins relative to this pair of conditions become a measure of stability. Phase margin is the additional open-loop phase shift necessary to give 180° excess phase when the open-loop gain is unity. All PLLs have $-90°$ phase shift due to the $1/s$ term, so excess phase is 90° even before any effect from the loop filter is considered. Gain margin is the additional gain necessary to give unity open-loop gain when the open-loop excess phase is 180°. While a loop will not achieve steady-state oscillation with a gain that exceeds unity at 180°, it may enter a condition wherein oscillations grow until average gain is reduced to unity due to saturation of components or the loop may lose lock due to excessive excursions of the phase error. Whether a loop will oscillate when its gain exceeds unity at 180° is more easily seen from the Nyquist plot or from the root locus plot than from the Bode plot. However, in most cases it will not operate properly under these conditions, and the Bode plot is adequate.

5.2.1   Bode Plot

Figure 5.2 illustrates a Bode plot. Open-loop gain and phase are plotted against modulation frequency, and the gain and phase margins can be seen in the figure. Tangential approximations for gain and phase are often used; we will use the tangential approximation for gain only, computing phase as accurately as necessary. When more accuracy is required exact calculations can be made, but much can be learned by the tangential approximations.

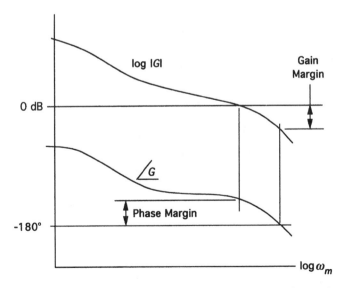

Fig. 5.2 Bode plot, open-loop gain on a log scale and open-loop phase plotted against modulation frequency on a log scale.

5.2.2 Nyquist Plot

Figure 5.3 illustrates a Nyquist plot [Kuo, 1987, pp. 556–590]. Normally, the shaded lines are not used, the information relative to the whole plot being easily discerned from the half shown in solid lines. However, we will consider the whole plot at this point.

To make the plot, the value of s is varied along the path shown at Fig. 5.3a. The object is to determine if the closed-loop transfer function $H(s)$ in Eq. (4.3) has any poles with positive real parts, corresponding to rising exponentials and instability. These would occur when the denominator $G(s) + 1$ would have one or more zeros with positive real parts, corresponding to a time response factor

$$e^{st} = e^{(j\omega + \sigma)t} = e^{j\omega t}e^{\sigma t}, \qquad \sigma > 0. \tag{5.1}$$

Assume that the open-loop function $G(s)$ is itself everywhere finite (stable) so that $G(s)$ [and so $(1 + G(s))$] has no poles in the right-half plane. The locus in the s plane is drawn in Fig. 5.3a in such a manner as to surround the nonallowed area, the right-half plane, which contains values of $s = \sigma + j\omega$ that have positive values of σ. As s traverses this locus, the corresponding value of $G(s)$ is plotted in Fig. 5.3b. If the $G(s)$ curve surrounds the point -1, where the value of $H(s)$ becomes infinite, n times, then the s-plane locus surrounds n zeros. But these are then in the right-half plane, so the

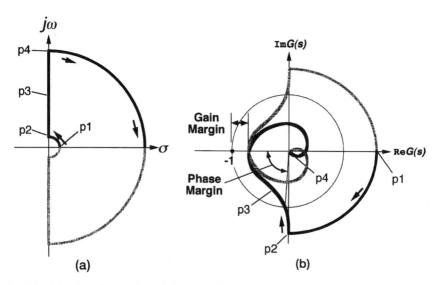

Fig. 5.3 Nyquist plot, a plot of the complex open-loop transfer function as the locus in the s plane encloses the right half of the plane. The shaded portion is a mirror image of the other half and is often not shown explicitly: (a) s plane and (b) $G(s)$ plane.

pole(s) of $H(s)$ [at $G(s) = -1$] would occur where σ, the real part of s, is positive, corresponding to a rising exponential and instability.

Let us study the illustration to see why it represents a PLL and why this diagram indicates stability. To avoid the allowable pole at zero, the locus in the s plane begins at an infinitesimal radius ($p1$). It moves from $\omega = 0$ to positive values of ω ($p2$) and then up the $j\omega$ axis. If any poles are encountered on the $j\omega$ axis, the locus passes these allowed singularities on the right. At the beginning of this locus, when s is very small, $|G(s)|$ is very large since, at low frequencies, $G(s) \Rightarrow K/s$, so it approaches infinity as the radius of the beginning circle in the s plane approaches zero. Thus the locus in the $G(s)$ plane begins at infinite radius. Also, since $G(s) \Rightarrow K/s$, $G(j\omega) \Rightarrow K/(j\omega) = -jK/\omega$ and the $G(s)$ locus reaches the negative imaginary axis when s reaches the positive imaginary axis ($p2$). As s increases along this axis, $|G(s)|$ decreases, as in the Bode plot, and the phase of $G(s)$ becomes more negative than $-90°$ ($p3$). For the function $G(s)$ represented here, the phase goes beyond $-360°$ as $|G(s)|$ continues to shrink ($p4$). With $|s|$ very large, the locus in the s plane is brought clockwise to the negative $j\omega$ axis, leaving $G(s)$ near zero. The s-plane locus is then completed and the second half of the $G(s)$ locus is symmetrical to the first because the Re $G(s)$ is an odd function of ω and Im $G(s)$ is an even function. The important fact is that the locus does not surround the point $-1 + j0$ so poles of $H(s)$ are in the left half of the s plane and that corresponds to the stable region. The gain and phase margin can also be seen from the Nyquist diagram and shown in Fig. 5.3b.

One of the benefits of the Nyquist diagram relative to the Bode plot is the ease of determining if a system where $|G(s)|$ exceeds unity at $180°$ is stable. Figure 5.4 represents a conditionally stable system. The locus of $G(s)$ does not surround -1 so it is stable. It is "conditionally stable" because a

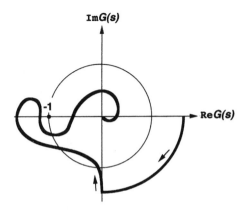

Fig. 5.4 Nyquist plot of a conditionally stable system. A reduction in the gain factor can cause the point -1, $j0$ to be enclosed, an indication of instability.

reduction of gain, such as might occur due to saturation or at power on, could cause it to enter the region of instability. Once there, it is likely to stay in a state where the waveforms are clipped due to limiting and the average gain is unity. In a phase-locked loop it may exceed the linear range of the phase detector and may lose lock. With complicated loci it may be beneficial to determine whether a line with a fulcrum at $-1 + j0$ rotates a net nonzero number of times (clockwise under the assumptions above) when its other end follows the locus around its path. If it does, the locus encloses -1.

One minor problem with the Nyquist plot is that it can be hard to draw in a manner that shows its features close to the critical -1 point as well as for large values. The representations that follow are only approximate. In many cases the radial distance has something like a logarithmic scale, giving an expanded presentation near the origin compared to further out. In fact we often use the greatest radius to represent infinity.

5.2.3   Evans Plot (Root Locus)

The third common method of stability analysis is to plot the poles of $H(s)$, which are the zeros of $[1 + KF(s)/s]$, as K increases to determine at what value of K they enter the right-half plane, taking on positive real parts σ [Kuo, 1987, pp. 384–456]. This is done according to a set of rules that permits the locations of the closed-loop poles to be determined from those of the open-loop poles and zeros. Figure 5.5 shows an Evans plot for a simple stable system (not for a practical PLL).

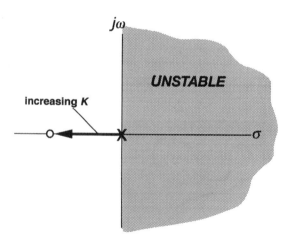

Fig. 5.5   Evans plot for a very simple system. The closed-loop poles are plotted in the s plane as the gain factor K increases.

5.3 STABILITY OF VARIOUS PLL CONFIGURATIONS

In this section we will consider the stability of PLLs, having various types of loop filters, using all three methods of analysis.

5.3.1 First-Order Loop

The simple loop described in Chapter 2 is represented in Fig. 5.6. In the Bode plot the gain falls in proportion to modulation frequency while the phase shift is constant at $-90°$, offering a fixed phase margin of 90°. The gain margin is not meaningful since the phase never reaches $-180°$. In any practical circuit, however, phase shift will continue to become more negative at sufficiently high frequencies due to parasitic poles and delays.

The Nyquist plot is also very simple.

The root locus begins at the only open-loop pole and moves along the real axis to infinity, moving further from the region of instability at increasing values of K.

5.3.2 Second-Order Loop

We will find that, like the first-order loop, all second-order loops are inherently stable because the phase shift only *approaches* $-180°$. Again, gain margin is therefore meaningless (infinite), but phase margin is important because it indicates the sluggishness or "ringiness" of the responses and because parasitic effects can turn a theoretically stable loop unstable.

The loop with a low-pass filter is represented in Fig. 5.7, that with an integrator and lead is in Fig. 5.8, and Fig. 5.9 represents the loop with a lag-lead filter. Phase margin ϕ_m is shown in each case.

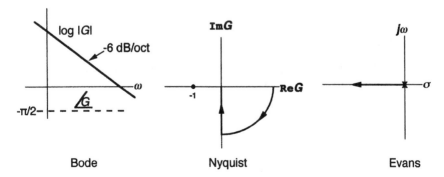

| Bode | Nyquist | Evans |

Fig. 5.6   Stability plots for a first-order loop.

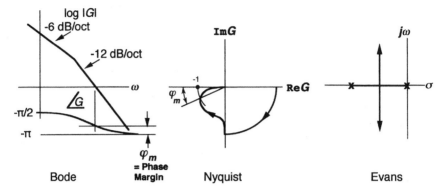

Fig. 5.7 Stability plots for a second-order type 1 loop with a low pass filter.

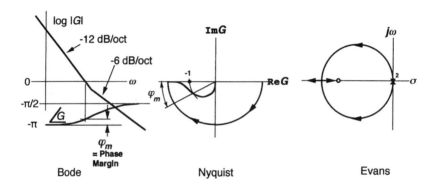

Fig. 5.8 Stability plots for a second-order type 2 loop.

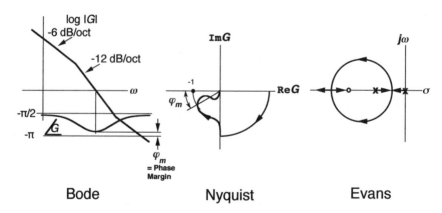

Fig. 5.9 Stability plots for a second-order type 1 loop with a lag-lead filter.

5.3.3   Third-Order Loop

We tend to stay away from higher-than-second-order loops in analysis because the number of variables is too large for compilation of standard
response curves. Often we approximate higher order loops as second order
by ignoring a pole that is further removed from the critical unity-gain region
than the other poles. However, higher order loops do occur in practice, and
the stability analyses, according to the methods here, are easily applied.
Figure 5.10 illustrates the plots for a third-order loop, showing gain and
phase margins for both a stable value of gain (curve b) and an unstable value
(curve a).

Figure 5.11 illustrates a conditionally stable third-order loop. Here a gain
factor between K_a and K_b results in instability.

5.3.4   Controlling Stability over Wide Gain Ranges

Sometimes the gain of a loop varies over wide ranges. If the gain is not
"rolled off," it may be too high at modulation frequencies where parasitic
phase shifts begin to appear. However, if it is rolled off too fast, there will be
excessive phase shift, also leading to instability or poor performance. A
solution is illustrated in Fig. 5.12. Here a number of poles and zeros are
alternated in the loop filter to give a roll-off rate faster than would be
obtained with a first-order loop while maintaining the phase margin. The
phase margin is never allowed to become too small, even though the gain can
vary, moving the curve vertically and moving ω_L along the frequency axis.
Loci are shown for the case where the highest frequency singularity is a pole
and for the case where it is a zero.

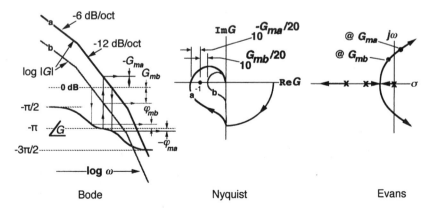

Fig. 5.10   Stability plots for a third-order type 1 loop at two values of gain factor K.
This could represent a low-pass filter with a parasitic pole at a higher frequency.

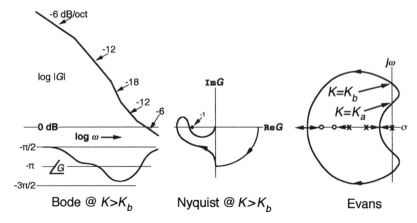

Fig. 5.11 Stability plots for a conditionally stable third-order loop. The loop is unstable for values of gain factor K between K_a and K_b.

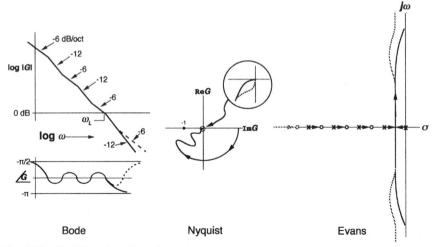

Fig. 5.12 Stability plots for a loop that is unconditionally stable over a wide range of frequencies.

5.3.5 Representing Delay

Usually any delay (transportation lag) in a loop will be much less than the reciprocal of the loop bandwidth and will therefore have little influence on stability or response.

A delay T_d is represented in Laplace transforms by

$$\exp(-sT_d) \Rightarrow \exp(-j\omega T_d), \tag{5.2}$$

a pure phase shift φ_T proportional to frequency. Phase, frequency, and voltage modulations, which apply to the loop's state variables, will be delayed by T_d and therefore

$$\varphi_T = -\omega_m T_d \qquad (5.3)$$

should be placed in the loop where appropriate. Note that, while Eq. (5.2) may give a very large phase shift at high carrier (signal) frequencies, that phase shift does not enter into the loop model. The modulation frequency is the parameter used in the loop [as in (5.3)], so it is wide loop bandwidth, rather than high carrier frequency, that aggravates the problem.

A phase shift proportional to ω_m can easily be introduced into the Bode phase plot. At the unity-gain frequency ω_L, the added phase shift will be $-\omega_L T_d$ and the phase margin will be correspondingly reduced. φ_T can also be added to the Nyquist diagram by shifting each point proportionally to the frequency represented. φ_T is more difficult to add to the Evans diagram; a number of poles and zeros must be used to approximate the frequency-dependent phase shift in regions of importance [Truxal, 1955, pp. 546–554].

5.4   COMPUTING OPEN-LOOP GAIN AND PHASE

We will use Figure 5.13 as an example to show how open-loop gain and phase can be computed at any frequency based on a knowledge of the critical frequencies and the known gain at some frequency [Kuo, 1987, pp. 693–706].

First note that a term that is proportional to the nth power of frequency (kf^n) is represented in decibels as

$$Q(f) = 20 \text{ dB} \log(kf^n) = 20 \text{ dB } n \log f + k'.$$

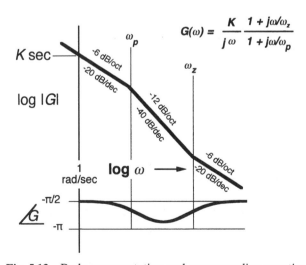

Fig. 5.13   Bode representation and corresponding equation.

Therefore, for each octave $(2:1)$ change in frequency

$$\Delta Q(f) = 20 \text{ dB } n \log 2 = 6n \text{ dB}$$

and for each decade $(10:1)$ change in frequency

$$\Delta Q(f) = 20 \text{ dB } n \log 10 = 20n \text{ dB}.$$

We begin with the tangential approximation to open-loop gain. The value at $\omega = 1$ rad/sec is given as K sec. The slope there is -6 dB/octave. That is, gain is inversely proportional to ω. Therefore, the gain at any frequency ω along this curve is

$$|G(\omega)| = K/\omega. \tag{5.4}$$

Notice that this gives the required gain at $\omega = 1$ rad/sec. If the gain is known instead to be K_a at some frequency ω_a, the equation is

$$|G(\omega)| = K_a \omega_a/\omega, \tag{5.5}$$

giving the known value at $\omega = \omega_a$. If the value is known instead along the -12 dB/octave slope, where the gain is therefore inversely proportional to ω^2, the gain is

$$|G(\omega)| = K_b(\omega_b/\omega)^2. \tag{5.6}$$

In each case the correct dependence on ω is provided (e.g., $1/\omega^2$), this is then multiplied by a constant (e.g., ω_b^2) to cause the resulting product to be unity at the known point, and the result is multiplied by the known gain at that point (e.g., K_b).

Suppose that the gain is desired at some frequency ω_c on the third segment of the curve. Then the gain is successively computed at each break point and finally at ω_c, basing each point upon the previous one. Gain at ω_p can be obtained from Eq. (5.4),

$$|G(\omega_p)| = K/\omega_p. \tag{5.7}$$

Then, using Eq. (5.6) with $\omega \Rightarrow \omega_z$, $\omega_b \Rightarrow \omega_p$, and $K_b \Rightarrow |G(\omega_p)|$, the value at ω_z can be found:

$$|G(\omega_z)| = |G(\omega_p)|(\omega_p/\omega_z)^2 = (K/\omega_p)(\omega_p/\omega_z)^2. \tag{5.8}$$

Finally the value at frequency ω_c is found using Eq. (5.5) with $\omega \Rightarrow \omega_c$, $\omega_a \Rightarrow \omega_z$, and $K_a \Rightarrow |G(\omega_z)|$:

$$|G(\omega_c)| = (K/\omega_p)(\omega_p/\omega_z)^2(\omega_z/\omega_c). \tag{5.9}$$

Notice also that this simplifies to

$$|G(\omega_c)| = (\omega_p/\omega_z)K/\omega_c, \qquad (5.10)$$

which is the same as for a continuous -6 dB/octave $(= -20$ dB/decade$)$ slope from $(\omega = 1, |G| = K$ sec$)$ except that the gain has been reduced by the ratio of the frequencies at the ends of the -12 dB/octave $(= -40$ dB/decade$)$ segment to take into account the additional $-20\log_{10}(\omega_p/\omega_z)$.

The exact gain can be calculated from the formula,

$$|G(\omega)| = \left|\frac{K}{j\omega}\right|\left|\frac{1 + j\omega/\omega_z}{1 + j\omega/\omega_p}\right| = \frac{K}{\omega}\sqrt{\frac{1 + (\omega/\omega_z)^2}{1 + (\omega/\omega_p)^2}}. \qquad (5.11)$$

Phase can be obtained from the formula as

$$\angle G(\omega) = \angle\frac{K}{j\omega} + \angle\left(\frac{1 + j\omega/\omega_z}{1 + j\omega/\omega_p}\right) \qquad (5.12a)$$

$$= -\frac{\pi}{2} + \tan^{-1}\frac{\omega}{\omega_z} - \tan^{-1}\frac{\omega}{\omega_p}. \qquad (5.12b)$$

The phase shift due to ω_z varies from $\tan^{-1}(0) = 0$ at $\omega = 0$ to $\tan^{-1}(1) = 45°$ at $\omega = \omega_z$ to $\tan^{-1}(\infty) = 90°$ at $\omega \Rightarrow \infty$. The phase shift is within about $18°$ of its final value, either $0°$ or $90°$, when ω/ω_z equals $\frac{1}{3}$ or 3, respectively. The same is true for every zero. Poles produce similar phase shifts except for a change of sign. Generally the phase shift due to singularities at frequencies that are many times the frequency at which it is being computed can be ignored.

Example 5.1   Bode Plot Open-loop gain at 10 Hz is 60 dB. The loop filter is an integrator-and-lead filter with the zero at 200 Hz. There are additional poles at 3 kHz and 10 kHz. Find the phase and gain margins.

The equation for open-loop gain has the form of (4.22) except there are two more poles. Writing it explicitly for the given singularities we obtain

$$G(s) = \frac{A(1 + s/\omega_z)}{s^2(1 + s/\omega_{p1})(1 + s/\omega_{p2})} \Rightarrow \frac{AF_b}{-(2\pi f_m/\text{cycle})^2 F_c F_d} \qquad (5.13a)$$

where

$$F_b \equiv 1 + j\frac{f_m}{200 \text{ Hz}}, \qquad F_c \equiv 1 + j\frac{f_m}{3000 \text{ Hz}}, \qquad F_d \equiv 1 + j\frac{f_m}{10,000 \text{ Hz}}.$$

$$(5.13b)$$

We begin by drawing the Bode tangential gain approximation plot as in Fig. Ex5.1. We start where the gain is given, at point a, and slope down at -40 dB/decade because of the $1/f^2$ dependency. In this frequency range, below the zero at b, F_b, F_c, and F_d are approximated as unity since the frequency-dependent term is smaller than the constant term in each factor.

As the modulation frequency increases, this approximation fails at the first singularity encountered at b, where the two terms in F_b have equal magnitude. At higher frequencies the second term is larger so we approximate F_b by the magnitude of the second term, $f_m/200$ Hz. Therefore, beyond 200 Hz the net frequency dependence is f^{-1} and the slope is reduced to -20 dB/decade. Similarly, at c, the approximation for F_c transitions from unity to a term proportional to frequency, returning the net frequency dependence to f^{-2} and giving a -40 dB/decade slope. The same thing occurs at d, giving an ultimate slope of -60 dB/decade.

The tangential approximation is poorest at the corners, being off by 3 dB at simple singularities. The error is slightly greater than 3 dB at c and d since

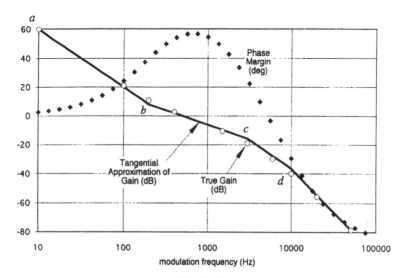

Fig. Ex5.1

these poles are not sufficiently separated for the effect of one not to be seen at the other.

According to the tangential approximation, the gain at b is [60 dB $-$ $40 \log_{10}(200/10) =$] 8 dB. It crosses unity gain at a frequency that is higher than the zero by a factor of 8 dB or $10^{8/20} = 2.5$, thus 500 Hz. The phase shift at 500 Hz, due to the poles at 3 and 10 kHz, is

$$-\tan^{-1}(500/3000) - \tan^{-1}(500/10{,}000) = -9.46° - 2.86° = -12.32°.$$
$$(5.14)$$

The zero at 200 Hz increases the phase by $\tan^{-1}(500/200) = 68.2°$. The net from these three singularities is $+55.9°$. This is also the phase margin because the two poles at the origin contribute $-180°$.

We can obtain the unity-gain frequency more accurately by trial and error using (5.13). A spreadsheet program can be very helpful in making such calculations. (The plots above were made from such a program.) Doing this, we find that the straight-line approximation is accurate, at 10 Hz, to one part in 1000 of $|G|$ and that the unity-gain frequency is actually 526 Hz rather than 500 Hz. At 526 Hz the phase margin is 0.3° greater than the approximate calculation showed. Thus the errors due to the tangential gain approximation are quite small.

By trial and error we find zero phase margin (180° excess phase shift) at 5235 Hz. The gain at that frequency is -27.5 dB. The gain margin is therefore $+27.5$ dB.

When we use the Bode tangential gain approximation to compute the unity-gain frequency, the inexactness of the computed frequency will cause an inaccuracy in the computed phase margin. To gain an idea of how large this error might be, we will consider some specific cases.

Figure 5.14 shows how large that error is in the vicinity of the pole of a lag filter or the zero of an integrator and lead filter. The errors are the same in the two cases when they are given as a function of the phase margin that is estimated from the tangential gain plot. The estimated margins are pessimistic. This plot is mainly of value for giving an indication of the magnitude of the inaccuracy because additional singularities are required for instability, and they would change the assumed conditions.

Figure 5.15 shows similar data in the vicinity of the double pole of a two-pole lag filter. Here the estimate is again pessimistic but by a larger amount. This is a severe case because the error at the corner is accentuated by the presence of two superimposed poles. However, it does represent a configuration that can be unstable without the addition of other singularities.

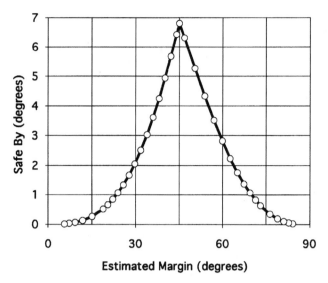

Fig. 5.14 Error in phase margin due to tangential gain approximation, single corner. This curve applies to both a lag filter (Fig. 5.7) and an integrator-and-lead filter (Fig. 5.8) in a PLL. (This applies specifically to a PLL because the additional $1/s$ term affects the result.)

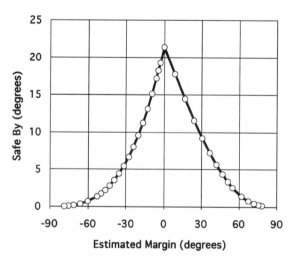

Fig. 5.15 Error in phase margin due to tangential gain approximation, double-pole loop filter. This curve applies to a PLL that has a lag filter with two superimposed poles.

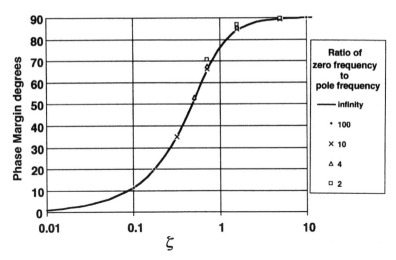

Fig. 5.16 Relationship between phase margin and damping factor, second-order loop. Solid curve applies to both integrator-and-lead and to low-pass loop filters. Data points are for lag-lead loop filters.

5.5 PHASE MARGIN VERSUS DAMPING FACTOR

Figures 4.5 and 4.6 indicate that small phase margin and small damping factor tend to go together. Figure 5.16 confirms this. It shows phase margin versus damping factor. The solid curve applies to a loop with either a low-pass or an integrator-and-lead filter. Separate points are shown for lag-lead filters having various ratios of zero to pole frequencies. For each of these ratios, a point is shown at the lowest ζ that can occur with that zero-pole ratio and at some higher values of ζ. The solid curve applies well for large ratios of zero to pole frequencies.

5.M APPENDIX: STABILITY PLOTS USING MATLAB

All of the stability plots in this chapter can easily be generated using MATLAB. In each case the open-loop transfer function is represented by a vector NUM consisting of the coefficients of its numerator and a vector DEN consisting of coefficients of its denominator. For example, the file `Bd2ordT1` produces a Bode plot for a second-order type 1 loop with the open-loop transfer function given by Eq. (4.2), which can be written

$$G(s) = \frac{s(K/\omega_z) + K}{s^2/\omega_p + s}.\tag{5.M.1}$$

The MATLAB representation is

$$NUM = [K/Wz \quad K]; \qquad (5.M.2)$$

$$DEN = [1/Wp \quad 1 \quad 0]. \qquad (5.M.3)$$

As can be seen from this example, the rightmost element of the vector is the coefficient of s^0, the next element is the coefficient of s^1, and so forth.

Individual files for various plots are available for downloading, and we can easily change their parameters in order to observe their effects on the stability plots. You can go through Bode, Evans, and Nyquist scripts and observe the effects of modifying the parameters. Note the simplicity of these scripts (programs).

The Nyquist plots in MATLAB are done only for real frequencies, just the $j\omega$ axis in Fig. 5.3a, so we can expect to be missing the corresponding parts of the plot (e.g., $p1$ to $p2$). Also, the standard `Nyquist` function draws arrows in the direction of increasing frequency on each half of the plot and shows a cross mark at $(-j + 0)$, but sometimes these marks become distorted. The arrows can be removed from the `Nyquist` script,[1] but the arrows tend to shrink or be in the undisplayed region in later plots.

One file, `ChsPlot` allows various loop configurations and various stability plots to be chosen dynamically, that is, without reexecuting the file. It is designed to permit this to be done with relative speed and ease to encourage rapid exploration and comparison. As the various loop types and orders are selected, parameters are automatically adjusted to keep the plots interesting —primarily to cause zero gain to occur at a frequency where phase is changing significantly—so, while the parameters can be modified, it may be best to do so on the simpler files first, leaving `ChsPlot` close to its original state. It might be instructive to compare Figs. 5.6 through 5.11 to similar plots generated using this program.

Since it is difficult to see the close-in (near $s = -1$) and far-out Nyquist pictures on the same plot, a zoom capability has been incorporated for the Nyquist plot in `ChsPlot`. It can be useful to compare the Nyquist plot to the Bode gain and phase plots to see what portion of the characteristic is being observed with a given magnification of the Nyquist plot.

In particular, observe the relationship between the Bode, Evans, and Nyquist plots when the third-order conditionally stable loop similar to Fig. 5.11 is selected. We can see how the Evans plot crosses into the right-half plane at some gains and, using the Bode plot, can observe what

---

[1]This procedure pertains in particular to MATLAB V.3.5. Find "nyquist" in the Signals_and_Systems_Toolbox. You may wish to copy it and rename the copy nyquistNoAr, for example, and then call that function when you want to get rid of the arrows. (Leave the modified function in the toolbox.) Find the comment "% Make arrows." Go down 12 lines to "plot(re,im,'r-',re,-im,'g--', real(xy),-imag(xy),'r-',real(xy2),imag(xy2),'g-');" delete the last part so it reads "plot(re,im,'r-',re,-im,'g--');."

these frequencies are, where the phase is below $-180°$. Using the Bode plot we can observe that the medium gain plot (obtained by choosing zm) has unity gain within the region of excessive phase but the high and low gain plots (zl and zh) do not. We can observe how the Nyquist locus surrounds the -1 point for the zm plot, confirming instability, but not for the others. Here we will need the zoom capability to verify that the -1 point is or is not surrounded in the sense of a vector from -1 to the locus rotating a full circle as the locus is followed.

Representative files are included in text form below and throughout the remainder of the book. These are written for MATLAB version 3.5. Scripts for versions 4 and 5 are available[2] on line, but the scripts given in the text usually require little or no modification to work with those later versions. Some necessary changes are shown as comments.

Note that data values can be obtained by deleting the ";" at the end of an appropriate line. This can be very useful but also tends to produce a large amount of output. The following resources are available to help you understand the MATLAB programs associated with this book:

- The comments in the programs
- The MATLAB Help utility
- The MATLAB manual
- The text of function programs (M-files)

Bd2ordT1

```
%Bode Plot from open- loop equation
%Second- Order, Type 1 (Lag- lead filter)
%Modify pole, zero, K from Eq. (5.M.1)
clg;
Wz = 1000;
Wp = .1;
K = 1000000;
num = [K/Wz  K];
den = [1/Wp  1  0];
bode(num, den);
```

Ev2ordT1

```
%Evans Plot
%Second- Order, Type 1 (Lag- lead filter)
%Input pole, zero, K (standard definitions)
clg;
```

---

[2]There is a difference in the clear graph command—"clg" becomes "clf"—and the recommended form for subplot(211), for example, is now subplot(2,1,1). However, these changes have been grandfathered so they are not currently necessary.

```
Wz = 1000;
Wp = 100;
K = 10000;
%f = logspace(2,4);
num = [K/Wz K];
den = [1/Wp 1 0];
rlocus(num,den)
```

NqLagLd

```
%Plots p / o nyquist for loop with lag- lead filter
%given K and pole and zero frequencies
%last statement could list output
clg;
Wz = 1000;
Wp = 10;
K = 10000;
f = logspace(2,4);
num = [K/Wz K];
den = [1/Wp 1 0];
[re,im,w] = nyquist(num,den,f);
nyquist(num,den,f)
%[w' re im]
```

ChsPlot

```
%Dynamic Choice of Bode, Evans, or Nyquist
%Dynamic Choice of Loop Types
%Nyquist plot zooming
%  Changing the following parameters could give poor scaling.
%      They will be modified automatcally as the various kinds
%      of loops are chosen to improve plots.
Wz = 1000;
Wp = 100;
K = 10;

f = logspace(0,5,200); % Frequency range
% v is axis vector to scale Nyquist plots depending on loop
                        type
% num and den are coefficients of open-loop transfer function

ny = 0; % nyquist looping control variable
go = 2; % main looping control variable
while (go ~= 0)
    order = input('enter order (1-3): ','s'); % SELECT LOOP
    if strcmp(order,'1'),
        disp('First Order');
        v = [-9 1 -9 1];
        num = [0 K];
        den = [1 0];
```

```
elseif strcmp(order,'2'),
   typ = input('enter type (1a [lag], 1b [lag-lead] or 2
                          [integrator & lead]): ','s');
   if strcmp(typ,'1a'),
      v = [-2 1 -5 1];
      disp('Second Order, Type 1a [lag] with K*10');
      num = [0 K*10];
      den = [1/Wp 1 0];
   elseif strcmp(typ,'1b'),
      disp('Second Order, Type 1b [lag-lead] with K*100');
      v = [-6 1 -10 1];
      Kx = 100*K;
      num = [Kx/Wz Kx];
      den = [1/Wp 1 0];
   else,
      disp('Second Order, Type 2 with K*1E5');
      v = [-40 1 -7 1];
      Kx = 100000*K;
      num = [Kx/Wz Kx];
      den = [1 0 0];
   end % if typ
else,
   disp('zp - Type 2 plus zero then pole');
   disp('ph - Type 1 plus pole, pole (Fig. 5.10), high
                          gain');
   disp('pl - Type 1 plus pole, pole (Fig. 5.10), low
                          gain');
   disp('zh - Type 1, 2 poles, 2 zeros (Fig. 5.11), high
                          gain');
   disp('zm - Type 1, 2 poles, 2 zeros (Fig. 5.11), medium
                          gain');
   disp('zl - Type 1, 2 poles, 2 zeros (Fig. 5.11), low
                          gain');;
   typ = input('Enter zp, ph, pl, zh, zm or zl: ','s');
   if strcmp(typ,'zp')
      disp('Third Order, Type 2 with pole<->zero, K*1E4');
      v = [-20 1 -20 1];
      Kx = K*10000;
      num = [Kx/Wp Kx];
      den = [1/Wz 1 0 0];
   elseif strcmp(typ,'ph') | strcmp(typ,'pl'),
      disp('Type 1 with poles at Wz and Wp');
      if strcmp(typ,'ph'),
         Kx = K*200;
         disp('high gain');
      else,
         Kx = K*50;
         disp('low gain');
      end % if p
```

```
            v = [-3.5 .5 -1 .5];
            num = [Kx];
            den = [1/(Wz*Wp) (1/Wz+1/Wp) 1 0];
        else,
            disp('Type 1 with poles at Wp/1.5 & Wp and zeros
                              higher by 5');
            v = [-3.5 .5 -1 .5];
            rpp = 1.5; % Wp2/Wp1 = Wz2/Wz1, Wp2 = Wp
            rzp = 5; % Wz1/Wp2
            rgg = 10; % gains, high/med = med/low
            Kmed = 1400;
            if strcmp(typ,'zh'),
                Kx = rgg*Kmed;
                disp('high gain');
            elseif strcmp(typ,'zl'),
                Kx = Kmed/rgg;
                disp('low gain');
            else % if strcmp(typ,'zm'),
                Kx = Kmed;
                disp('medium gain');
            end % if z
            num = Kx*[1/(rzp^2*rpp*Wp^2) (1+1/rpp)/(rzp*Wp) 1];
            den = [rpp/Wp^2 (1+rpp)/Wp 1 0];
        end % if typ
    end % if order
    go = 1;
    while(go == 1) % SELECT PLOT TYPE
        disp('bode, nyquist, evans, change loop or quit?');
        ctrl = input('Enter b, n, e, c, or q: ', 's');
        clg;
        if strcmp(ctrl,'e')|strcmp(ctrl,'evans'),
            rlocus(num,den);
        elseif strcmp(ctrl,'b')|strcmp(ctrl,'bode'),
            bode(num,den,f);
        elseif strcmp(ctrl,'n')|strcmp(ctrl,'nyquist'),
            ny = 1;
            vnow = v;
            while(ny == 1) % NYQUEST LOOP TO PERMIT ZOOMING
% ###########################  OK for MATLAB v. 3.5.
                axis(vnow);              % #  For MATLAB v. 4, 5
                nyquist(num,den,f); % #  swap these 2 lines
% ###########################
                xpnd = input('zoom or continue? ','s');
                if strcmp(xpnd,'z')|strcmp(xpnd,'zoom')
                    xp = input('enter expansion factor ');
                    vnow = vnow / xp;
                else
                    ny = 0;
                end % if xpnd
            end % while ny
```

```
% ############ replace next line by "axis auto" for MATLAB v.
                        4, 5
        axis; axis;
    elseif strcmp(ctrl,'q')|strcmp(ctrl,'quit'),
        go = 0;
    elseif strcmp(ctrl,'c')|strcmp(ctrl,'change'),
        go = 2;
    end % if ctrl
  end % while go = 1
end % while go~=0
```

When entering scripts into a computer, type any line that is indented 2″ as a continuation of the previous line.

PROBLEMS

5.1 (a) Sketch a Bode gain and phase plot for a loop where $K_p = 1$ V/cycle, $K_V = 3.5$ kHz/V, and the filter is shown in Fig. 3.19 where $R_1 = 1$ kΩ, $R_2 = 100$ Ω, and $C = 1$ μF.

(b) What are the gain and phase margins?

5.2 Give gain and phase margins and the frequencies at which they occur for a PLL with $K = 10^4$ sec^{-1} and whose filter has a pole at 1 kHz, two poles at 8 kHz, and a zero at 2 kHz.

5.3 Pick the value of R_1 in the filter of Fig. 3.19, with $R_2 = 2$ kΩ and $C = 0.2$ μF, and the value of K for the loop to give 45° phase margin and a unity-gain bandwidth frequency $f_L = 200$ Hz.

5.4 What is the gain margin for a loop with a gain of 10^6 at 1 Hz and filter poles at 100 kHz and 10 MHz?

CHAPTER 6

TRANSIENT RESPONSE

One important aspect of phase-locked loop (PLL) performance is the manner in which it responds to a change in input frequency or phase, and one of the most important measures is its response to a step change of the input. Does the output reach its final value quickly or slowly? Does it overshoot and, if so, how long does it take to come finally within a specified offset from the final value? How great is the phase error during a frequency step? Does it surpass the range of the phase detector? To answer such questions we need a set of curves that show time responses under various conditions.

6.1 STEP RESPONSE

For a linear system, the step response can be obtained by use of Laplace transforms. The transform of a unit step, $1/s$, multiplied by $H(s)$ from Eq. (4.10) or (4.20), can be separated, by use of the Heaviside (partial fraction) expansion theorem [Thomas and Finney, 1984], into parts for which transform pairs are available. The error response, $[1 - H(s)]/s$, can be similarly obtained or can be simply taken as $[1/s - H(s)/s]$. Step responses can then be plotted for various combinations of loop parameters. Either output or error can be plotted since one is easily obtainable from the other by Eqs. (2.28).

6.1.1 Form of the Equations

The challenge is to find a reasonably small set of curves that covers a large number of cases of practical importance. If we restrict our attention to linear performance, as we will in this chapter, the response to any size input step will be proportional to that step, so a curve of unit step response will suffice to indicate the response to all step sizes. Another device for limiting the number of curves is to restrict our study to loops of order 1 or 2. We have already shown the step response for a first-order loop (Figs. 2.2 and 2.6). There, just one curve suffices if its abscissa is normalized to the open-loop gain at $\omega = 1$ rad/sec, that is, if $x = Kt$. For second-order loops, curves can be plotted against normalized time, $\omega_n t$, for various damping factors ζ. However, Eq. (4.10) has an additional variable, ω_z, that threatens to add an additional, potentially intractable, dimension. As a first step, for integrator-and-lead filters ω_z can be eliminated from Eq. (4.10). This is done using Eq. (4.16) to give

$$H(s) = \omega_n^2 \frac{1 + s(2\zeta/\omega_n)}{s^2 + 2\zeta\omega_n s + \omega_n^2}. \tag{6.1}$$

The expression now contains a larger number of individual variables, but it is characterized entirely by two parameters, ω_n and ζ. Thus two sets of step response curves can be generated, one for a lag filter, based on Eq. (4.20), and the other for the integrator and lead, based on Eq. (6.1). Fortunately, responses for the more general lag-lead filter can be obtained from a combination of these, as we shall now see.

From Eq. (4.9) $1/\omega_z$ can be written in terms of ζ:

$$\frac{1}{\omega_z} = \frac{2\zeta\omega_n - \omega_p}{\omega_n^2}, \tag{6.2}$$

and ω_p can be eliminated by using Eq. (4.7) to give

$$\frac{1}{\omega_z} = \frac{2\zeta}{\omega_n} - \frac{1}{K}. \tag{6.3}$$

Substituting this expression into Eq. (4.10) we obtain

$$H(s) = \omega_n^2 \frac{1 + s(2\alpha\zeta/\omega_n)}{s^2 + 2\zeta\omega_n s + \omega_n^2} \tag{6.4a}$$

and

$$1 - H(s) = \frac{s^2 + 2\zeta\omega_n(1 - \alpha)s}{s^2 + 2\zeta\omega_n s + \omega_n^2}, \tag{6.4b}$$

where

$$\alpha = 1 - \frac{\omega_n}{2\zeta K} = \frac{1}{1 + \omega_z/K}. \tag{6.5}$$

For a lag filter, $\omega_z \Rightarrow \infty$, so $\alpha \Rightarrow 0$ and Eq. (6.4a) is the same as Eq. (4.20). For an integrator and lead filter, $K \Rightarrow \infty$, so $\alpha \Rightarrow 1$ and Eq. (6.4a) is the same as Eq. (6.1).

Lag filter	$\alpha \Rightarrow 0,$
Integrator-and-lead filter	$\alpha \Rightarrow 1.$

But Eq. (6.4a) can also be written

$$H(s)|_{\text{lag-lead}} = (1 - \alpha)H(s)|_{\text{lag}} + \alpha H(s)|_{\text{integrator-and-lead}} , \qquad (6.6)$$

which can be easily verified by substituting Eq. (6.4a) with $\alpha = 0$ for $H(s)|_{\text{lag}}$ and Eq. (6.4a) with $\alpha = 1$ for $H(s)|_{\text{integrator-and-lead}}$ in Eq. (6.6) to obtain Eq. (6.4a). Therefore the step response with any lag-lead filter can be obtained from curves for a loop with a lag filter and curves for a loop with an integrator-and-lead filter, each for the same ζ as the desired curve, by combining their values at each (normalized) time according to Eq. (6.6). The same relationship holds for the error response, as can be seen by subtracting Eq. (6.6) from 1:

$$1 - H(s)|_{\text{lag-lead}} = (1 - \alpha) + \alpha - H(s)|_{\text{lag-lead}}$$

$$= (1 - \alpha)[1 - H(s)]_{\text{lag}} + \alpha[1 - H(s)]_{\text{integrator-and-lead}} \cdot$$
$$(6.7)$$

Equations (6.6) and (6.7) are impulse responses; they can be multiplied by $1/s$ to give step response, or by $1/s^2$ to give the response to ramp inputs, and so on. Thus a relatively small set of curves can indicate the transient response for common second-order loops regardless of filter type.

6.1.2 Step Response Equations

The step error response of the PLL in the time domain, that is, the phase error $\varphi_e(t)$ in response to a unit phase step at the input or the frequency error $\omega_e(t)$ in response to a unit frequency step at the input, is obtained from the inverse Laplace transform of $(1/s)[1 - H(s)]$, giving (for $t > 0$):

$$Y_{u,\,\text{error}} = \frac{1}{2}\left[\left(\zeta\frac{2\alpha - 1}{\sqrt{\zeta^2 - 1}} + 1\right)e^{-\omega_n t(\zeta + \sqrt{\zeta^2 - 1})}\right.$$

$$\left. - \left(\zeta\frac{2\alpha - 1}{\sqrt{\zeta^2 - 1}} - 1\right)e^{-\omega_n t(\zeta - \sqrt{\zeta^2 - 1})}\right]. \qquad (6.8)$$

This form is most useful for $\zeta > 1$, for which an alternate form is

$$Y_{u,\,\text{error}}(t) = e^{-\zeta\omega_n t}\left[\cosh\left(\omega_n t\sqrt{\zeta^2 - 1}\right) + \zeta\frac{1 - 2\alpha}{\sqrt{\zeta^2 - 1}}\sinh\left(\omega_n t\sqrt{\zeta^2 - 1}\right)\right].$$

$$(6.9)$$

For $\zeta = 1$ Eq. (6.8) becomes

$$Y_{u,\,\text{error}}(t) = e^{-\omega_n t}\left[1 + (1 - 2\alpha)\omega_n t\right], \qquad (6.10)$$

while for $\zeta < 1$ it is

$$Y_{u,\,\text{error}}(t) = e^{-\zeta\omega_n t}\left[\cos\left(\omega_n t\sqrt{1 - \zeta^2}\right) + \zeta\frac{1 - 2\alpha}{\sqrt{1 - \zeta^2}}\sin\left(\omega_n t\sqrt{1 - \zeta^2}\right)\right].$$

$$(6.11)$$

In each case the output response, that is, the phase $\varphi_{\text{out}}(t)$ at the voltage-controlled oscillator (VCO) output after a unit step in $\varphi_{\text{in}}(t)$ at the loop input or the frequency $\omega_{\text{out}}(t)$ at the VCO output after a unit frequency step in $\omega_{\text{in}}(t)$ at the loop input, is

$$Y_{u,\,\text{out}} = 1 - Y_{u,\,\text{error}}. \qquad (6.12)$$

Figure 6.1 shows the output response to a step for $\zeta = 1$ and five values of α. It is apparent from this how each curve is a weighted average of the curves for $\alpha = 0$ and $\alpha = 1$, according to Eq. (6.6).

Example 6.1   Step Response, $0 < \alpha < 1$

a. The frequency input to a loop steps 10 kHz. What will be the change in output frequency 5 msec after the input step if $\omega_n = 600$ rad/sec, $\zeta = 1$, and $\alpha = 0.25$, where $\omega_n t$ equals 3?

Look at Fig. 6.1 at $\omega_n t = 3$ and $\alpha = 0.25$, the second curve from the bottom. The unit step response there is 0.88 so the output equals that number times the magnitude of the input step, thus *8.8 kHz.* If we use (6.10), for improved accuracy compared to reading the curves, we obtain 8.753 kHz.

b. Repeat for $\alpha = 0.9$.

Again use Fig. 6.1. For $\alpha = 0$, the response is 0.8. For $\alpha = 1$, the response is 1.1. From (6.6), the step response for $\alpha = 0.9$ is

$$Y_{u,\,\text{out}} = 0.9 \times 1.1 + 0.1 \times 0.8 = 1.07.$$

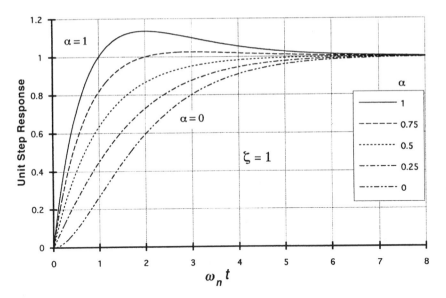

Fig. 6.1   Output response to a step input with $\zeta = 1$ for five values of α.

Thus the output would overshoot to *10.7 kHz* at the specified time. The equation gives 10.697 kHz.

c. Repeat for $\alpha = 0.4$ and $\zeta = 0.707$.

We will first obtain the error response from Fig. 6.3, since no output response is shown for this case. The responses at $\omega_n t = 3$ for α equal to zero and one are, respectively, 0.04 and -0.16. Multiplying the latter by 0.4 and the former by $(1 - 0.4)$ we obtain a sum of -0.040 so the error at that time is -0.040×10 kHz or -400 Hz. Since the output at zero error would be 10 kHz, the output has stepped *10.40 kHz* at 5 msec. Equation (6.9) gives 10.423 kHz.

    d. Using the same α and ζ, what is the output phase change 5 msec after a 20° phase step at the input?

As before, the step output is 1.042 times the step input, that is, 20.84°.

Figure 6.2 shows the step error response for $\zeta \geq 1$ and Fig. 6.3 shows it for $\zeta < 1$. It is desirable to plot the curves against a variable that is normalized in a way that keeps the curves close together. Not only does this make more feasible the plotting of multiple curves on the same graph, but it facilitates interpolation between curves. This goal is not satisfied especially well in

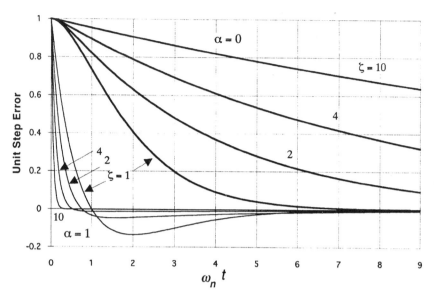

Fig. 6.2 Error response to a step input with $\zeta \geq 1$ versus $\omega_n t$.

Fig. 6.2. For $\zeta \geq 1$, it is better to plot the curves against the unity-gain frequency ω_L, as can be seen in Fig. 6.4. (ω_L' is the unity-gain frequency of the tangential approximation of the Bode plot and is approximately equal to ω_L, especially for large ζ.) Here the curves become almost coincident for large ζ. With this normalization, a single curve is sufficient for ζ greater than about 2. In fact, that curve is given, for both $\alpha = 0$ and $\alpha = 1$, by $\exp(-\omega_L t)$, the equation for the step response of a first-order loop. [This is identical to Eq. (2.12a).] We might have expected this because, for $\zeta > 1$, ω_L' occurs where the slope of the Bode plot is -6 dB/octave, as for a first-order loop, and, as ζ increases, the -12 dB/octave region moves further from ω_L (curve 3 in Fig. 4.5 and 2 in 4.6). As noted in Section 4.1, when two loops have similar $G(\omega)$ in the region of ω_L, their gross performance is similar.

The approximate value for ω_L for $\zeta \geq 1$ can be obtained from the geometry illustrated in Figs. 4.5 and 4.6. From Fig. 4.5, for the low-pass filter,

$$\omega_L \approx \omega_L' = \omega_n/(2\zeta), \tag{6.13}$$

while, from Fig. 4.6, for the integrator-and-lead filter,

$$\omega_L \approx \omega_L' = 2\zeta\omega_n. \tag{6.14}$$

For $\zeta < 0.5$, $\omega_L \approx \omega_n$ so Fig. 6.3 is already plotted approximately against the unity-gain frequency.

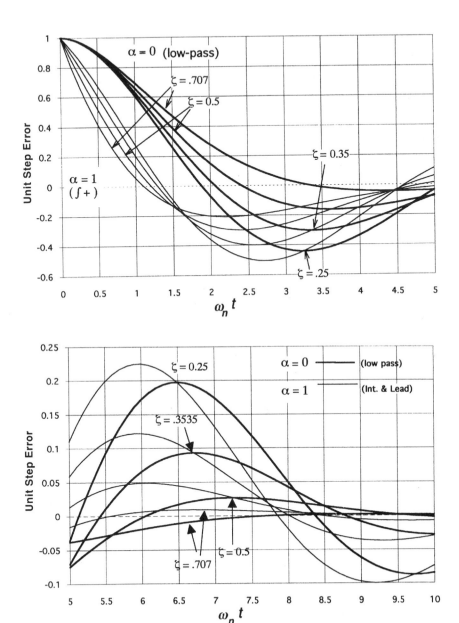

Fig. 6.3 Error response to a step input with $\zeta < 1$.

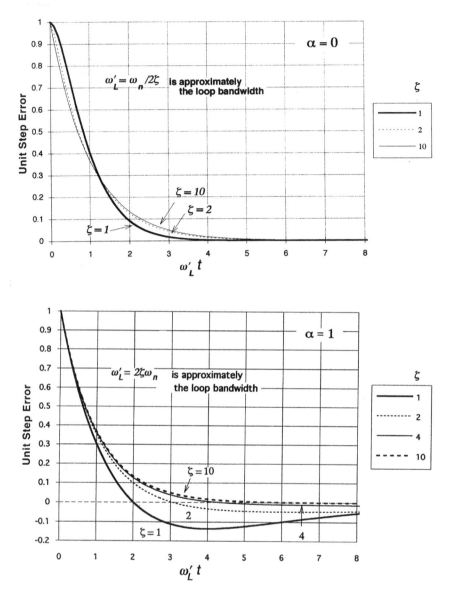

Fig. 6.4   Error response to a step input with $\zeta \geq 1$ versus $\omega_L' t$.

6.2 ENVELOPE OF THE LONG-TERM STEP RESPONSE

Figures 6.5 and 6.6 are log plots of the error response that are used to show the error, as the value becomes small, after many time constants. Figure 6.5 is for $\zeta \leq 1$. Since the response for $\zeta < 1$ is oscillatory, it is only practical to show the envelope of the response. This is obtained as the magnitude of the vector sum of two normal phasors having the amplitudes of the sine and cosine components in Eq. (6.11). For small ζ the response depends little on α, so only one curve is shown in those cases.

Example 6.2   Envelope of Step Response   How long will it take for the error after a frequency step to be reduced to 10^{-6} of the step size if $\zeta = 0.5$ and $\omega_n = 10^4$ rad/sec?

Figure 6.5 shows that $\omega_n t \approx 27.7$ for $\zeta = 0.5$, regardless of the value of α, when the envelope has been reduced to 10^{-6} of the step size. This occurs at 2.77 msec. The error may be smaller than its envelope, depending on where the peaks of the overshoots occur, but cannot be larger.

6.3   RESPONSE TO RAMP INPUT

The integral of Eq. (6.8) is an important quantity because it gives the phase error in response to a frequency step, and this is important in determining whether operation is confined to the linear portion of the phase detector curve, as is assumed in developing the linear equations. It can be obtained by using $1/s^2$ as the forcing function, (that is, taking the inverse Laplace transform of $(1/s^2)[1 - H(s)]$) or by integration in the time domain. (Since integration is a linear function, the integral of the response to a given input is the same as the response to the integral of that input.) The same response

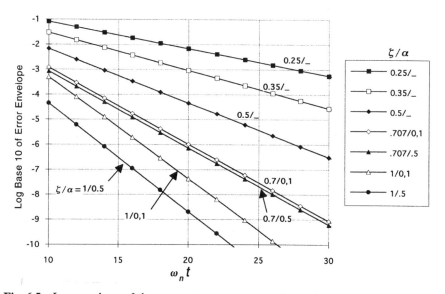

Fig. 6.5   Log envelope of the error response to a step input for large $\omega_n t$ with $\zeta \le 1$. Curves are labeled with ζ and α, separated by a slash.

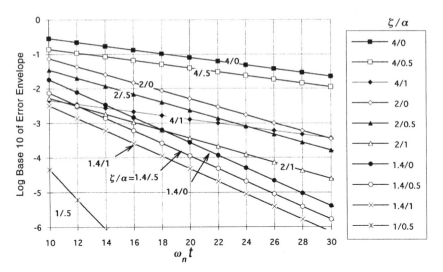

Fig. 6.6 Log envelope of the error response to a step input for large $\omega_n t$ with $\zeta > 1$. Curves are labeled with ζ and α, separated by a slash.

applies to either frequency or phase when the input is a ramp of either frequency or phase, respectively. The phase response to a phase ramp is also the phase response to a frequency step.

By integrating Eq. (6.8) we obtain the response to a unit ramp (t). The forms for $\zeta > 1$, $\zeta = 1$, and $\zeta < 1$, respectively, are:

$$Y_{r,\,\text{error}}(t) = 2\zeta \frac{1 - \alpha}{\omega_n} - \frac{e^{-\zeta \omega_n t}}{\omega_n}$$

$$\times \left[2\zeta(1 - \alpha)\cosh\left(\sqrt{\zeta^2 - 1}\,\omega_n t\right) + \frac{2\zeta^2(1 - \alpha) - 1}{\sqrt{\zeta^2 - 1}}\sinh\left(\sqrt{\zeta^2 - 1}\,\omega_n t\right) \right],$$

$$(6.15)$$

$$Y_{r,\,\text{error}}(t) = \frac{1}{\omega_n}\left\{ 2(1 - \alpha) - e^{-\omega_n t}[2(1 - \alpha) + (1 - 2\alpha)\omega_n t] \right\}, \qquad (6.16)$$

$$Y_{r,\,\text{error}}(t) = 2\zeta \frac{1 - \alpha}{\omega_n} - \frac{e^{-\zeta \omega_n t}}{\omega_n}$$

$$\times \left[2\zeta(1 - \alpha)\cos\left(\sqrt{1 - \zeta^2}\,\omega_n t\right) + \frac{2\zeta^2(1 - \alpha) - 1}{\sqrt{1 - \zeta^2}}\sin\left(\sqrt{1 - \zeta^2}\,\omega_n t\right) \right].$$

$$(6.17)$$

It is convenient to normalize these equations to show the response to a ramp of slope ω_n (by multiplying the equations by ω_n). These responses are shown in Figs. 6.7 through 6.9.

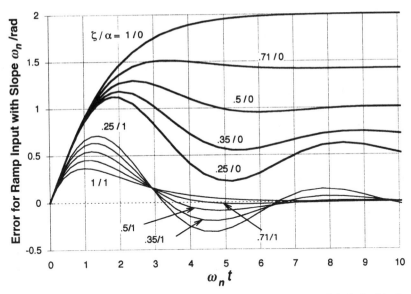

Fig. 6.7 Error response to a ramp input with $\zeta \leq 1$. Curves are labeled with ζ and α, separated by a slash.

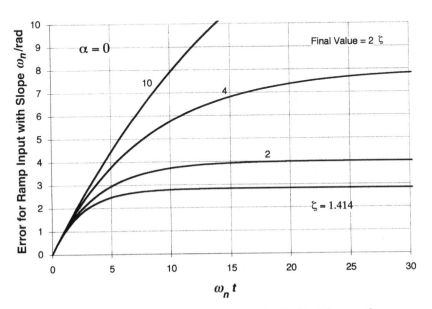

Fig. 6.8 Error response to a ramp input with $\zeta > 1$ for $\alpha = 0$.

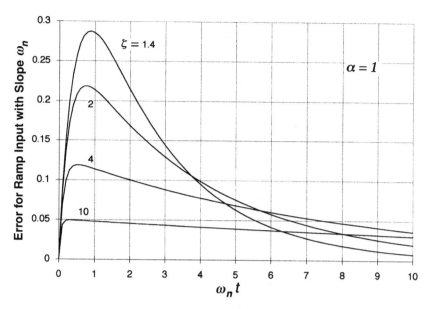

Fig. 6.9 Error response to a ramp input with $\zeta > 1$ for $\alpha = 1$.

Example 6.3 Ramp Response

a. The input frequency to a PLL is changing at a uniform rate of 1 kHz/sec. What is the error at the output 20 msec after the ramp starts if $\zeta = 0.5$, $\alpha = 0$, and $\omega_n = 100$ rad/sec?

Use Fig. 6.7 at $\zeta = 0.5$, $\omega_n t = 2$. There, according to the graph, the output error will be 1.26 if the input slope is $\omega_n = 100/\text{sec}$. We scale the input and response, multiplying them by 10 Hz, to obtain the given input ramp. The output error is then 12.6 Hz. Thus the output will lag the input ramp by 12.6 Hz at 20 msec. Equation (6.17) gives 12.687 Hz.

b. Using the same parameters, what will be the output phase error at 20 msec if the input frequency steps 1 kHz?

A 1-kHz frequency step is a phase ramp of 1000 cycles/sec (by definition). Therefore, the problem is basically the same as the previous problem except we scale by 10 cycles, rather than 10 Hz, this time. As a result, the output phase will lag the input phase by 12.6 cycles at 20 msec.

6.4 RESPONSE TO PARABOLIC INPUT

The response to a parabolic input $t^2/2$ can be obtained by integrating the above responses to a ramp input (or by taking the inverse Laplace transform of $(1/s^3)[1 - H(s)]$). We will write the response to $t^2/2$ since it corresponds to a unit frequency ramp. If the input frequency is represented by a unit ramp t, then the input phase is the integral of that frequency with respect to time, $t^2/2$. The phase output for a unit frequency ramp input is, for $\zeta > 1$, $\zeta = 1$ and $\zeta < 1$, respectively:

$$Y_{p,\text{error}}(t) = \frac{1 - 4\zeta^2(1 - \alpha)}{\omega_n^2} + 2\zeta \frac{1 - \alpha}{\omega_n} t$$

$$+ \frac{e^{-\zeta \omega_n t}}{\omega_n^2}\left[\zeta \frac{2(1 - \alpha)(2\zeta^2 - 1) - 1}{\sqrt{\zeta^2 - 1}}\sinh\left(\sqrt{\zeta^2 - 1}\,\omega_n t\right)\right.$$

$$\left. + \left[4\zeta^2(1 - \alpha) - 1\right]\cosh\left(\sqrt{\zeta^2 - 1}\,\omega_n t\right)\right], \quad (6.18)$$

$$Y_{p,\text{error}}(t) = \frac{t}{\omega_n}\left[2(1 - \alpha) - e^{-\omega_n t}(1 - 2\alpha)\right] + \frac{3 - 4\alpha}{\omega_n^2}(e^{-\omega_n t} - 1),$$

$$(6.19)$$

$$Y_{p,\text{error}}(t) = \frac{1 - 4\zeta^2(1 - \alpha)}{\omega_n^2} + 2\zeta \frac{(1 - \alpha)}{\omega_n} t$$

$$+ \frac{e^{-\zeta \omega_n t}}{\omega_n^2}\left[\zeta \frac{2(1 - \alpha)(2\zeta^2 - 1) - 1}{\sqrt{1 - \zeta^2}}\sin\left(\sqrt{1 - \zeta^2}\,\omega_n t\right)\right.$$

$$\left. + \left[4\zeta^2(1 - \alpha) - 1\right]\cos\left(\sqrt{1 - \zeta^2}\,\omega_n t\right)\right]. \quad (6.20)$$

Figure 6.10 shows the error in response to a parabolic input of $[0.5(\omega_n t)^2]$ with $\alpha = 1$. Figure 6.11 shows responses for $\alpha < 1$. Note that, unless $\alpha = 1$, the response eventually climbs.

Example 6.4 Response to a Parabolic Input   The frequency into a loop is increasing at a rate of 1 kHz/sec. What is the output phase error 1 msec after the start of this frequency ramp if $\zeta = 0.5$, $\alpha = 0.5$, and $\omega_n = 7$ krad/sec?

We are seeking the phase error when the input phase has a time derivative (frequency) that is proportional to time so we use Fig. 6.11. At $\omega_n t = 7$ the

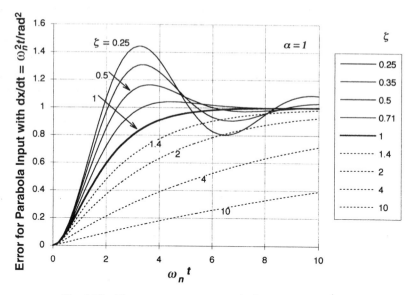

Fig. 6.10   Error response to a parabolic input with $\alpha = 1$.

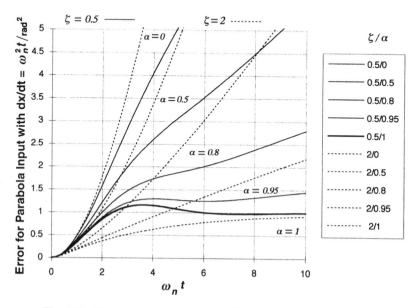

Fig. 6.11   Error response to a parabolic input with $\zeta = 0.5$ and 2.

error is 4. This is in response to an input with a time derivative of $\omega_n^2 t/\text{rad}^2 = 49 \times 10^6 t/\text{sec}^2$. We multiply the input and output by 10^3 cycles/49×10^6 to scale the graph to our problem. (The input now is a frequency ramp with a slope of t kHz/sec, which is also a phase parabola with a second derivative 10^3 cycles/sec.) The output, 1 msec after the start of the frequency ramp, is thus

$$(10^3 \text{ cycles}/49 \times 10^6)(4) = 8.2 \times 10^{-5} \text{ cycles}.$$

Equation (6.20) gives 8.1×10^{-5} cycles.

If the loop filter had been an integrator-and-lead ($\alpha = 1$, Figs. 6.10 and 6.11), the phase error would have settled to 2×10^{-5} cycles whereas it will continually increase otherwise.

6.5 OTHER RESPONSES

Function $-H(s)$ is the closed-loop transfer function from any point of signal injection to a point just prior to the injection. Function $1 - H(s)$ is the closed-loop transfer function from any point of signal injection to a point just after the injection. The usual point of injection is to the summer that represents part of the phase detector, as shown in Fig. 6.12a. However, another important injection point is just after the VCO, as shown in Fig. 6.12b.

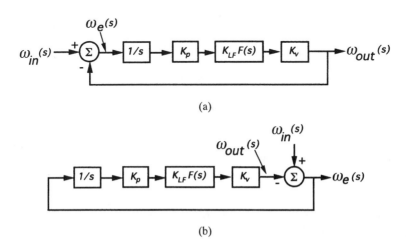

(a)

(b)

Fig. 6.12 Two different injection points. (a) Represents excitation at the reference input. (b) Represents a change in the open-loop VCO frequency. The observable frequency is shown as output in both cases.

This might represent, for example, a change in the open-loop frequency of the VCO (i.e., the center frequency) due to some undesired influence such a power supply voltage change. Then, with the loop closed, the VCO's output is represented by the error response, $1 - H(s)$, to the disturbance, whereas $-H(s)$ represents the ratio of the tuning voltage (in terms of equivalent frequency) to the injected disturbance. See Sections 7.4 and 7.3 for further discussion.

6.6 NOTE ON UNITS FOR GRAPHS

The graphs are normalized. Most are for a unit step so the ordinate must be multiplied by the magnitude of the step that is actually encountered. For example, if the step is 0.1 rad, the value given for the response on the curve must be multiplied by 0.1 rad. If the step is 10 Hz and the response value given by the graph at the appropriate value of $\omega_n t$ is 0.4, that actually represents 0.4×10 Hz $= 4$ Hz. In the case of the response to an input ramp (e.g., Fig. 6.9), the graph is given for an input slope of ω_n. If the input slope is 0.1 rad/sec and the graph value at the desired $\omega_n t$ is 0.6, then the actual response is $0.6 \times (0.1$ rad/sec) per ω_n. Since this would have no units, the units of radians for the phase error would be added. If $\omega_n = 10$ rad/sec and the ramp were a frequency ramp with a slope of 10 Hz/sec, then 0.6 would represent $0.6 \times (10$ Hz/sec) per $(10$ rad/sec$) = 0.6$ Hz, a frequency error in response to the frequency ramp. Similarly, a parabolic input with a slope kt must be multiplied by k rad$^2/\omega_n^2$ (see Figs. 6.10 and 6.11).

6.7 EQUIVALENT CIRCUIT

As was done for the first-order loop in Chapter 2, an analogous circuit can be drawn. Figure 6.13 shows a circuit whose voltages are proportional to the frequencies or phases of the second-order PLL. In order that the same output can be used to show both $H(s)$ and $1 - H(s)$, two different injection points are used and are labeled with subscripts, a and b, corresponding to the part of Fig. 6.12 that shows the same injection point. The voltage at the bottom of the resistor represents the output for a PLL with a lag filter, that at the top of the resistor represents the output from a PLL with an integrator-and-lead filter, and the tap has a voltage representative of the output from a PLL with a lag-lead filter.

6.8 GENERAL LONG-TERM (STEADY-STATE) RESPONSE CHARACTERISTICS

These eventual, or steady-state, changes generally depend on the loop type (see Section 4.1). When the phase changes to a new value at the input to any

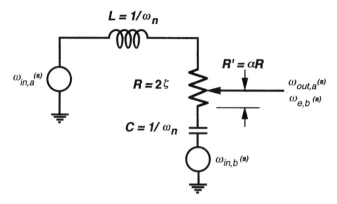

Fig. 6.13 Equivalent circuit for a second-order loop. $H(s) = \omega_{\text{out},\,a}/\omega_{\text{in},\,a}$; $1 - H(s)$ $= \omega_{e,\,b}/\omega_{\text{in},\,b}$.

PLL, the output eventually follows exactly (in the limit, i.e., at infinite time); see Figs. 6.1 through 6.6. Since there is no absolute phase reference, there is really no way for the loop to differentiate between different phases on a long-term basis. And all PLLs must have equal input and output frequencies.

A step of input frequency, which is equivalent to a phase ramp, causes the phase error to change in order to drive the VCO to a new frequency. As the DC zero-frequency gain approaches infinity, the long-term phase error approaches zero, so the only exception to the requirement for a change in steady-state phase error is with the integrator-and-lead filter (type 2 loop), which, theoretically, has infinite zero-frequency gain; see Figs. 6.9 and 6.7 for $\alpha = 1$. With a low-pass filter, the steady-state phase error in following a unit frequency step is $2\zeta/\omega_n$; see Fig. 6.8. This is easily understood since, by Eqs. (4.7) and (4.15b), $2\zeta/\omega_n = 1/K \equiv d\varphi_e/d\omega_{\text{out}}$.

When driven by an input frequency ramp, the phase error in a type 2 loop takes on a constant value that causes the integrator output, and thus the VCO, to ramp and maintain the constant phase offset from the input ramp; see Fig. 6.10. Since the phase error becomes constant, the frequency error must become zero. With a type 1 loop the phase error increases continuously, implying that the output takes on a constant frequency offset. See Fig. 6.11.

6.9 OPEN-LOOP EQUATIONS IN TERMS OF CLOSED-LOOP PARAMETERS

Now that we have defined closed-loop parameters ω_n, α, and ζ, it might be useful to have our open-loop equations available in those terms. Solving Eq. (4.4) for $G(s)$ in terms of $H(s)$, we obtain

$$G(s) = \frac{1}{1 - H(s)} - 1 = \frac{H(s)}{1 - H(s)}. \tag{6.21}$$

Using Eqs. 6.4a and 6.4b, this becomes

$$G(s) = \frac{2\alpha\zeta\omega_n}{s} \frac{s + \omega_n/(2\alpha\zeta)}{s + 2(1-\alpha)\zeta\omega_n}. \qquad (6.22)$$

6.10   STATE-SPACE ANALYSIS

In Chapter 2 we obtained transient responses for a simple loop using classical differential equations. In previous sections of this chapter we used Laplace transform techniques to obtain responses for more complex loops. The solution of $H(s)$ in terms of its poles becomes generally quite difficult for more complex loops. Adding even one more pole to $G(s)$ can create this problem. Therefore, as the complexity increases, it becomes more desirable to use computer-aided solutions employing the state-space method [Dorf, 1965; Chen, 1970]. To illustrate how this method is applied to the PLL problem, we will consider the solution of the transient response for the second-order system, which we have already solved using Laplace transforms, but the application of the method to more complex loops should be apparent. Hopefully we can get a computer to do most of the work for us, so increasing the complexity is not of such great concern.

6.10.1   Basic Equations

We write the closed-loop transfer function for a general second-order system as

$$H(s) = \frac{\varphi_{\text{out}}(s)}{\varphi_{\text{in}}(s)} = \frac{n_1 s^2 + n_2 s^1 + n_3 s^0}{s^2 + d_2 s^1 + d_3 s^0}. \qquad (6.23)$$

[For simplicity we have set $d_1 = 1$. This is always possible without restricting the value of $H(s)$.] This is consistent with the output and input being written as

$$\varphi_{\text{out}}(s) = \left[n_1 s^2 + n_2 s + n_3\right] X_3(s) \qquad (6.24)$$

and

$$\varphi_{\text{in}}(s) = \left[s^2 + d_2 s^1 + d_3\right] X_3(s). \qquad (6.25)$$

Interpreting the Laplace operator s as time differentiation, we can write the output as a function of time as

$$\varphi_{\text{out}}(t) = n_1 \ddot{x}_3 + n_2 \dot{x}_3 + n_3 x_3, \qquad (6.26)$$

where $x_i \equiv x_i(t)$ and $\dot{x}_i = dx_i/dt$.

However, since the values of each of the orders of derivative are necessary to define the state of the system at any time, we consider each separately and write

$$\varphi_{\text{out}}(t) = n_1 x_1 + n_2 x_2 + n_3 x_3 \tag{6.27}$$

with

$$x_2 \triangleq \dot{x}_3 \tag{6.28}$$

and

$$x_1 \triangleq \dot{x}_2 = \ddot{x}_3, \tag{6.29}$$

or, in general,

$$x_i \triangleq \dot{x}_{i+1}. \tag{6.30}$$

Following the same procedure we can obtain

$$\varphi_{\text{in}}(t) = x_1 + d_2 x_2 + d_3 x_3 \tag{6.31}$$

or, writing this in terms of the highest derivative,

$$x_1 = \varphi_{\text{in}}(t) - d_2 x_2 - d_3 x_3. \tag{6.32}$$

We can now write the vector

$$\mathbf{X} = \begin{bmatrix} x_2 \\ x_3 \end{bmatrix} \tag{6.33}$$

and its derivative

$$\dot{\mathbf{X}} = \begin{bmatrix} \dot{x}_2 \\ \dot{x}_3 \end{bmatrix} = \begin{bmatrix} x_1 \\ x_2 \end{bmatrix} = \begin{bmatrix} \varphi_{\text{in}}(t) - d_2 x_2 - d_3 x_3 \\ x_2 \end{bmatrix} \tag{6.34}$$

and incorporate these in a state-vector equation,

$$\dot{\mathbf{X}} = \mathbf{A}\mathbf{X} + \mathbf{B}u_{\text{in}}, \tag{6.35}$$

with

$$\mathbf{A} = \begin{bmatrix} -d_2 & -d_3 \\ 1 & 0 \end{bmatrix}, \tag{6.36}$$

$$\mathbf{B} = \begin{bmatrix} 1 \\ 0 \end{bmatrix}, \tag{6.37}$$

and

$$u_{in} = \varphi_{in}(t). \tag{6.38}$$

In order to more easily see its equivalence to Eq. (6.34), let us expand (6.35) as

$$\begin{bmatrix} \dot{x}_2 \\ \dot{x}_3 \end{bmatrix} = \begin{bmatrix} -d_2 & -d_3 \\ 1 & 0 \end{bmatrix} \begin{bmatrix} x_2 \\ x_3 \end{bmatrix} + \begin{bmatrix} 1 \\ 0 \end{bmatrix} \varphi_{in}(t). \tag{6.39}$$

This matrix equation is then solved (by a computer program, preferably) to give the state variables in **X** as a function of time.

We still must obtain the desired output as a function of **X**. We will do this for $n_1 = 0$, which is true for the loop of interest, and for the outputs φ_{out} and $\dot{\varphi}_{out}$. Equation (6.27) then becomes

$$\varphi_{out}(t) = n_2 x_2 + n_3 x_3, \tag{6.40}$$

and its derivative is

$$\dot{\varphi}_{out}(t) = n_2 \dot{x}_2 + n_3 \dot{x}_3 = n_2 x_1 + n_3 x_2 \tag{6.41}$$

$$= n_2 [\varphi_{in}(t) - d_2 x_2 - d_3 x_3] + n_3 x_2 \tag{6.42}$$

$$= n_2 \varphi_{in} + [n_3 - n_2 d_2] x_2 - n_2 d_3 x_3. \tag{6.43}$$

Here we have used Eq. (6.32) to eliminate x_1.

The output is now expressed in the form of a second vector equation,

$$\mathbf{Y} = \mathbf{CX} + \mathbf{D}u_{in} \tag{6.44}$$

where

$$\mathbf{Y} = \begin{bmatrix} \dot{\varphi}_{out} \\ \varphi_{out} \end{bmatrix}, \tag{6.45}$$

$$\mathbf{C} = \begin{bmatrix} (n_3 - n_2 d_2) & -n_2 d_3 \\ n_2 & n_3 \end{bmatrix}, \tag{6.46}$$

and

$$\mathbf{D} = \begin{bmatrix} n_2 \\ 0 \end{bmatrix}. \tag{6.47}$$

As before, we can write this out as

$$\begin{bmatrix} \dot{\varphi}_{\text{out}}(t) \\ \varphi_{\text{out}}(t) \end{bmatrix} = \begin{bmatrix} (n_3 - n_2 d_2) & n_2 d_3 \\ n_2 & n_3 \end{bmatrix} \begin{bmatrix} x_2 \\ x_3 \end{bmatrix} + \begin{bmatrix} n_2 \\ 0 \end{bmatrix} \varphi_{\text{in}}(t), \quad (6.48)$$

which we can compare to (6.40) and (6.43). If only one output [e.g., $\varphi_{\text{out}}(t)$] were desired, $\mathbf{C}$ and $\mathbf{D}$ would have only one row each.

Now we can apply these equations more specifically to $H(s)$ as given by Eq. (6.4a). Comparing Eq. (6.4a) to Eq. (6.23), the following equivalencies become apparent:

$$n_1 = 0; \quad n_2 = 2\alpha\zeta\omega_n; \quad n_3 = \omega_n^2; \quad d_1 = 1; \quad d_2 = 2\zeta\omega_n; \quad d_3 = \omega_n^2.$$
$$(6.49)$$

With these equations we can not only obtain $\varphi_{\text{out}}(t)$ and $\omega_{\text{out}}(t) = \dot{\varphi}_{\text{out}}(t)$ but also a phase-plane plot $\omega_{\text{out}}(\varphi_{\text{out}})$, which we will meet later in Chapter 8.

Fortunately, there are programs, such as MATLAB, that make the process described above easy, not only solving the matrix equations but converting from numerator and denominator coefficients of $H(s)$ to the required matrix form.

6.10.2   Initial Conditions

The transient responses we have worked with up to now have had zero initial conditions. That is, the state variables are initially zero because we begin at steady state. Sometimes we may know the frequency and phase at some time and want to know what they are at some later time. Initial conditions can be accommodated by the state-space method, but we must know how to get the values of the initial state variables from the initial phase and frequency.

We can solve (6.40) and (6.43) to get

$$x_2 = \frac{\varphi_{\text{out}} n_2 d_3 + \dot{\varphi}_{\text{out}} n_3 - \varphi_{\text{in}} n_2 n_3}{\text{denom}} \quad (6.50)$$

and

$$x_3 = \frac{\varphi_{\text{out}}(n_3 - n_2 d_2) - \dot{\varphi}_{\text{out}} n_2 + \varphi_{\text{in}} n_2^2}{\text{denom}} \quad (6.51)$$

where

$$\text{denom} = n_3^2 - n_2 n_3 d_2 + n_2^2 d_3. \quad (6.52)$$

Therefore, if we put the initial values $\omega_{\text{out}}(t = 0+)$, $\varphi_{\text{out}}(t = 0+)$, and $\varphi_{\text{in}}(t = 0+)$ in the above equations and begin with the resulting state variables, we obtain the response starting with those initial conditions.

Something interesting that we find by doing this is that the step response, for a given ω_n and ζ, starting with a given set of initial conditions, is independent of α. If a phase step is applied, α determines the initial value of frequency, but the response from then is determined by that initial frequency and the initial phase $[\varphi_{out}(t = 0+) = 0]$.

We may be able to obtain the initial value of **X** more easily if we have access to a computer program that will perform inverses. If **Y** contains n independent variables, where n is the order of the transfer function [2 in Eq. (6.23)], then, if we multiply (6.44) by $\mathbf{C}^{-1}$, the inverse of **C** (which will be square), we obtain

$$\mathbf{C}^{-1}\mathbf{Y} = \mathbf{X} + \mathbf{C}^{-1}\mathbf{D}u_{in}. \tag{6.53}$$

Rearranging and applying to the initial condition, we obtain

$$\mathbf{X}(0) = \mathbf{C}^{-1}[\mathbf{Y}(0) - \mathbf{D}u_{in}(0)] \tag{6.54}$$

or

$$\begin{bmatrix} x_2(0) \\ x_3(0) \end{bmatrix} = \mathbf{C}^{-1}\left[\begin{bmatrix} \dot{\varphi}_{out}(0) \\ \varphi_{out}(0) \end{bmatrix} - \begin{bmatrix} n_2 \\ 0 \end{bmatrix}\varphi_{in}(0)\right]. \tag{6.55}$$

The inverse for the case we are studying is

$$\mathbf{C}^{-1} = \frac{1}{\text{denom}}\begin{bmatrix} n_3 & n_2 d_3 \\ -n_2 & (n_3 - n_2 d_2) \end{bmatrix}, \tag{6.56}$$

which we can verify, or discover, by using (6.56) in (6.55) and comparing the result to Eq. (6.50) and (6.51). We can also verify it by computing $\mathbf{C}^{-1}\mathbf{C}$ and observing that it equals the unity matrix.

6.11   AN APPROXIMATE SOLUTION USING STATE-SPACE VARIABLES

Another method [Gill, 1981], produces an approximate, but often adequate, solution that does not require complex matrix computations. New values for the state variables are computed at the end of a time interval based on the values at the beginning of that interval (end of the previous interval). Each computation interval starts with the solution of Eq. (6.32), based on the values of the state variables (e.g., x_2, x_3) and input (e.g., φ_{in}) at the end of the previous interval. Then the other state variables are found, based on

Eq. (6.30), by integration:

$$x_{i+1}(t_j) = x_{i+1}(t_{j-1}) + \int_{t_{j-1}}^{t_j} x_i \, dx_i \approx x_{i+1}(t_{j-1}) + \frac{x_i(t_{j-1}) + x_i(t_j)}{2}(t_j - t_{j-1}).$$

$$(6.57)$$

This is done in sequence, starting with x_2, until all of the state variables have been computed. Then the output is computed using (6.48). Accuracy gets better as the time interval $t_j - t_{j-1}$ gets smaller.

6.12   EFFECT OF AN ADDED POLE

Figure 6.14 illustrates the effect of a pole added to what would otherwise be a type 2, second-order, loop with $\zeta = 0.5$. We see little effect on the transient response when the frequency ω_a of the added pole is 10 times ω_n, which approximately equals ω_L when $\zeta = 0.5$ (they are equal in the tangential approximation). As ω_a comes closer to ω_n, the effect grows until an undamped oscillation occurs when they coincide. At that point the added pole cancels the zero, giving a constant 180° excess phase and producing instability.

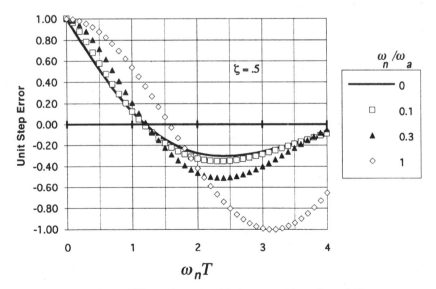

Fig. 6.14   Effect of a pole added to type 2 loop ($\zeta = 0.5$).

6.M APPENDIX: TRANSIENT RESPONSES USING MATLAB

The simplest way to generate a step response with MATLAB is to use the step function, step (NUM, DEN). Here NUM and DEN are vectors as discussed in Section 5.M, but this time they are part of the closed-loop transfer functions, $H(s)$ or $1 - H(s)$, whereas they were part of $G(s)$ in Section 5.M.

6.M.1 step Program with Simple Vectors

If, in Eqs. (6.4), we let $\zeta = 0.5$, $\omega_n = 1$, and set α either to zero or one, then all of the coefficients of $H(s)$ and $1 - H(s)$ are either 0 or 1. This is of no fundamental importance but it does lead to the simple vectors used in the ErTrn... and OutTrn... files as illustrated in the following program:

ErTrn1a

```
% ERROR TRANSIENT RESPONSE FOR
% alpha = 1, zeta = 0.5, Wn = 1
% Step, Ramp, or Parabolic inputs
% Choose response type by changing "den" below
num = [1 0 0];
denstep = [1 1 1];
denramp = [1 1 1 0];
denparab = [1 1 1 0 0];
t = linspace(0, 10, 100);
den = denstep; %Set den to denstep, denramp, or denparab
step(num,den,t);
grid;
```

If we set $\alpha = 1$ in Eq. (6.4b), only the s^2 term exists in the numerator, corresponding to num from ErTrn1a. The denominator coefficients are all 1, as in denstep above. When we execute this file, with den = denstep, we obtain results as shown in Fig. 6.3 for $\alpha = 1$, $\zeta = 0.5$, $\omega_n = 1$.

If we integrate the step input to produce a ramp, the resulting output will be the integral of the step response. We can find it by multiplying $1 - H(s)$ by $1/s$, corresponding to a shifting of all the 1s in denstep to the left, producing denramp. When we execute the file with den = denramp, therefore, we obtain the ramp error response of Fig. 6.7 with $\alpha = 1$, $\zeta = 0.5$, $\omega_n = 1$.

Similarly, if we integrate again, the output will be the response to a parabola $t^2/2$, as shown in Fig. 6.10 with $\alpha = 1$, $\zeta = 0.5$, $\omega_n = 1$. Of course, we can obtain responses for other values of α, ζ, and ω_n by changing NUM and DEN appropriately to represent Eqs. (6.4).

6.M.2 Effect of Added Pole

The curves of Fig. 6.14 can be generated with a slight modification of the program given above. The open-loop transfer function for $\alpha = 1$ is Eq. (4.23). We are using $\omega_n = 1$ for simplicity and to normalize the time ($\omega_n t \Rightarrow t$). Since $\zeta = 0.5$ also, by Eq. (4.16) $\omega_z = 1$ and the open loop transfer function is

$$G(s) = \frac{1}{s^2}(1 + s). \tag{6.M.1}$$

With an added pole at

$$\omega_+ = \omega_n / r_n, \tag{6.M.2}$$

the open-loop transfer function becomes

$$G_+(s) = \frac{1}{s^2}\frac{1 + s}{1 + r_n s}, \tag{6.M.3}$$

leading to a close-loop transfer function of

$$H_+(s) = \frac{1}{1 + G_+(s)} = \frac{r_n s^3 + s^2}{r_n s^3 + s^2 + s + 1}. \tag{6.M.4}$$

The corresponding NUM and DEN vectors can be seen in the program that follows.

ExtrPole

```
% ERROR STEP RESPONSE WITH ADDED POLE
% Compare to Fig. 6.14
% alpha = 1, zeta = 0.5, Wn = 1
% rn: ratio of natural frequency to added pole

t = linspace (0, 10, 100);

clg; m = 0
while (rn> = 0)
num = [rn 1 0 0];
den = [rn 1 1 1];
step (num, den, t);
grid;
rn = input ('enter ratio of Wn to added pole (negative to quit)')
end % while
```

6.M.3 Transient Response Program with Dynamic Entry

Transnts is a program that uses the techniques above to plot error responses to step, ramp, and parabolic inputs to a second-order loop while permitting the user to select parameters dynamically, that is, the parameters (except for ω_n) or the input can be altered and a new plot obtained without starting the program over. Results will be as shown in the responses in Figs. 6.2 to 6.4 and 6.7 to 6.11, but α and ζ can be selected arbitrarily, not limited to the common values in the plots.

Transnts

```
% This program plots the error response for an input
% unit step or
% a ramp with slope Wn or
% a parabola with slope (Wn^2)t

% MODIFY THESE PARAMETERS
% ***********************
Wn = 1    % <- Natural Frequency
zeta = 1; % <- Damping Factor
% ***********************
% Then Go

zeta
go = 2;

while (go ~= 0)
   if (go == 2)
      alpha = 2;
      while ((alpha < 0) | (alpha >1))
         alpha = input('input alpha (0 ≤ alpha ≤ 1): ');
      end; % while alpha
   else
      zeta = -1;
      while (zeta <= 0)
         zeta = input('input zeta (>0): ');
      end; % while zeta
   end; % if
   x2 = 2*zeta*Wn;
   x3 = Wn*Wn;
   N0 = 2*(1-alpha)*zeta*Wn;
   num = [1 N0 0]; % <- numerator
   denstep = [1 x2 x3]; % <- denominator for step in
   denramp = [1 x2 x3 0]; % <- denominator for ramp in
   denparab = [1 x2 x3 0 0]; % <- denominator for parabola in
   t = linspace(0,10,50); % <- 0 to 10 with 50 steps
```

```
   go = 1;
   while(go == 1)
      ctrl = input('step, ramp, parabola, alpha or zeta
                             change, or quit? Enter first
                             letter: ', 's');
      clg;
      if strcmp(ctrl,'s'),
         step(num,denstep,t);
         grid;
      elseif strcmp(ctrl,'r'),
         step(num,denramp,t);
         grid;
      elseif strcmp(ctrl,'p'),
         t = linspace(0,5,20);
         step(num,denparab,t);
         grid;
      elseif strcmp(ctrl,'q'),
         go = 0;
      elseif strcmp(ctrl,'a'),
         go = 2;
      elseif strcmp(ctrl,'z'),
         go = 3;
      end % if
   end % while go == 1
end % while go ~= 0
```

6.M.4 State Space

SSstep produces the same step response as does ErTrn1a above but employs MATLAB's ordinary state-space matrix method that is hidden by the (step) function used in the previous program.

SSstep

```
%Develop error step response by standard matrix procedure
% alpha = 1, zeta = 0.5, Wn = 1
% Step Input

% Generate vector of computation times:
inc = .05; %sampling period
ending = 10; % Student MATLAB limits ending/inc to to 1024 in
                         v.3.5, 8192 in v.4, 16384 in v.5
last = 1 + ending/inc; % index of last data points
t = linspace(0,ending,last); % linear sequence of times
dt = t(2) - t(1); % time increment
```

```
% Generate input step vector:
```

```
u = ones(1,last); % unit step input; column vector with input
                                at each time
```

```
clg; % clear previous graphs
```

```
NUM = [1 0 0]; % for error response, as before
DEN = [1 1 1]; % denstep from ErTrnla
[a,b,c,d] = tf2ss(NUM,DEN); % get elements of dynamic matrix
                             equation (c is bottom row of C and
                             d is bottom row of D in Eq. (6.44)
[Phi, Gamma] = C2D(a,b,dt); % get matrices for use in solution
x = ltitr(Phi,Gamma,u'); % state variable solution for each
                           value of t
y = c*  +d*u; % outputs. x is transpose of X in Eq. (6.33); y
               is lower element in Eq. (6.45)
            % x', y and u have a column for each
              time increment
```

```
plot(t,y)
grid
title('Time Response, Error, Step Input')
xlabel('seconds')
ylabel('radians')
```

To generate a ramp response we could change DEN, as we did in ErTrnla but the more usual way is to change the driving function, which is enclosed in a box above. For a ramp, it would be replaced by the following. (last and inc were defined above in the program.)

```
for i = 1:last,
        u(i) = (i-1)*inc;
end % generation of input vector
```

This generates a vector whose elements are proportional to time. However, it represents a stepped (sampled and held) approximation to a ramp whereas, previously, a stepped version of a step, which is still a step, was integrated within the response function to produce a time-continuous ramp, so the previous method is more accurate. (The labels on the graph should also be changed appropriately.)

The reader might wish to generate a vector representing a parabolic input, $t^2/2$.

6.M.5   Adding Phase-Plane Output

SSaddedP performs the same function as ExtrPole but using the basic state-space equations. We have also added a phase-plane plot, a plot of frequency versus phase. To get frequency, we must output the derivative of the phase so **Y** has two elements, φ and $\dot{\varphi}$. To accomplish this we give NUM two rows, first the numerator that will produce φ_{out} and second its derivative. Since taking the derivative is equivalent to multiplying by s, the orders of all of the powers of s are increased by 1. This amounts to shifting the coefficients in the second row to the left, each to apply to a one-higher order of derivative (or of s). If we had retained the numerator for $1 - H$, as in the previous programs, the left-most coefficient, which was 1, would have shifted left, increasing the order of the vector and making it of higher order than DEN. This is not permitted so we changed from representing the error transfer function, $1 - H$, to the output transfer function, H. We did this by subtracting the previous NUM from DEN, equivalent to subtracting $1 - H$ from 1. We will obtain the error response later by subtracting the output from the input.

SSaddedP

```
%ERROR STEP RESPONSE WITH ADDED POLE
% Develop output time response and
% linear phase plane by ordinary matrix procedure

% Compare to Fig. 6.14
% alpha = 1, zeta = 0.5, Wn = 1
% rn: ratio of natural frequency to added pole

% Generate vector of computation times
inc = .05; %sampling period
ending = 10; % Student MATLAB limits ending/inc to to 1024 in
                       v.3.5, 8192 in v.4, 16384 in v.5
last = 1 + ending/inc; % index of last data points
t = linspace(0,ending,last); % linear sequence of times

u = ones(1,last); % unit step input

rn = 0
while(rn>=0)
   NUM = [0 0 1 1           % first row is output (phase)
          0 1 1 0];         % second row is derivative of
                            output (frequency)
   DEN = [rn 1 1 1];
   [a,b,c,d] = tf2ss(NUM,DEN); % elements of dynamic matrix
                              equation
   [Phi, Gamma] = C2D(a,b,inc); % values for use in solution
                              eqaution
   x = ltitr(Phi,Gamma,u'); % solutions for each value of t
   y = c*x'+d*u; % outputs
```

```
clg;
er = u - y(1,:);
  subplot(211), plot(t,er,'g')
  grid
  title('Time Response, Error (linear)')
  xlabel('seconds')
  ylabel('radians')
  subplot(212), plot(y(1,:),y(2,:),'r')
  grid
  title('Phase Plane, Output (linear)')
  xlabel('phase (radians)')
  ylabel('freq (rad/sec)')


  rn = input('enter ratio of Wn to added pole (negative to
                         quit)')
end %while
```

The resulting graph for $r = 0.5$ is shown in Fig. 6.M.1. Note how the damped oscillations of the time response appear in the phase-plane plot. The phase-plane plot with $r = 1$ is especially interesting.

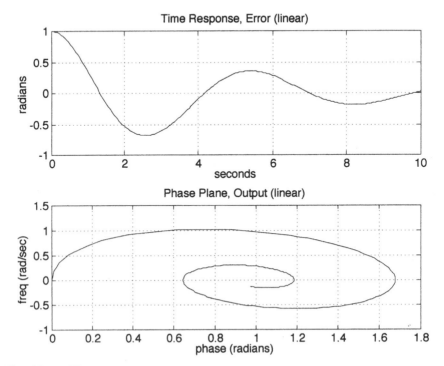

Fig. 6.M.1 Time response and phase-plane plot for step input with added pole, $r = 0.5$.

6.M.6   Adding Initial Conditions

LMatPhPn is a step response program for second-order loops that shows both the error response versus time and the phase plane and permits entry of damping factor and alpha, converting these values to the inputs required for NUM and DEN. It also allows the specification of initial phase and frequency.

To obtain an ordinary step response, set SR to 1 and provide a value for A, the input phase step amplitude. In that case the state variables are set to zero (initial steady state). Otherwise the transient starts with the specified initial output phase Phinit and radian frequency Winit, the state variables being set to whatever is necessary to produce them. The presence of a phase with magnitude A at the input still influences the results, however—it establishes the final value.

Enter parameter changes before executing the program. Try several values of α with other parameters constant and initial conditions specified. Note how the response is independent of α under these conditions.

A second program LMPhPnHz differs only in that frequency is given in hertz.

LMatPhPn

```
%TIME RESPONSE & PHASE PLANE PLOT
%Develop output time response and
% linear phase plane by ordinary matrix procedure

% MODIFY PARAMETERS BELOW & GO
%******************************
SR = 0 ; % 1 for step response, otherwise start at initial
                             conditions
Phinit = 0 ; % radian initial phase
Winit = 0 ; % radian/sec initial frequency
A = 1 % radian phase step amplitude
Wn = 1 % rad/sec' % natural frequency
z = .75 % damping factor
alpha = 0
ending = 10/Wn; % Student MATLAB limits ending/inc to1024 in
                          v.3.5, 8192 in v.4, 16384 in v.5
inc = .05/Wn %sampling period
%******************************

% Generate vector of computation times
last = 1 + ending/inc; % index of last data points
t = linspace(0,ending,last); % linear sequence of times
dt = t(2) - t(1); % time increment
u = A * ones(1,last); % step input, amplitude A
```

```
%Polynominals for closed-loop gain:
   N = [ 0        2*z*Wn*alpha    Wn^2     % numerator for phase
        2*z*Wn*alpha     Wn^2         0 ] ;  % numerator for
                            frequency
   D = [ 1         2*z*Wn        Wn^2 ] ;   % denominator
[a,b,c,d] = tf2ss(N,D) ;    % elements of dynamic matrix
                            equation

if (SR == 1)&(A~=0),
   x0 = [0,0]; % step response starts with zero state
                            variables;
   disp('STEP RESPONSE');
else % state variables to give required initial conditions
     % reference Eq. (6.50)-(6.52)
   denom = N(1,2)^2*D(3)-N(1,3)*N(1,2)*D(2)+N(1,3)^2;
   x0(1)=(Winit*N(1,3)+Phinit*N(1,2)*D(3)-
                            u(1)*N(1,2)*N(1,3))/denom;
   x0(2)=(Phinit*(N(1,3)-N(1,2)*D(2))-
                            Winit*N(1,2)+u(1)*N(1,2)^2)/denom;
   disp('START AT INITIAL CONDITIONS');
end % if

[Phi, Gamma] = C2D(a,b,dt) ;    % values for use in solution
                            equation
x = ltitr(Phi,Gamma,u',x0) ;    % solutions for each value of
                            t
y = c*x'+d*u ; % outputs

clg; % clear previous graphs
er = u - y(1,:);
subplot(211), plot(t,er,'g')
grid
title('Time Response, Error (linear)')
xlabel('seconds')
ylabel('radians')
subplot(212), plot(y(1,:),y(2,:),'r')
grid
title('Phase Plane, Output (linear)')
xlabel('phase (radians)')
ylabel('freq (rad/sec)')
```

PROBLEMS

6.1 A PLL has an integrator-and-lead filter, Fig. 3.23c with $R_1 = 2 \text{ k}\Omega$ and $G_a \Rightarrow \infty$. $K_p = 1$ V/rad and $K_v = 1$ MHz/V. The loop is locked and following an input frequency ramp of 100 MHz/sec. The steady voltage u_1 at the input to the loop filter is 1 V. The answers to the following questions should not involve many equations. If you think about them properly, they are simple.

(a) What is the size of C in the loop filter?

(b) What is the value of the phase error?

6.2 A PLL has the loop filter in Fig. 3.21 with $R_1 = 1$ kΩ and $C = 0.014$ μF. The VCO tuning curve is linear with 999 kHz at 3 V and 996 kHz at 2.5 V. The phase detector is a flip-flop with output states of 0.5 and 3.5 V. Give the following (including units of course):

(a) Minimum input frequency at which lock can occur

(b) Maximum input frequency at which lock can occur

(c) K_p

(d) K_{LF}

(e) K_v

(f) ω_p

(g) ω_n

(h) ζ

(i) ω_L (unity open-loop-gain frequency)

(j) Absolute difference between input and output phases, $\varphi_{in} - \varphi_{out}$, with 1 MHz steady input

(k) Time for the output to achieve 90% of its ultimate change after an input step (e.g., output and input phases)

6.3 For the locked loop shown below, pick R such that the first occurrence of zero frequency error is 47 msec after a frequency step input.

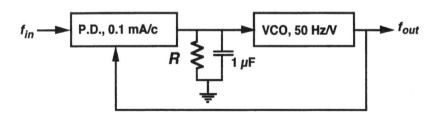

6.4 A PLL has a filter shown in Fig. 3.19 with $R_1 = 3$ kΩ, $R_2 = 1$ kΩ, and $C = 0.16$ μF. $K_p = 0.5$ V/rad and $K_v = 10,000$ (rad/sec)/V. What is the phase error in the locked loop 160 μsec after a 0.2-rad input phase step?

6.5 For a loop with $\omega_n = 300$ rad/sec, $\alpha = 1$, $\zeta = 0.5$, $K_v = 1$ kHz/V, what is the peak phase error φ_e in response to a 1-V step added to the tuning voltage and at what time does it occur?

6.6 For a loop with $\omega_n = 1000$ rad/sec, $\alpha = 1$, $\zeta = 1$, the VCO center frequency ramps at 10 Hz/msec due to temperature changes. What is the resulting phase error φ_e?

CHAPTER 7

MODULATION RESPONSE

Since phase-locked loops (PLLs) can be used as phase and frequency modu-
lators and demodulators, the transfer functions from phase or frequency to
voltage and from voltage to phase or frequency are of obvious importance.
We will want to know how the loop filters the modulation. We will also be
interested in how the phase error responds since we will need to know
whether the loop is remaining linear (and locked). As with the transient
response, the equations that we need have already been developed in
Chapter 4, but we will here interpret them in terms of their implications for
modulation. As was done for the transient response, the modulation response
will be computed for the output and the error. And, as was done for the
transient, important related responses will also be considered.

7.1 PHASE AND FREQUENCY MODULATION

A sinusoid can be written

$$v = V \sin[\psi(t)], \qquad (7.1)$$

where we have chosen to use symbols appropriate to a voltage. The instantaneous phase is

$$\psi(t) = \omega_c t + \varphi(t), \qquad (7.2)$$

where the phase consists of a steadily advancing component, $\omega_c t$, and a
modulation component,

$$\varphi(t) = m \sin(\omega_m t + \theta_m). \qquad (7.3)$$

139

Here m is the peak phase deviation, also called the modulation index. This implies an instantaneous frequency

$$\omega(t) \equiv \frac{d\psi(t)}{dt} = \omega_c + \frac{d\varphi(t)}{dt} \qquad (7.4)$$

$$= \omega_c + m\omega_m \cos(\omega_m t + \theta_m) \qquad (7.5)$$

$$= \omega_c + \Delta\omega \sin(\omega_m t + \theta_m + \pi/2), \qquad (7.6)$$

where $\Delta\omega = m\omega_m$. Thus the peak phase and frequency deviations are related by

$$m = \Delta\varphi = \frac{\Delta\omega}{\omega_m} = \frac{\Delta f}{f_m} \qquad (7.7)$$

while, from Eqs. (7.3) and (7.6), the frequency deviation leads the phase deviation by 90°.

Note that, if the signal in Eq. (7.1) is phase modulated, it is also frequency modulated. So, what is the difference? Why do we have two names for the same process? The difference is only significant when the modulation represents information at multiple frequencies. For example, with phase modulation, a modulating signal that has constant amplitude as its frequency changes will produce frequency-independent phase deviation, but by Eq. (7.7), the frequency deviation will be proportional to modulation frequency f_m. Conversely, with frequency modulation, the same signal would produce constant frequency deviation Δf, but the phase deviation $m = \Delta\varphi$ would decrease with modulation frequency. This is illustrated in Fig. 7.1.

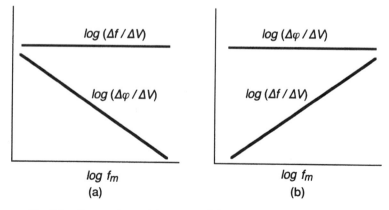

Fig. 7.1 Comparison of frequency (a) and phase (b) modulations.

7.2 MODULATION RESPONSES

Suppose Eqs. (7.1) through (7.7) represent the reference input to a PLL. A similar set of equations could then be written to describe its output. With phase modulation, the phase change at the output is related to the phase change at the input by [see Eq. (4.3)]

$$\varphi_{\text{out}}(\omega_m) = \varphi_{\text{in}}(\omega_m)H(\omega_m), \tag{7.8}$$

which also implies

$$|\varphi_{\text{out}}(\omega_m)| = |\varphi_{\text{in}}(\omega_m)||H(\omega_m)| \tag{7.9}$$

and

$$\angle\,\varphi_{\text{out}}(\omega_m) = \angle\,\varphi_{\text{in}}(\omega_m) + \angle\,H(\omega_m). \tag{7.10}$$

Then the equation like (7.3) that describes the phase modulation at the output would be

$$\varphi_{\text{out}}(t) = m|H(\omega_m)|\sin[\omega_m t + \theta_m + \angle H(\omega_m)]. \tag{7.11}$$

Similarly, the output frequency modulation would be described by an equation like (7.6) [see Eq. (2.27)]

$$\omega_{\text{out}}(t) = \omega_c + \Delta\omega|H(\omega_m)|\sin[\omega_m t + \theta_m + \angle H(\omega_m) + \pi/2]. \tag{7.12}$$

The error responses could be similarly written as

$$\varphi_e(t) = m|1 - H(\omega_m)|\sin\{\omega_m t + \theta_m + \angle[1 - H(\omega_m)]\} \tag{7.13}$$

and

$$\omega_e(t) = \Delta\omega|1 - H(\omega_m)|\sin\{\omega_m t + \theta_m + \angle[1 - H(\omega_m)] + \pi/2\}. \tag{7.14}$$

7.3 RESPONSES IN A FIRST-ORDER LOOP

We begin the study of modulation and demodulation with the first-order loop shown in Fig. 7.2. We have previously shown loops with phase change input (Fig. 2.4) and with frequency change input (Fig. 2.7). Figure 7.2 shows both types of inputs simultaneously. Since we are considering the loop to be linear in this chapter, superposition applies.

   In this case the only voltage variable v appears at the output of the phase detector, which is also the input to the voltage-controlled oscillator (VCO). Phase or frequency modulation, represented by φ_{in} or ω_{in}, respectively, is

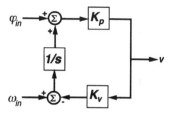

Fig. 7.2 Demodulation in a first-order loop. Comment about where the phase reversal ($-$) is placed: since the output of K_v is VCO frequency, it is *subtracted* from the input frequency.

introduced into the phase detector. The phase appearing after the summer is the error response, so

$$v(\omega_m) = \varphi_e(\omega_m)K_p = \varphi_{in}(\omega_m)[1 - H(\omega_m)]K_p. \tag{7.15}$$

Using Eq. (2.26b) for φ_e, with $s \Rightarrow j\omega$, we obtain

$$v(\omega_m) = \varphi_{in}(\omega_m)j\omega_m K_p/(j\omega_m + K_p K_v). \tag{7.16}$$

Voltage v is also related to ω_{in} by

$$v(\omega_m) = \omega_{out}(\omega_m)/K_v = \omega_{in}(\omega_m)H(\omega_m)/K_v. \tag{7.17}$$

Using Eq. (2.25b), this is

$$v(\omega_m) = \omega_{in}(\omega_m)K_p/(j\omega_m + K_p K_v). \tag{7.18}$$

Since the two inputs are connected by $1/s$, it is not surprising that the two responses differ by a factor of $j\omega_m$. These responses are illustrated in Fig. 7.3.

We see that at low frequencies Eq. (7.18) becomes

$$v(\omega_m \ll \omega_L) = \omega_{in}/K_v \tag{7.19}$$

so that v is proportional to the input frequency and provides frequency demodulation. At low frequencies the gain is high enough to force the

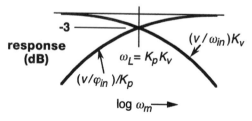

Fig. 7.3 Responses of the first-order loop.

frequency of the VCO to equal that of the input, and thus the VCO's tuning voltage is proportional to the input frequency.

At high frequencies Eq. (7.16) becomes

$$v(\omega_m \gg \omega_L) = \varphi_i(\omega_m)K_p, \tag{7.20}$$

providing phase demodulation. At high frequencies the loop is essentially open so the VCO's phase does not vary. Thus the phase error equals the input phase, and the voltage at the output of the phase detector is therefore proportional to input phase.

Moreover, if, as in Fig. 7.4, a voltage is introduced at the phase detector output,

$$\omega_{\text{out}} = v[1 - H(\omega_m)]K_v \tag{7.21}$$

$$= vj\omega_m K_v/[j\omega_m + K_p K_v] \tag{7.22}$$

and

$$\varphi_{\text{out}}(\omega_m) = vH(\omega_m)/K_p \tag{7.23}$$

$$= vK_v/(j\omega_m + K_p K_v). \tag{7.24}$$

At low frequencies Eq. (7.24) becomes

$$\varphi_{\text{out}}(\omega_m \ll \omega_L) = v/K_p \tag{7.25}$$

so the phase is proportional to v, providing phase modulation. Because the gain is high, the phase is forced to follow v.

At high frequencies Eq. (7.22) becomes

$$\omega_{\text{out}}(\omega_m \gg \omega_L) = vK_v, \tag{7.26}$$

enabling us to obtain frequency modulation. Here the low gain permits the VCO to be modulated as if the loop were open, but the loop keeps the average [direct current] value of the frequency equal to the reference frequency (which is not shown because it is not relevant to the discussion of the modulator).

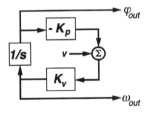

Fig. 7.4 Modulation in a first-order loop. The K_p block includes the *usual* summer, which contains the minus sign required to produce phase reversal in the loop. The reference input is not shown because it is constant.

Example 7.1   Modulation in a Simple Loop

a. Choose an appropriate bandwidth for a first-order loop to act as a frequency modulator and for a second such loop to serve as demodulator. The tones to be transmitted have modulation frequencies between 1 and 5 kHz.

We apply the tones at v in Fig. 7.4 and obtain the demodulated tones at v in Fig. 7.2. We must choose a bandwidth for the loop in Fig. 7.4 that is low compared to the tone frequencies so the feedback does not attempt to counter v and its full value will be impressed on the VCO. Conversely, we must choose a bandwidth in Fig. 7.2 that is high enough so the VCO will follow the input and v will be accurately proportional to frequency. Thus we might select the frequency at unity loop gain f_L to be *200 Hz in Fig. 7.4* and *10 kHz in Fig. 7.2*, for example.

b. What will be the effect on the communication link if the bandwidths are interchanged between the two loops?

Then the existence of high gains at the modulation frequencies in Fig. 7.4 will cause the second input to the summer to follow v, implying that φ_{out} is also proportional (by $1/K_p$) to v. Thus we will produce phase modulation rather than frequency modulation. Similarly, the narrow bandwidth in Fig. 7.2 will cause v to be proportional to φ_{in}. We will thus have constructed a *phase-modulated (PM) link rather than a frequency modulated (FM) link*.

7.4   TRANSFER FUNCTIONS IN A SECOND-ORDER LOOP

The responses in a second-order loop are generally similar to those in the first-order loop, but, because of the filter block, there are more isolated points to be considered and the response shape is more variable.

The closed-loop transfer function $H(s)$ and the error function $[1 - H(s)]$ in Table 4.1 describe the response of the loop's output phase φ_{out} or phase error φ_e, respectively, to the input phase φ_{in}. Figure 4.1 illustrates these relationships. However, the same functions describe other responses in the loop. Figure 7.5 illustrates all of these combinations. Here a disturbance $\Delta x(s)$ is introduced into the loop through a summer at any point where two of the loop's functional blocks are joined. The disturbance could be noise or an intended signal. All the variables are functions of s, even though the dependency is not shown explicitly.

The variable x has been primed to differentiate it from the output that precedes the summer. When $\Delta x = 0$, $x = x'$. Since, in what we have studied so far, Δx has been zero for u_1 and u_2, we have not used the primed

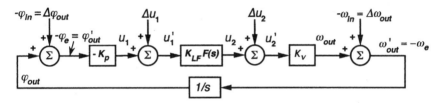

Fig. 7.5   Loop with the various injection points for disturbances shown.

variables. Since we have concentrated on φ and ω (see Fig. 4.1) as points of disturbance, the associated variables were given special names for the disturbance (e.g., φ_{in}) and the error (e.g., φ_e). Rather than invent new special names for all the other disturbances and errors, we will use the convention shown, and we may refer to the φ and ω disturbances and errors in the same manner to keep the presentation symmetrical and to point out the similarity in the treatment of the different variables. There is one awkward difference between the errors in Fig. 7.5 and those considered before, however. A 180° phase shift, or minus sign, must be place somewhere in the loop and is customarily placed before the summer. We could do that for any one of the summers in Fig. 7.5 but only for one. So we place the minus sign approximately in its usual location, before K_p but not before the summer. To maintain the usual definitions for φ_{in}, ω_{in}, φ_e, and ω_e, we then place minus signs before them.

There are two ways to relate a variable to a disturbance. We can multiply Δx by $(1 - H)$ and then by the gain to the point of interest or we can multiply Δx by $-H$ and then divide by the gain from the point of interest to the point of disturbance. For example, the value of u_2 in response to a disturbance Δu_1 can be obtained as

$$u_2 = u_1' K_{\text{LF}} F(s) = \Delta u_1 [1 - H(s)] K_{\text{LF}} F(s) \qquad (7.27)$$

or as

$$u_2 = u_1 /(-K_p K_v /s) = \Delta u_1 H(s) s /(K_p K_v). \qquad (7.28)$$

7.4.1   Output Response

The output response, which precedes the summer, can be obtained by multiplication of the disturbance by $-H(s)$:

$$x(s) = -\Delta x(s) H(s). \qquad (7.29)$$

Here $x(s)$ may equal u_1, u_2, ω_{out}, or φ_{out}.

At modulation frequencies that are low compared to ω_L, $H(\omega_m) \approx 1$, so it then may be convenient to express various responses in terms of $H(\omega_m)$. For example, a power supply transient might cause a change $\Delta\omega_{\text{out}}$ in the VCO frequency when the loop is open. The disturbance can be represented by the injection of $\Delta\omega_{\text{out}}$ as shown in Fig. 7.5, and the closed-loop response at ω_{out} can be obtained by multiplying $-\Delta\omega_{\text{out}}(\omega_m)$ by $H(\omega_m)$, which is conveniently unity at low frequencies. Then u_2 can be obtained by dividing by K_v so $u_2(\omega_m) \approx -\Delta\omega_{\text{out}}(\omega_m)/K_v$. If also $\omega_m \ll \omega_p$, then $u_1(\omega_m) \approx -\Delta\omega_{\text{out}}(\omega_m)/(K_v K_{\text{LF}})$.

At modulation frequencies that are high compared to ω_L, $H(\omega_m) \approx G(\omega_m)$. Therefore a similar procedure can be followed except that the response includes the open-loop $G(\omega_m)$ and, where it is used, $K_{\text{LF}} \Rightarrow K_{\text{LF}} F(\omega_m)$. If $\omega_m \gg \omega_z$, then $F(\omega_m) \Rightarrow \omega_p/\omega_z$. For example, we may wish to know the effect, upon the output frequency, of a high-frequency signal coupled into the circuit at the output of the phase detector. If $\omega_m \gg \omega_L$, ω_z, then $\omega_{\text{out}}(\omega_m) \approx \Delta u_1(\omega_p/\omega_z)K_{\text{LF}}K_v$.

7.4.2 Error Response

The response that follows the disturbance is the error and can be obtained by multiplying $\Delta x(s)$ by $[1 - H(s)]$.

$$x'(s) = \Delta x(s)[1 - H(s)]. \tag{7.30}$$

At modulation frequencies that are high compared to ω_L, $[1 - H(\omega_m)] \approx 1$, so it may be convenient to express responses in terms of $[1 - H(\omega_m)]$. For example, the phase error φ_e in response to an input φ_{in} can be obtained by multiplying $\varphi_{\text{in}}(\omega_m)$ by $[1 - H(\omega_m)]$, which is unity at high frequencies. Then u_1 can be obtained by multiplying by K_p to give $u_1(\omega_m) \approx \varphi_{\text{in}}(\omega_m)K_p$. Then u_2 can be obtained by multiplication by $K_{\text{LF}} F(\omega_m)$ and, if $\omega_m \gg \omega_z$, we have $F(\omega_m) \approx \omega_p/\omega_z$.

At modulation frequencies that are low compared to ω_L, $[1 - H(\omega_m)] \approx 1/G(\omega_m)$. Therefore a similar procedure can be followed except that the response includes $1/G(\omega_m)$ and, where it is used, $K_{\text{LF}} F(\omega_m) \Rightarrow K_{\text{LF}}$ if $\omega_m \ll \omega_p$.

7.4.3 Responses Near ω_L

Where the approximations above cannot be used because ω_m is commensurate with ω_L, it is necessary to know $H(\omega_m)$ or $[1 - H(\omega_m)]$. Table 7.1 is a compilation of transfer functions and error functions relating the variables in Fig. 7.5. The disturbances are shown at the bottom, and the transfer functions to the primed (error) and unprimed (output) variables are shown in the table on the same line as the variable. The diagonal split separates transfer functions that relate to the errors from those that relate to the outputs.

TABLE 7.1. Transfer Functions Between Loop Variables[a]

	$[y/x](s)$				
u'_1	$-(1-H)K_pK_v/s$	$-(1-H)K_p/s$	$-(1-H)K_p$	$(1-H)$	⇓ y
$\varphi'_{out}(-\varphi_e)$	$(1-H)K_v/s$	$(1-H)/s$	$(1-H)$	$-H$	u_1
ω'_{out}	$(1-H)K_v$	$(1-H)$	$-H$	H/K_p	φ_{out}
u'_2	$(1-H)$	$-H$	$-Hs$	Hs/K_p	ω_{out}
y ⇑	$-H$	$-H/K_v$	$-Hs/K_v$	$Hs/(K_pK_v)$	u_2
	Δu_2	$\Delta\omega_{out}(-\omega_{in})$	$\Delta\varphi_{out}(-\varphi_{in})$	Δu_1	
		x			

[a]Those above the diagonal refer to the leftmost column. Those below refer to the rightmost column.

Of the two ways to link the response to the disturbance, the function that avoids $F(s)$ has been chosen for Table 7.1 because $F(s)$ is highly variable and relatively complex in form. Given responses for $H(s)$ and $[1 - H(s)]$, it is a simple matter to multiply or divide them by constants. It is somewhat more complicated to multiply or divide them by s, but it is still simpler than taking into account the details of the filter (which are, nevertheless, embedded in the form of H or $[1 - H]$).

Example 7.2   Use of Table 7.1   What is the transfer function between the input frequency and (a) the phase error? (b) the output frequency?

  a. Using the second column for $x = -\omega_{in}$, we go up to the second row where $y = -\varphi_e$ appears on the left. At the intersection we see the transfer function $(1 - H)/s$.
  b. Since ω_{out} appears in the fourth row on the right, the transfer function appears in the fourth row of column two, $-H$. However, this is the transfer function from $-\omega_{in}$ to $+\omega_{out}$ so the desired function is H.

7.4.4 Phase or Frequency at Inputs and Outputs

As mentioned above, $\Delta\omega_{out}$ can represent the value of a disturbance in the open-loop frequency of the VCO. However, $\Delta\omega_{out}$ might also represent an input frequency transient or modulation. In that case $\omega'_{out} = -\omega_e$ is the frequency error at the input and ω_{out} is the loop's output frequency. Similarly, while $\Delta\varphi_{out}$ may represent $-\varphi_{in}$, it may also represent an open-loop transient in the VCO. In the former case, the summer output is $-\varphi_e$

whereas, in the latter case, it is the phase of the output of the loop. The reason that these symbols can represent either input or output is that there is no physical separation between the input and output. The term $1/s$ is a mathematical rather than a physical entity. It represents an integration that converts frequency to phase, and this can be considered to occur at input or output, whichever is convenient. If we want to know the output phase, we consider the integration to occur between the VCO and the output, converting output frequency to output phase. If we want to consider the input to be a frequency, we consider the integration to occur after the input summer, converting input frequency difference to input phase difference before entry into the phase detector.

7.5 TRANSIENT RESPONSES BETWEEN VARIOUS POINTS

While we have been discussing the frequency response between various points in the loop, Table 7.1 also applies to transient responses. We will discuss this briefly before continuing with our discussion of frequency response.

We have developed equations and graphs for error response to steps, ramps, and parabolas. These are represented in Laplace transforms by $[1 - H(s)]/s$, $[1 - H(s)]/s^2$, and $[1 - H(s)]/s^3$. From these we can also obtain the output responses, related to the driving functions by $H(s)$. Therefore we can easily find $y(t)$ in Table 7.1 when it is related to the $x(t)$ by $[1 - H(s)]$ or $H(s)$ and $x(t)$ is a step, ramp, or parabola.

However, we can also find values of $y(t)$ that are related to $x(t)$ by $(1 - H)/s$ if $x(t)$ is a step or a ramp by using the ramp or parabola response, respectively. This is because $1/s$ represents integration, and the integral of a response to some input equals the response to the integral of that input. For example, if $\Delta\omega_{\text{out}}$ is a step, from the table,

$$u_1'(s) = -\Delta\omega_{\text{out}}(s)\{[1 - H(s)]K_p/s\} = -(1/s)\{[1 - H(s)]K_p/s\} \tag{7.31}$$

$$= -\{(1/s^2)[1 - H(s)]\}K_p. \tag{7.32}$$

But the expression in { } in Eq. (7.32) is the error response to a unit ramp and that is shown in Figs. 6.7 through 6.9. Thus we can find the voltage at the output of the phase detector in response to a frequency step by multiplying the value given by one of those figures by $-K_p$.

Responses in Table 7.1 containing $[H(s)s]$ represent differentiation of responses given by $H(s)$. Therefore we can obtain such a response to a ramp or parabola by using the responses developed for a step or ramp, respectively. For example, to find u_2 in response to a ramp at φ_{in}, we can use the step response from Figs. 6.2 through 6.4, subtracting them from one to get $H(s)$ and dividing by K_v. Since we have not developed graphs for impulse responses, however, we must differentiate step responses to find $[H(s)s]$ in that case.

7.6   MAGNITUDE AND PHASE OF THE TRANSFER FUNCTIONS

The variables can be written in complex form as

$$x = |x|\exp(j\angle x) \tag{7.33}$$

so

$$y = Hx \Rightarrow |y|\exp(j\angle y) = \{|H|\exp(j\angle H)\}\{|x|\exp(j\angle x)\} \tag{7.34}$$

As a result, we can write separate equation for the magnitudes and the phases:

$$|y| = |H||x|, \tag{7.35}$$

$$\angle y = \angle H + \angle x. \tag{7.36}$$

Similarly,

$$|y'| = |(1 - H)||x|, \tag{7.37}$$

$$\angle y' = \angle(1 - H) + \angle x. \tag{7.38}$$

7.6.1   Output Responses

We write $H(\omega_m)$ from Eq. (6.4a) with $s \Rightarrow j\omega_m$, normalizing the modulation frequency to the natural frequency. [See Hgeneric Excel file.]

$$H(\Omega) = \frac{1 + j2\alpha\zeta\Omega}{(1 - \Omega^2) + j2\zeta\Omega}, \tag{7.39}$$

where

$$\Omega = \omega_m/\omega_n. \tag{7.40}$$

The magnitude and phase are then

$$|H(\Omega)| = \sqrt{\frac{1 + (2\alpha\zeta\Omega)^2}{(1 - \Omega^2)^2 + (2\zeta\Omega)^2}}, \tag{7.41}$$

$$\angle H(\Omega) = \tan^{-1}(2\alpha\zeta\Omega) - \tan^{-1}[2\zeta\Omega/(1 - \Omega^2)]. \tag{7.42}$$

7.6.2 Error Responses

The error response can be written [see 1minHgen Excel file]

$$1 - H(\Omega) = \frac{-\Omega^2 + j2(1 - \alpha)\zeta\Omega}{(1 - \Omega^2) + (j2\zeta\Omega)}, \tag{7.43}$$

$$|1 - H(\Omega)| = \sqrt{\frac{\Omega^4 + [2(1 - \alpha)\zeta\Omega]^2}{(1 - \Omega^2)^2 + (2\zeta\Omega)^2}}, \tag{7.44}$$

$$\angle[1 - H(\Omega)] = -\tan^{-1}[2(1 - \alpha)\zeta/\Omega] - \tan^{-1}[2\zeta\Omega/(1 - \Omega^2)]. \tag{7.45}$$

7.6.3 Effect of α

Figure 7.6 shows both the output and error amplitude responses for $\zeta = 1$ at various values of α. Theoretically, we can add a weighted response for $\alpha = 0$ to one for $\alpha = 1$ to produce the response for any desired value of α, according to Eq. (6.6), as we did for transient responses. However, we would require that the responses be given separately for real and imaginary parts so that we could add the two real parts and the two imaginary parts separately and then combine them to get magnitude and phase. It is probably easier to compute the values from the equations above. However, knowing the way in which these responses combine can help us to predict the results in a rough way from analysis of the curves that will be given.

Figure 7.7 shows the corresponding phases. Note that the phase of the output response $H(\Omega)$ with a low-pass filter ($\alpha = 0$) is identical to that of the error response $[1 - H(\Omega)]$ with an integrator-and-lead filter ($\alpha = 1$).

7.6.4 Responses for $\zeta \le 1$

Figures 7.8 and 7.9 show amplitude responses for output and error, respectively. Figures 7.10 and 7.11 show corresponding phases.

According to Eq. (4.20) or (6.4), when $\zeta = 1/\sqrt{2} \approx 0.71$, the two closed-loop poles are at $(-1 \pm j)\omega_n/\sqrt{2}$ and therefore lie on lines that are 45° from the real axis. When also $\alpha = 0$, these are the only singularities (there are no zeros). Such a set produces a two-pole "maximally flat" (Butterworth) response, which has zero-valued first and second derivatives of magnitude relative to ω^2 at $\omega = 0$. The flatness of this response is evident in Fig. 7.8. Note also that the 3-dB bandwidth equals ω_n.

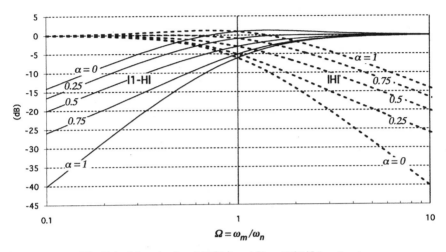

Fig. 7.6   Magnitude of $H(\Omega)$ and $[1 - H(\Omega)]$ for $\zeta = 1$.

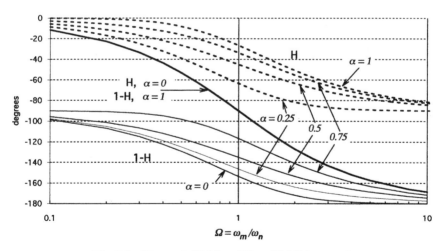

Fig. 7.7   Phase of $H(\Omega)$ and $[1 - H(\Omega)]$ for $\zeta = 1$.

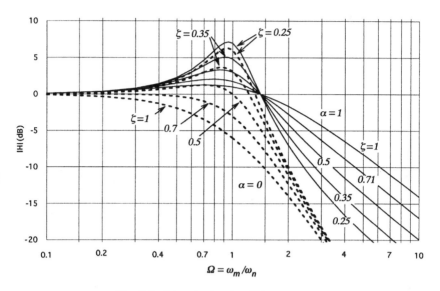

Fig. 7.8 Magnitude of $H(\Omega)$ for $\zeta \leq 1$.

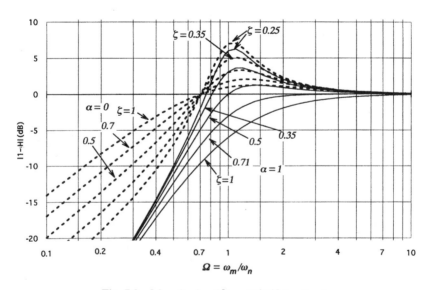

Fig. 7.9 Magnitude of $[1 - H(\Omega)]$ for $\zeta \leq 1$.

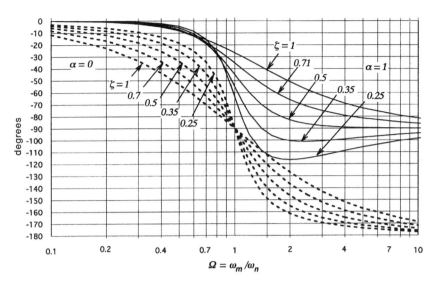

Fig. 7.10 Phase of $H(\Omega)$ for $\zeta \le 1$.

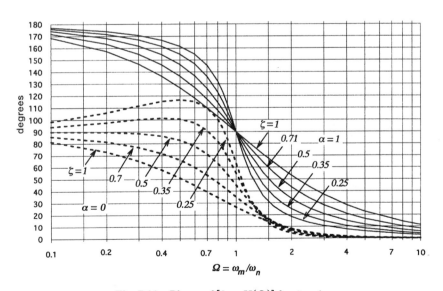

Fig. 7.11 Phase of $[1 - H(\Omega)]$ for $\zeta \le 1$.

7.6.5 Responses for $\zeta \geq 1$

Figures 7.12 and 7.13 show amplitude responses for output and error, respectively, and Figs. 7.14 and 7.15 show the corresponding phases. These are normalized to ω'_L [see Eqs. (6.13) and (6.14)]. As with the step response, we find that the modulation responses are more closely related to the unity open-loop gain frequency ω_L than to ω_n, and we therefore plot the responses versus ω'_L, an approximation to ω_L.

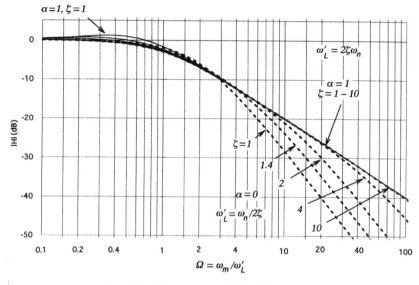

Fig. 7.12 Magnitude of $H(\Omega)$ for $\zeta \geq 1$.

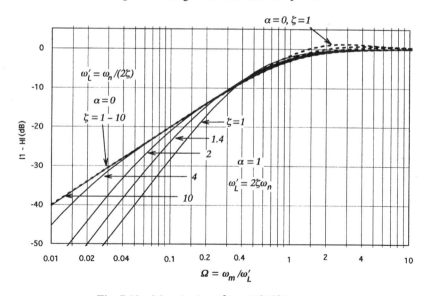

Fig. 7.13 Magnitude of $[1 - H(\Omega)]$ for $\zeta \geq 1$.

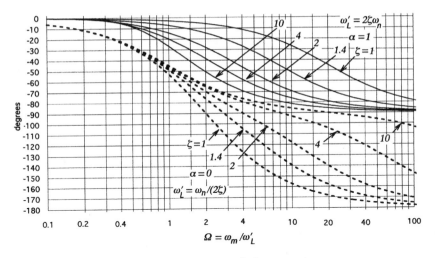

Fig. 7.14   Phase of $H(\Omega)$ for $\zeta \geq 1$.

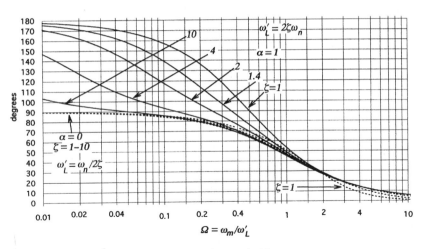

Fig. 7.15   Phase of $[1 - H(\Omega)]$ for $\zeta \geq 1$.

Example 7.3   Modulation Response Curves   A sinusoidal phase modulation is impressed on the reference signal to a PLL with $\alpha = 1$, $\zeta = 0.5$, and $\omega_n = 2\pi\ 300$ rad/sec. The input has a peak phase deviation of 40° and a modulation frequency of 600 Hz. Describe the modulation of the loop's output signal.

From Fig. 7.8, at $\Omega = 600$ Hz/300 Hz = 2 and $\zeta = 0.5$, $|H| = -4$ dB. The output modulation amplitude is therefore 40° × $10^{-4/20}$ = 25.2°. From Fig. 7.10, this modulation lags the input modulation by 82°. Here 25.2° is the amplitude of the phase modulation and $-82°$ is its phase. Both are parameters of the state variable, output phase.

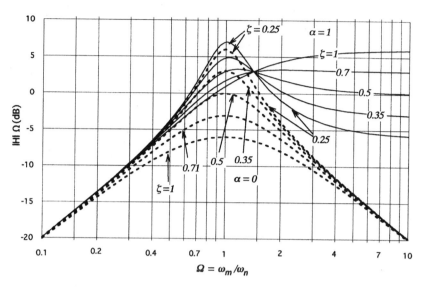

Fig. 7.16 Magnitude of $sH(\Omega)$ for $\zeta \le 1$. Note: The peak value, for $\alpha = 0$, is $|H|\Omega = -20\log_{10}(2\zeta)$.

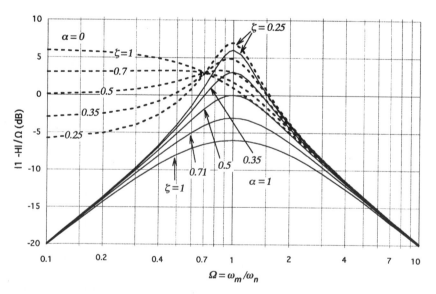

Fig. 7.17 Magnitude of $[1 - H(\Omega)]/s$ for $\zeta \le 1$. Note: The peak value, for $\alpha = 1$, is $|1 - H|/\Omega = -20\log_{10}(2\zeta)$ [Blanchard, 1976, p. 125].

7.7 RELATED RESPONSES

The two other amplitude responses required for Table 7.1 are shown in Figs. 7.16 and 7.17. Phase is not shown since the phase of Hs is just the phase of H plus 90° and the phase of $(1 - H)/s$ is the same as that of $(1 - H)$ less 90°. Therefore the phase curves previously given can be used again.

Example 7.4   Use of Related Modulation Response Curves   Describe the frequency of the loop's output signal in Example 7.3.

Since frequency is the time derivative of phase, the frequency is $j\omega_m$ times the phase (in radians). Using Fig. 7.16, we find $|H|\Omega = 2$ dB higher than the input amplitude, or

$$|H|\left(\frac{\omega_m}{\omega_n}\right) = 10^{2/20} = 1.26,$$

$$|H|\omega_m = 1.26\,\omega_n \Rightarrow 1.26(2\pi)(300 \text{ rad/sec})\left(\frac{\text{cycle}}{2\pi \text{ rad}}\right) = 378 \text{ Hz.}$$

The magnitude of the frequency modulation is *264 Hz* the product of the input, $[40°(\pi \text{ rad}/180°) =]0.7$ rad, and the response. The phase angle of the frequency modulation is 90° plus the phase angle of the phase modulation (obtained in Example 7.3) or $+8°$.

7.8 MODULATION AND DEMODULATION IN THE SECOND-ORDER LOOP

Modulation and demodulation are performed in a manner similar to what was described for the first-order loop.

7.8.1 Frequency Demodulation

By Table 7.1,

$$u_2 = \omega_{in} H(\omega_m)/K_v, \tag{7.46}$$

which is like Eq. (7.17) and, as for the first-order loop, leads to Eq. (7.19). Thus the voltage at the input to the VCO equals the input frequency divided by K_v at low frequencies, and the shape of the magnitude response is given by the curves in Figs. 7.8 and 7.12 while the phase is shown in Figs. 7.10 and 7.14.

The usable range of modulation frequencies can be extended beyond the limits indicated by Eq. (7.46) by compensation of the demodulated output to counter the effects of the loop response near and beyond ω_L. Thus a circuit, external to the loop, with response $C/H(\omega_m)$, over a sufficient range of ω_m, could be driven by the tuning voltage, producing an output $\omega_{in}C/K_v$, where C is a constant. Of course, the signal level u_2 becomes increasingly weaker at higher ω_m, so it will eventually drop into thermal noise, or other circuit noise, and the compensated signal will not be useful for ω_m/ω_L sufficiently large.

Again according to Table 7.1,

$$|\varphi_e(\omega_m)| = \omega_{in}|1 - H(\omega_m)|/\omega_m \tag{7.47}$$

$$= (\omega_{in}/\omega_n)|1 - H(\omega_m)|/\Omega. \tag{7.48}$$

The magnitude of the phase error must be kept small enough to be within the range of the phase detector and, if the loop parameters that affect H are to be approximately constant, within its more or less linear portion (say $\pm \pi/4$ rad for a balanced mixer). Figure 7.17 can be used to predict this error.

Example 7.5   Frequency Demodulator   Let $\alpha = 1$, $\zeta = 0.35$, $\omega_n = 2\pi(10^4)$ rad/sec, $K_v = 10$ kHz/V. At low frequencies, Eq. (7.19) gives

$$u_2 = \omega_{in}/(10^4 \text{ Hz/V}) = f_{in}10^{-4} \text{ V/Hz}.$$

Thus an FM signal with peak deviation 1 kHz will produce a waveform with a peak deviation of 0.1 V at low frequencies. At higher frequencies the response is modified according to the response curve shown in Fig. 7.8 for the given parameters and thus peaks 5 dB higher at approximately $\omega_m = 0.82 \omega_n \Rightarrow f_m = 0.82 f_n = 8200$ Hz. The response is 3 dB below the low-frequency response at

$$\omega_m = 1.7\omega_n \Rightarrow f_m = 1.7f_n = 17{,}200 \text{ Hz}.$$

By Eq. (7.47) and Fig. 7.17, the maximum phase deviation will occur at $\omega_m = \omega_n$ and will be 3 dB above (ω_{in}/ω_n) at that frequency. Thus Eq. (7.48) leads to

$$|\varphi_e| \leq (\omega_{in}/\omega_n)\sqrt{2}.$$

To keep this error below $\pi/4$, we must have

$$(\omega_{in}/\omega_n)\sqrt{2} < \pi/4.$$

Using the given ω_n, f_{in} must therefore be kept below $[\pi \times 10^4/(4\sqrt{2})]$Hz $= 5550$ Hz to keep the phase detector characteristic in its more linear range.

When the loop has a low-pass filter ($\alpha = 0$), $u_2 = K_{LF}u_1$ at low frequencies, so a frequency demodulation response can be obtained from u_1 also, as shown by Eq. (7.48) and Fig. 7.17. There are similarities between this response at u_1 when $\alpha = 0$ and the response at u_2 when $\alpha = 1$ (Fig. 7.8).

Both are flat at low frequencies and fall off -6 dB/octave at high frequencies. We can obtain a stronger filtering action by passing either response through a series low-pass filter to produce a -12 dB/octave fall-off at high frequencies. This can be done quite naturally when $\alpha = 0$ by simply taking the response from u_2. Then the responses with the steeper high-frequency skirts, shown in Fig. 7.8 for $\alpha = 0$, including the maximally flat response (see Section 7.6.4), are obtained. It is also possible to obtain these same responses from loops were $\alpha \neq 0$ by adding the proper low-frequency filter, as we will now show.

It is apparent from Table 4.1 that $H(s)$ for $\alpha \neq 0$ (Eq. 4.10) can be changed to $H(s)$ for $\alpha = 0$ [Eq. (4.20)] through multiplication by $1/(1 + s/\omega_z)$, the response for a low-pass filter whose pole frequency is the same as the zero frequency of the loop. While this can be done by adding such a low-pass filter in series with the demodulated output at u_2, we can obtain the desired response without adding a filter if the loop is using a passive lag-lead filter. From Fig. 3.19 and Eq. (3.15) we see that the desired filtering of u_2 is available across C in that loop filter. Thus the demodulated voltage taken there is

$$v = u_2 \frac{1}{1 + s/\omega_z} = \frac{\omega_{\text{in}}}{K_v} \left\{ \frac{H(\omega_m)_{\alpha \neq 0}}{1 + s/\omega_z} \right\} = \frac{\omega_{\text{in}}}{K_v} H(\omega_m)_{\alpha = 0}. \qquad (7.49)$$

7.8.2 Phase Demodulation

By Table 7.1,

$$u_1 = \varphi_{\text{in}}[1 - H(\omega)] K_p, \qquad (7.50)$$

which is like Eq. (7.15) and leads to Eq. (7.20). Thus the voltage at the phase detector output equals the input phase multiplied by K_p at high frequencies, and the response shape is given by Figs. 7.9 and 7.13. As with frequency demodulation, the fall-off of the response beyond ω_L can be compensated.

By Table 7.1, the output frequency that accompanies the input phase, and which must be accommodated by the VCO, is

$$|\omega_{\text{out}}| = |\varphi_{\text{in}} H(\omega_m) \omega_m| = |\varphi_{\text{in}}| \omega_n [|H(\Omega)| \Omega] \qquad (7.51)$$

and $|H(\omega_m)|\Omega$ can be obtained from Fig. 7.16.

7.8.3 Frequency Modulation

By Table 7.1,

$$\omega_{\text{out}} = \Delta u_2 [1 - H(\omega_m)] K_v, \qquad (7.52)$$

which is like Eq. (7.21) and leads to Eq. (7.26). Thus the output frequency equals Δu_2 multiplied by K_v at high frequencies and the response shape is given by Figs. 7.9 and 7.13.

The accompanying phase error is given by Table 7.1 as

$$- \varphi_e = \varphi'_{\text{out}} = -j \Delta u_2 K_v [1 - H(\omega_m)]/\omega_m$$
$$= -j(\Delta u_2 K_v/\omega_n)[1 - H(\omega_m)]/\Omega, \qquad (7.53)$$

and $[1 - H(\omega_m)]/\Omega$ can be obtained from Fig. 7.17.

7.8.4 Phase Modulation

By Table 7.1,

$$\varphi_{\text{out}} = \Delta u_1 H(\omega_m)/K_p, \qquad (7.54)$$

which is like Eq. (7.23) and leads to Eq. (7.25). Thus the output phase equals Δu_1 divided by K_p at low frequencies, and the response is given by Figs. 7.8 and 7.12.

The tuning range of the VCO would be of concern. The required range can be obtained from Table 7.1, where the output frequency is given by

$$\omega_{\text{out}} = \Delta u_1 H(\omega_m) j \omega_m/K_p = j \Delta u_1(\omega_n/K_p)|H(\omega_m)|\Omega, \qquad (7.55)$$

and $|H(\omega_m)|\Omega$ can be obtained from Fig. 7.16.

7.8.5 Extending the Modulation Frequency Range

By introducing properly related modulations at both Δu_1 and Δu_2, a combined response can be obtained that has bandwidth much wider than would normally be permitted by the loop bandwidth.

7.8.5.1 Frequency Modulation. If modulation is introduced at both Δu_1 and Δu_2, the combined response will be the sum of Eqs. (7.52) and (7.55), giving

$$\omega_{\text{out}} = \Delta u_2 [1 - H(\omega_m)] K_v + j \Delta u_1 H(\omega_m) \omega_m/K_p. \qquad (7.56)$$

If Δu_1 is processed such that

$$\Delta u_1 = -j \Delta u_2 K_p K_v/\omega_m, \qquad (7.57)$$

Eq. (7.57) becomes

$$\omega_{\text{out}} = \Delta u_2 K_v \qquad (7.58)$$

independently of modulation frequency. The process indicated by Eq. (7.57) would involve integration of Δu_2, as illustrated in Fig. 7.18a. Comparing Eqs. (7.57) and (3.35), we see that

$$RC = \frac{1}{K_p K_v}. \qquad (7.59)$$

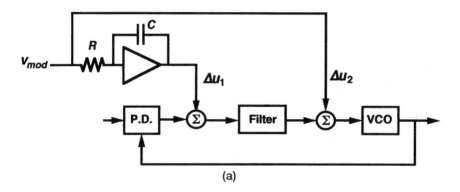

(a)

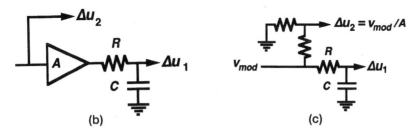

(b)   (c)

Fig. 7.18   Expanding the bandwidth of a frequency modulator.

The integration could be obtained approximately by a low-pass filter, operating beyond its cutoff frequency, over a limited range of frequencies. This is illustrated in Fig. 7.18b where

$$\Delta u_1 = \Delta u_2 \, A/[1 + j\omega_m RC] \Rightarrow -j\,\Delta u_2 \, A/[\omega_m RC] \qquad (7.60)$$

for

$$\omega_m \gg 1/(RC). \qquad (7.61)$$

Equation (7.57) will be satisfied if also

$$RCK_p K_v = A. \qquad (7.62)$$

An amplifier is not needed if the last two equations can be satisfied without it or if A is achieved effectively by attenuating the modulation source to obtain Δu_2, as shown in Fig. 7.18c.

We must be aware that φ_e will continually increase as ω_m falls [Eq. (7.7)], assuming constant frequency deviation, so that we will run into linearity problems at low enough modulation frequencies.

7.8.5.2 Phase Modulation.   A similar extension can be performed for phase modulation. Combining Eqs. (7.54) and (7.53) we obtain

$$\varphi_{\text{out}} = \Delta u_1 \, H(\omega_m)/K_p - j\,\Delta u_2 \, K_v[1 - H(\omega_m)]/\omega_m. \qquad (7.63)$$

Use

$$\Delta u_2 = j\,\Delta u_1 \, \omega_m/(K_p K_v) \qquad (7.64)$$

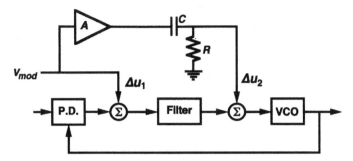

Fig. 7.19   Expanding the bandwidth of a phase modulator.

to give a composite of

$$\varphi_{\text{out}} = \Delta u_1 / K_p. \tag{7.65}$$

Equation (7.64) requires a differentiator but can be realized by a high-pass circuit with cutoff frequency sufficiently higher than the modulation frequencies of interest. This is illustrated in Fig. 7.19 where

$$\Delta u_2 = A / [1 + 1/(RCj\omega_m)] \Delta u_1 \Rightarrow j \Delta u_1 \, \omega_m \, ARC \tag{7.66}$$

for

$$\omega_m \ll 1/(RC). \tag{7.67}$$

Equation (7.65) will be satisfied also if

$$ARCK_p K_v = 1. \tag{7.68}$$

Comments in the previous section concerning the need for an amplifier apply here also.

We must also be aware that the required VCO tuning range will continually increase as ω_m increases [Eq. (7.7)], assuming constant phase deviation.

7.9   MEASUREMENT OF LOOP PARAMETERS FOR $\alpha = 0$ OR 1 FROM MODULATION RESPONSES

One can verify the design loop parameters by comparing measured responses to the expected responses using the curves in this chapter. A more direct method is available for determining ω_n and ζ for a loop when α is 0 or 1. With an integrator-and-lead filter, we can frequency modulate the input, sweep the modulation frequency (constant frequency deviation), and observe the phase detector output. By Table 7.1 the response shape will be like the curves shown in Fig. 7.17 for $\alpha = 1$. It can be shown, by algebraic manipulation of Eq. (4.23) that [Blanchard, 1976, pp. 124–127], if ω_- and ω_+ are the two values of ω_m where the response is 3 dB below the peak response, then

$$\omega_n^2 = \omega_+ \omega_- \qquad (7.69)$$

and

$$\omega_+ - \omega_- = 2\zeta\omega_n. \qquad (7.70)$$

Comparing Eqs. (4.25) and (4.20) (see Table 4.1), it is apparent that $|[1 - H(\omega_m)]/\omega_m|$ for $\alpha = 1$ has the same shape as does $|H(\omega_m)\omega_m|$ for $\alpha = 0$. This can also be seen by comparing Figs. 7.16 for $\alpha = 0$ and 7.17 for $\alpha = 1$. Therefore, by Table 7.1, Eqs. (7.69) and (7.70) also apply to a loop with a low-pass filter if we phase modulate the input, sweep the modulation frequency (constant phase deviation), and determine the modulation frequencies, ω_+ and ω_-, at which the deviation of the tuning voltage is 3 dB below its peak value.

7.10   EFFECT OF AN ADDED POLE

Figure 7.20 illustrates the effect of a pole added to what would otherwise be a type 2, second-order, loop with $\zeta = 0.5$. We see little effect on the modulation response when the frequency ω_a of the added pole is 10 times ω_n, which approximately equals ω_L when $\zeta = 0.5$ (they are equal in the tangential approximation). As ω_a comes closer to ω_n, the effect grows until the response becomes infinite at the natural frequency when $\omega_a = \omega_n$. At that point the added pole cancels the zero, giving a constant 180° excess phase and producing instability at the frequency where open-loop gain is one. Compare the effects of the added pole on the modulation response to its effect on the transient response (see Section 6.12).

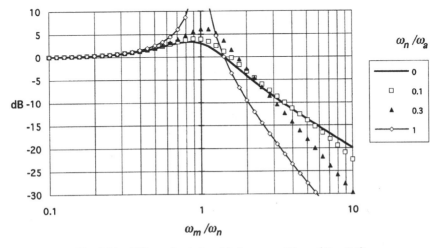

Fig. 7.20   Effect of a pole added to type 2 loop ($\zeta = 0.5$).

7.M APPENDIX: MODULATION RESPONSE USING MATLAB

FreqRsp draws the magnitude and phase of the output and error responses versus log of frequency, normalized (to ω_n) in most cases. It can draw curves like those in Figs. 7.6 through 7.15 or like Fig. 7.20 as well as first-order loop response. We can select the kind of loop and its parameters dynamically. While the multiple curves presented in the text have some advantages, the advantage of the program is that it gives us more freedom to choose parameters, especially α. We have shown how transient responses for any α can be obtained by combining response for $\alpha = 0$ and $\alpha = 1$ in the correct proportion, but this is not easily done with the phase or magnitude of the responses.

FreqRsp

```
%FREQUENCY RESPONSE
%  Choose loop and parameters dynamically.

r2deg = 180/pi; % for use in changing radians to degrees
w = logspace(-1,1,51); % Frequency range
Wn = 1 % Natural frequency normalized

go = 2;
while (go ~= 0)
   order = input('Enter order (1-3), 0 to quit: ','s');
   if strcmp(order,'0'),
      go = 0;
   elseif strcmp(order,'1'),
      disp('First Order');
      K = input('First Order.  Input K: ');
      out = [0 K];          % from Eq. (2.25b)
      er = [1 0]; % from Eq. (2.26b)
      den = [1 K];
    elseif strcmp(order,'3'),
      rn = input('Third Order, Type 2, zeta = 0.5.  Enter
                        Wn/W+: ');
      out = [0 0 1 1];      % as in SS Added Pole , paragraph
                             6.M.5
      er = [rn 1 0 0];      % from Eq. (6.M.4)
      den = [rn 1 1 1];
   elseif (go~=0),
      alpha = input('Second Order. Input alpha: ');
      zeta = input('Input zeta: ');
      out = [2*alpha*zeta 1];  % from Eq. (6.4a)
      er = [1 2*zeta*(1-alpha) 0]; % from Eq. (6.4b)
      den = [1 2*zeta 1];
   end
```

```
hout = freqs(out, den, w);
magout =  20*log10(abs(hout));
phout = r2deg*angle(hout);
her = freqs(er, den, w);
mager = 20*log10(abs(her));
pher = r2deg*angle(her);

clg
subplot(211), semilogx(w, magout, 'r', w, mager, '--r')
grid
title('Frequency Response Magnitude.  solid = output    dash
                      = error')
xlabel('Omega / Omega-n')
ylabel('dB')
subplot(212), semilogx(w, phout, 'g', w, pher, '--g')
grid
title('Frequency Response Phase.  solid = output    dash =
                     error')
xlabel('Omega / Omega-n')
ylabel('degrees')
end % while go
```

In FRspDorI, the second-order response is modified to produce the derivatives and integrals, as shown in Figs. 7.16 and 7.17, by shifting the numerator vector elements (see Section 6.M.1).

PROBLEMS

7.1 $f = 10$ MHz $+ 1$ kHz $\cos(200$ rad/sec $t + \pi/4$ rad)

(a) What is the carrier frequency?

(b) What is the peak frequency deviation?

(c) What is the modulation frequency?

(d) What is the peak phase deviation?

(e) What is the earliest time $t \geq 0$ when the frequency deviation is 0?

(f) What is the earliest time $t \geq 0$ when the phase deviation is 0?

7.2 Given a first-order loop and $K_p = 0.1$ V/rad:

(a) When the reference has a peak deviation of 0.1 rad, what will be the peak deviation of the phase detector output at high modulation frequencies?

(b) What is the value of K_v for a frequency demodulation response at $f_m = 1$ kHz that is 6 dB below the DC response.

(c) If $K_v = 1$ kHz/V, at what modulation frequency will the phase demodulation output be 20 dB below its high frequency value?

(d) What peak voltage injected after the phase detector will produce 0.01-rad peak phase deviation at low frequencies?

7.3 The reference frequency input to a second-order loop with a low-pass filter is swept in modulation frequency. The frequency deviation of the output (VCO) equals that of the input at low modulation frequencies but peaks 5 dB higher than the input deviation when a modulation frequency of 1 kHz occurs ("peaks" meaning that is the highest response).

(a) What is the damping factor?

(b) As the modulation frequency increases, at what frequency will the output deviation again equal the input deviation?

7.4 A loop is used as a phase modulator. The modulating voltage, which is added at the phase detector output, is swept in frequency with a 1-V peak sine wave. At low frequencies the peak phase deviation produced is 0.1 rad. It is a maximally flat response with -3 dB at 10 kHz.

(a) What is K_p?

(b) What are α and ζ.

(c) What is ω_n?

7.5 A loop has an integrator-and-lead filter with unity open-loop gain (straight-line gain approximation) at 10 kHz and a zero at 5 kHz. What is the peak phase error when the input reference signal is frequency modulated at a peak deviation $\Delta f = 4$ kHz? At what frequency f_m does it occur?

7.6 A frequency modulator with a frequency range that extends well below the loop bandwidth is represented by the block diagram as in Fig. 7.18b. The amplifier gain is 100. $K_p K_v = 1000$ sec^{-1}. $R = 10$ kΩ.

(a) What should be the value of C.

(b) The loop has an integrator-and-lead filter with $\zeta = 1$. If the input Δu_1 were removed, by how much would the frequency deviation drop at a modulation frequency equal to half of the natural frequency?

(c) If both ω_m and the peak frequency deviation equal $\omega_n/2$ when Δu_1 is connected, what will be the peak phase error after the input at Δu_1 is disconnected [same loop as in (b)]?

7.7 Using Figs. 7.8 to 7.11, confirm the theory of Section 7.8.5.1 by showing equal responses at $\omega_m = 0$, $\omega_m = \omega_n$, and $\omega_m = 10\omega_n$.

CHAPTER 8

ACQUISITION

So far we have considered linear performance almost exclusively. Of course, we normally seek to ensure that the loop will operate in this fashion. However, it is essential to know under what conditions the loop will not lock, and it is valuable to understand what is involved in the acquisition of phase lock [Blanchard, 1976, Chapter 10; Gardner, 1979, Chapter 5].

8.1 OVERVIEW

Refer to Fig. 8.1. The frequency of the voltage-controlled oscillator (VCO) when the phase detector is in the center of its range will be called the center frequency, ω_c. This can be varied by adjusting the fixed frequency-determining elements in the oscillator or perhaps some bias that is added to the tuning voltage or to the phase detector output. We will call the difference between the input frequency and the center frequency the mistuning, Ω:

$$\Omega = \omega_{in} - \omega_c. \tag{8.1}$$

The initial value of frequency error, when the signal is applied, is

$$\omega_e(0) = \Omega. \tag{8.2}$$

In general, there is a range of mistuning, called the seize or lock-in range, $|\Omega| \leq \Omega_S$, over which the ω_{out} will immediately move to the steady-state lock frequency, ω_{in}. In other words, Ω will immediately begin decreasing toward zero when the input switch is closed.

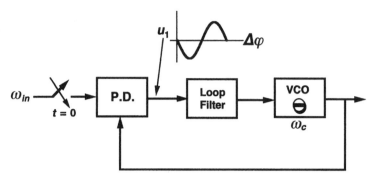

Fig. 8.1   Loop for acquisition.

There is a range of mistuning, called the pull-in or acquisition range, $|\Omega| \le \Omega_{PI}$ over which the loop will eventually lock but may skip cycles, sometimes many cycles, before doing so. By skipping cycles is meant that the difference between the input and the output frequency causes the phase detector characteristic to be traversed past the phases where the voltage is minimum and maximum.

There is range of mistuning, called the hold-in or synchronization range, $|\Omega| \le \Omega_H$, over which the loop will maintain lock but over which it will not necessarily acquire lock.

The Ω_S and Ω_{PI} ranges are guaranteed maximum values where the action described above will occur regardless of initial phase. These actions may occur at wider mistunings if the phase happens to have a fortuitous value.

These ranges are illustrated in Fig. 8.2. We will sometimes use notation such as $\pm\Omega_{PI}$ to emphasize that we are considering the deviation of ω_{in} from ω_c rather than the total range over which ω_{in} may vary, which is twice as large.

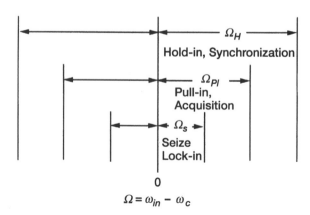

Fig. 8.2   Mistuning.

The relationship

$$\Omega_S \leq \Omega_{PI} \leq \Omega_H \qquad (8.3)$$

must always hold, just because of the definitions. The equalities in this relationship apply to first-order loops and highly damped second-order loops, which act like first-order loops. First-order loops move toward steady-state whenever it is possible to tune to it. The phase does not overshoot but moves directly toward its final value. In a first-order loop, the value of the phase error is instantly related to the value of the frequency; there can be only one frequency for a given phase error. In a second-order loop, due to the filter, this is not so and various total states can exist at a given frequency. Thus the second-order loop might move directly toward steady state when it is at a given frequency or it might skip cycles at the same frequency, depending on the mix of state variables under each condition—we might also say, depending on its history.

We will now describe a possible experiment with a second-order loop to illustrate the various ranges. The loop is locked with f_{in} = 9 MHz. The input frequency is *slowly* moved higher, past the 10-MHz center frequency, which is the frequency of the VCO when the phase detector is at midrange. At 14 MHz, lock is lost because the phase detector cannot put out enough voltage to tune the VCO to 14 MHz. (Generally we assume that the phase detector is the only nonlinear element, since its nonlinearity is inherent.) This is an edge of the total hold-in range. Now the f_{in} is slowly lowered until lock is reestablished at 12 MHz. This is an edge of the total acquisition range. When the input frequency is adjusted to 12 MHz, with the input switch open, and the switch is then closed, the loop skips cycles for a long time but finally comes into lock. But, if the frequency is a little higher when the switch is closed, no lock occurs. The experiment is repeated at lower and lower frequencies until a frequency is found, say 11 MHz, where, every time, the loop goes to steady state, after the switch is closed, without skipping cycles. This is one edge of the seize range. We can repeat the experiment on the low-frequency side and, assuming we have an ideal (linear except for the balanced-mixer phase detector) loop, the deviations from 10 MHz will be equal, for each range, on the low side to what they were on the high side. In that case we would conclude that Ω_H = ±4 MHz, Ω_{PI} = ±2 MHz, and Ω_S = ±1 MHz. Or we might be more informative by saying that the hold-in range extends from 6 to 14 MHz, the pull-in range extends from 8 to 12 MHz, and the seize range extends from 9 to 11 MHz.

In contrast to the above, if the loop were first order and we tuned higher, past f_c = 10 MHz, the results might be as follows. At 13 MHz, for example, the loop breaks lock. As we tune slowly back, it will again acquire lock, with no cycle skipping, at 13 MHz. Thus, for this first-order loop, $\Omega_S = \Omega_{PI} = \Omega_H$ = 3 MHz.

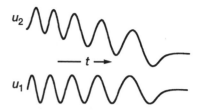

Fig. 8.3   Waveforms during acquisitions.

The time required to get from an initial frequency error $\omega_e = \Omega$ to $\omega_e = \Omega_S$ is the acquisition or pull-in time, T_{PI}. During this time the phase error and frequency (or, more easily observed, u_1 and u_2) might look as pictured in Fig. 8.3. The difference between ω_{in} and ω_{out} causes the phase to move continuously, leading to approximately a sinusoid at u_1 as the phase-detector characteristic is repeatedly traced out. However, this is not a perfect sinusoid. Due to the small feedback that does exist, it spends slightly more time at low voltages than at high voltages, leading to a negative average value. As shown, u_2 is approximately sinusoidal, following u_1, but the negative average value in u_1 causes the average output of the loop filter to fall. As it does so, the average output frequency comes closer to the input frequency. As this happens, $\omega_e (= \omega_{in} - \omega_{out})$ is lowered so the loop gain increases, causing more feedback and greater distortion of the sinusoid and accelerating the pull-in process.

Let us look at the same process from a slightly different perspective. Fig. 8.4 shows, on a Bode plot, some of the parameters of importance for a "high-gain" ($\omega_z \ll \omega_L$) second-order loop. It turns out that the acquisition frequency falls approximately at the geometric mean of the hold-in range K and ω_L (i.e., $\Omega_{PI} \approx \sqrt{K\omega_L}$) while the seize frequency is approximately equal to ω_L (i.e., $\Omega_S \approx \omega_L$). When the initial mistuning is between these frequencies, the loop skips cycles while the DC content of u_1 is integrated and slowly reduces the mistuning. The process is accelerated as the seize frequency is approached and, when it is finally reached, the loop locks rapidly. During the last part of the process, when the loop is in a linear region (gains are approximately constant), the frequency changes in a manner given by the

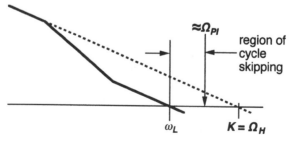

Fig. 8.4   Acquisition parameters on a Bode plot, sine PD high gain.

transient response curves previously studied. The appropriate response is a combination of a phase step response and a frequency step response, the combination giving the appropriate initial phase and frequency.

The developments that follow are for a sinusoidal phase detector characteristic. Somewhat improved performance should be expected (for the same K_p) when the phase detector characteristic becomes more triangular, say as a result of near equal signal levels at the two inputs (see Section 3.1.4.2).

8.2   ACQUISITION AND LOCK IN A FIRST-ORDER LOOP

We will study the process of acquisition in a first-order loop because the relatively easy analysis will show effects that have similarities to those in higher order loops.

From Fig. 8.5, the output frequency is

$$\omega_{\text{out}} = \omega_c + K_v u_1 \tag{8.4}$$

while u_1 is given by

$$u_1 = K_p \cos[\Delta\varphi(t)]. \tag{8.5}$$

We can subtract Eq. (8.4) from ω_{in} and employ the definition of Ω in Eq. (8.1) to give

$$\omega_e = \omega_{\text{in}} - \omega_{\text{out}} = \omega_{\text{in}} - \omega_c - K_v u_1 = \Omega - K_v u_1. \tag{8.6}$$

Since $\omega_{\text{in}} - \omega_{\text{out}} = d\Delta\varphi/dt$, this can also be written

$$d\Delta\varphi/dt = \Omega - K_v u_1. \tag{8.7}$$

Thus if u_1 is greater than Ω/K_v, the value at which ω_{out} equals ω_{in}, $\Delta\varphi(t)$ will decrease. If this occurs on the positive slope of $\cos\Delta\varphi$ ($-\pi \leq \Delta\varphi \leq 0$), the decreasing $\Delta\varphi$ will cause u_1 to decrease and thus reduce the value of ω_{out}. This will continue until $\omega_{\text{out}} = \omega_{\text{in}}$ and $u_1 = \Omega/K_v$ and then cease (Fig. 8.6). The smaller the frequency error ω_e the smaller the absolute value

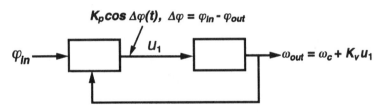

Fig. 8.5   First-order loop.

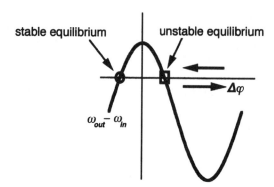

Fig. 8.6 First-order loop characteristic.

of Eq. (8.7) and the more slowly moves $\Delta\varphi$. Thus the pull-in is eventually exponential. If this occurs on the negative slope of the phase detector characteristic ($0 \leq \Delta\varphi \leq \pi$), the decreasing $\Delta\varphi$ will lead to higher values of u_1, which will cause the frequency error to increase and $\Delta\varphi$ to move even faster until $\Delta\varphi = 0$, at which point the process described before will occur.

There is a second equilibrium point on the opposite slope of the characteristic, but it is unstable. No error signal is generated at that point, but any deviation from that point will cause the phase to move away from it, as indicated by the arrows in Fig. 8.6, whereas any deviation from the stable point causes the phase to move back toward it. If the initial phase is near the value for unstable equilibrium, the phase error will be near zero so the phase will move very slowly initially. This is called "hang up" [Gardner, 1977] and can be a problem when the time required to attain steady state is important.

Another way to look at the process is illustrated in Fig. 8.7, which shows a phasor ox of magnitude K_p at an angle $\Delta\varphi$. The horizontal projection equals u_1. The negative of the rate of change of $\Delta\varphi$ is, by Eq. (8.7), proportional to the deviation of u_1 from the line $u_1 = \Omega/K_v$, the value that, by Eq. (8.6), produces $\omega_{out} = \omega_{in}$. Thus the outside arrows in Fig. 8.7 show $\Delta\varphi$ decreasing when u_1 is higher than this value and increasing otherwise; a is the point of unstable equilibrium and c is the point of stable equilibrium. If x is between b and c, u_1 will decrease continuously as c is approached. If x is between a and b, u_1 will peak at b before decreasing to its final value at c. If x is between a and b', u_1 will reach a negative peak before settling at c. These various possibilities are illustrated in Fig. 8.8 where an attempt has been made to show the effect of the varying $d\Delta\varphi/dt$.

Figure 8.8 covers all the response shapes because the change in $\Delta\varphi$ is completely determined by u_1. If the initial $\Delta\varphi$ is closer to final value than the initial condition in Fig. 8.8, we merely start further to the right on the appropriate curve. Surprisingly, the time to get from a phase near the unstable equilibrium point to a small offset from steady state is the same for

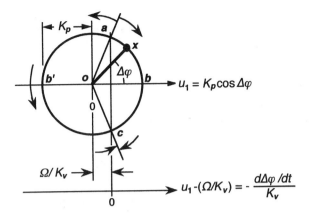

Fig. 8.7   Phasor representation of first-order loop pull-in.

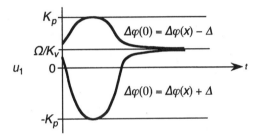

Fig. 8.8 u_1 during pull-in.

both trajectories as long as the small initial and final offsets are identical, which we will now show.

8.2.1   Transient Time

Combining Eqs. (8.5) and (8.7) we obtain

$$dt = \frac{d\Delta\varphi(t)}{\Omega - K\cos\Delta\varphi(t)}. \tag{8.8}$$

This can be integrated to give the time for $\Delta\varphi(t)$ to change from an initial value $\Delta\varphi(0)$ at $t = 0$ to a value $\Delta\varphi(t)$ at time t:

$$t = \int_0^t dt = \frac{1}{\sqrt{K^2 - \Omega^2}} \ln\left| \frac{\tan[\Delta\varphi(t)/2] - \beta}{\tan[\Delta\varphi(t)/2] + \beta} \frac{\tan[\Delta\varphi(0)/2] + \beta}{\tan[\Delta\varphi(0)/2] - \beta} \right|, \tag{8.9}$$

where

$$\beta = \sqrt{\frac{K - \Omega}{K + \Omega}} \,. \tag{8.10}$$

As t approaches infinity, the denominator inside the $|\;|$ must approach zero, implying that

$$\beta = -\tan\frac{\Delta\varphi(\infty)}{2} \,. \tag{8.11}$$

The time t also becomes infinite when $\tan[\Delta\varphi(0)/2] = \beta$, consistent with this being a point of unstable equilibrium (hang-up). Comparing β for the two cases, we can see that unstable equilibrium occurs when the initial phase is the negative of the final phase [i.e., $\Delta\varphi(0) = -\Delta\varphi(\infty)$], as can be seen from Fig. 8.7.

To improve our understanding of the process, let us address this question: If $\Delta\varphi(0)$ is offset by a small angle ε from the point of unstable equilibrium, how will the time required, to come within the small angular offset μ of final value, when the initial offset is positive compare to the time required when it is negative? A small offset in angle causes a small offset in the tangent related by

$$d \tan x = \frac{dx}{\cos^2 x} \,. \tag{8.12}$$

Thus we can write, for the point of unstable equilibrium,

$$\beta = \tan\frac{\Delta\varphi(0) + \varepsilon}{2} \approx \tan\frac{\Delta\varphi(0)}{2} + \frac{\varepsilon}{2\cos^2[\Delta\varphi(0)/2]} \tag{8.13}$$

and, for the point of stable equilibrium,

$$-\beta = \tan\frac{\Delta\varphi(t) - \mu}{2} \approx \tan\frac{\Delta\varphi(t)}{2} - \frac{\mu}{2\cos^2[\Delta\varphi(t)/2]} \,. \tag{8.14}$$

The terms $\Delta\varphi(0)$ and $\Delta\varphi(t)$ represent an initial angle slightly less than the angle of unstable equilibrium and a final angle slightly more than the angle of stable equilibrium and, thus, a transient that moves around the right side of the circle in Fig. 8.7. Substituting these into Eq. (8.9) we obtain

$$t \approx \frac{1}{\sqrt{K^2 - \Omega^2}} \ln\left| \frac{\dfrac{(-2\beta)}{\mu}}{2\cos^2(\Delta\varphi(t)/2)} \cdot \frac{\dfrac{2\beta}{(-\varepsilon)}}{2\cos^2(\Delta\varphi(0)/2)} \right|$$

$$= \frac{1}{\sqrt{K^2 - \Omega^2}} \ln\left\{ \frac{16\beta^2}{\mu\varepsilon} \cos^4\!\left(\frac{\Delta\varphi(0)}{2}\right) \right\} \,. \tag{8.15}$$

Since μ and ε appear only as a product, if we repeat this process with the signs changed, to represent a locus around the left side of the circle, we obtain the same answer, which is what we were to demonstrate.

8.2.2 Acquisition

We know that, if $|\Omega| > K = K_p K_v$, the loop cannot be locked because the mistuning is outside the hold-in range. Graphically, when Ω/K_v exceeds K_p, a and c no longer exist in Fig. 8.7 because the vertical line on which they lie no longer intersects the circle. Then $d\Delta\varphi/dt$ is always positive and the phase continuously traverses the circle, slowing as it approaches the line and speeding as it moves away. Of course, similar statements apply to mistuning in the opposite direction. We can analyze the out-of-lock waveforms to determine their shape more exactly in a manner similar to that used in the last section. Now, however, β in Eq. (8.10) will be imaginary so we define

$$\gamma = \sqrt{\frac{\Omega + K}{\Omega - K}} \qquad (8.16)$$

and employ it in another form of the solution of Eq. (8.8) to obtain

$$t \approx \frac{2}{\sqrt{\Omega^2 - K^2}} \tan^{-1}\left\{ \gamma \tan\left(\frac{\Delta\varphi(t)}{2} \right) \right\} \qquad (8.17)$$

or, equivalently,

$$\tan \alpha t = \gamma \tan\left(\frac{\Delta\varphi(t)}{2} \right), \qquad (8.18)$$

where

$$\alpha = \frac{\sqrt{\Omega^2 - K^2}}{2}. \qquad (8.19)$$

We can interpret this with the help of Fig. 8.9, which is drawn for $\gamma < 1$ ($\Omega < 0$).

As αt increases steadily with time, $\Delta\varphi(t)$ increases to maintain the relationship between tangents. When αt is small,[1] $\Delta\varphi(t)/2$ must increase at

---

[1] As in the typical tangent representation, when αt is outside this range, the back projection of the radius line is used to intersect the vertical line.

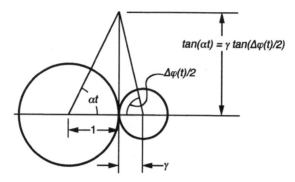

Fig. 8.9 Aid for interpreting cycle skipping waveshape.

a rate $1/\gamma$ times faster than αt because $\tan(x) \approx x$ and the radius of the right circle is that much smaller. However, both angles reach $\pi/2$ at the same time so the rate of change of $\Delta\varphi(t)/2$ must slow as it approaches that value. This leads to the waveforms shown in Fig. 8.10. Notice the typical behavior in which more time is spent at negative values of u_1. In a second-order loop this tends to bring ω_{out} down to meet ω_{in}, which is lower, as implied by $\Omega < 0$. If $\Omega > 0$, the right circle would be larger than the left circle and more time would be spent at positive values to bring the VCO frequency up to the input frequency.

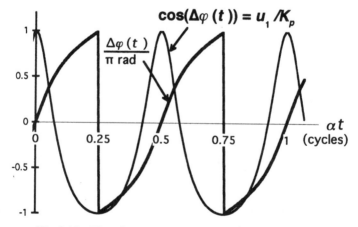

Fig. 8.10 Waveforms in first-order loop skipping cycles.

8.3 ACQUISITION FORMULAS FOR SECOND-ORDER LOOPS WITH SINE PHASE DETECTORS

Before developing our understanding of the acquisition further through analysis, some equations relating to acquisition in second-order loops will be presented. These have been developed through analysis or experiment and often involve approximations so it is essential to understand the limits of applicability.

All of these equations are for a sinusoidal phase detector characteristic. Results for other types [Egan, 1981, pp. 211–220] are summarized in Appendix 8A. To reiterate,

For a first-order loop,

$$\Omega_H = \Omega_{PI} = \Omega_S = K \text{ rad.} \tag{8.20}$$

For all orders,

$$\Omega_H = K \text{ rad.} \tag{8.21}$$

With an ideal integrator-and-lead filter ($\alpha = 1$), K is infinite, so the average component produced by the nonsymmetrical u_1 waveform during cycle skipping, no matter how small, will eventually be integrated to a large enough value to produce acquisition.

For an integrator-and-lead filter ($\alpha = 1$), with ideal components,

$$\Omega_H = \Omega_{PI} = \infty. \tag{8.22}$$

Even if K could be infinite, real phase detectors do not produce exactly zero average value in the absence of a signal and real op-amps do not have zero offset. There will always be some offset and it may overcome the tiny average component. In the presence of finite offset, excessive DC gain can do more harm than good. However, we defer consideration of optimization in the presence of offsets and here give equations that apply to the degree that such offsets can be neglected.

For a lag-lead filter ($0 < \alpha < 1$), with ideal components (see Fig. 8.11),

$$\Omega_{PI} \approx 2K\sqrt{x - x^2}, \qquad K \gg \omega_z, \tag{8.23}$$

$$x \equiv \omega_p/(2\omega_z), \qquad \text{[Greenstein, 1974]} \tag{8.24}$$

$$\Omega_{PI} \approx 2K\sqrt{x} \approx 2\sqrt{K\zeta\omega_n}, \qquad K \gg \omega_z \gg \omega_p, \tag{8.25}$$

$$\Omega_S = \omega_n^2/\omega_z = \omega_L, \qquad \omega_z \ll \omega_L. \text{ [Gardner, 1979, pp. 68–70;}$$

$$\text{Kroupa, 1973, p. 177] } \tag{8.26}$$

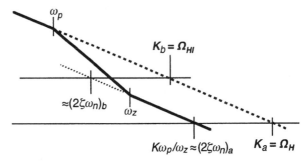

Fig. 8.11   Some parameters if $\omega_z \gg \omega_p$.

Under the conditions of Eqs. (8.25) and (8.26), the time to go from $\omega_e = \Omega$ to $\omega_e = \Omega_S$ is [Richman, 1954a]

$$T_{PI} = \Omega^2/(2\zeta\omega_n^3), \qquad \Omega_{PI} \gg \Omega \gg \Omega_S. \tag{8.27}$$

For a low-pass filter ($\alpha = 0$) [Rey, 1960],

$$\Omega_H \geq \Omega_{PI} \approx 3\zeta K \sqrt{\sqrt{0.423 + 1.19\zeta^4} - 1.092\zeta^2}. \tag{8.28}$$

Here the inequality indicates that the pull-in range cannot exceed the hold-in range, even if the formula gives a larger value.

The value given by Eq. (8.26) for seize frequency can be justified as follows. The loop cannot skip a whole cycle without producing the peak output from the phase detector, $\pm K_p'$. If that peak output appears immediately upon connection to the reference, a worst case in which the initial phase is at the edge of the hold-in range, the resulting step will be amplified by the high-frequency gain, ω_p/ω_z, of the loop filter. This filter output will be amplified by K_v to produce an output frequency step of magnitude,

$$|\Delta\omega| = K_p'(\omega_p/\omega_z)K_v = K'\omega_p/\omega_z = \omega_n^2/\omega_z. \tag{8.29}$$

If this is bigger than the mistuning ($|\Omega| < \omega_n^2/\omega_z$), the mistuning will be overcome by the frequency step. The net frequency error will then cause the phase error to move toward lock. Therefore, locking can occur without cycle skipping with a mistuning as high as ω_n^2/ω_z.

A loop with a low-pass filter has no seize range because the filter prevents an initial phase step from being seen immediately so any mistuning can send the phase in the wrong direction initially.

Example 8.1   Acquisition Formulas   Find Ω_H, Ω_S and Ω_{PI}, for a loop with a sinusoidal phase detector and the following parameters:

$$\omega_p = 100 \text{ rad/sec}; \qquad \omega_z = 500 \text{ rad/sec}; \qquad K = 10,000 \text{ sec}^{-1}.$$

From (8.21),

$$\Omega_H = K = 10,000 \text{ rad/sec} = 1592 \text{ Hz}.$$

From (8.24),

$$x = 100/(2 \times 500) = 0.1.$$

From (8.25),

$$10,000 \gg 500 > ? > 100$$

($> ? >$ indicates inequality satisfied weakly or questionably);

$$\Omega_{PI} \approx 2(10,000 \text{ rad/sec})(0.1)^{0.5} = 6325 \text{ rad/sec} = 1007 \text{ Hz}.$$

Equation (8.23) does not require the latter inequality needed for (8.25). For comparison, from (8.23),

$$\Omega_{PI} \approx 2(10,000 \text{ rad/sec})(0.1 - 0.01)^{0.5} = 6000 \text{ rad/sec}.$$
$$\omega_L = K\omega_p/\omega_z = 2000 \text{ rad/sec} > ? > \omega_z = 500 \text{ rad/sec}.$$

From (8.26),

$$\Omega_S = \omega_L = 2000 \text{ rad/sec} = 318 \text{ Hz}.$$

8.4   APPROXIMATE PULL-IN ANALYSIS

During acquisition of lock, the spectrum of the VCO might transition through a set of states such as shown in Fig. 8.12. If the loop were slow enough for us to observe this on a spectrum analyzer, we might first see the VCO output at its initial offset from the reference signal. It would slowly move closer to the reference and, as it did, we would begin to notice two frequency-modulation (FM) sidebands. These would be caused by the beat frequency produced by the mixing of the reference and VCO output in the phase detector. Since FM sidebands are offset from the central spectral line by the modulation frequency ω_e, one of the sidebands must be at the same frequency as the reference. As the VCO frequency moves closer to the reference, the frequency of that sideband remains fixed at the reference frequency. As ω_e becomes smaller and the modulation index (and possibly the gain in the loop filter) at that frequency is therefore increased, the sidebands grow in amplitude, and more sidebands begin to appear while the VCO output closes ever more quickly on the reference. Since one sideband is always at the input frequency, it will produce a zero-frequency, or DC, signal

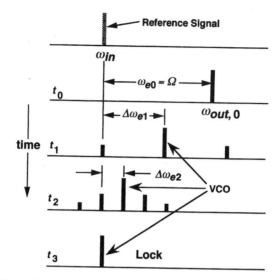

Fig. 8.12 Sequence of spectral pictures during pull-in.

from the phase detector. We will see that this is the signal that re-tunes the VCO to produce pull-in. In the time domain it appears as distortion of a near-sinusoidal beat note (Fig. 8.10).

More can be learned about the acquisition performance of the second-order loop through an approximate analysis that computes this bias (DC) component of u_1. The approximation involved is the assumption, implicit in the computation of the sideband level, that Ω is constant. That cannot really be true because we use the computed value to determine how Ω changes with time. Thus the approximation is best when the change in bias is slow so it is poorest when Ω approaches Ω_S. When we have developed an expression for the bias we will first consider it in the most general terms. That will permit us to predict certain performance features, such as false lock, that can occur when G is more complex than what occurs in second-order loops. Then we will again restrict our considerations to second-order loops and use the computed bias value to obtain some of the results listed in the last section.

8.4.1 Basic Equations

We will now determine the amount ω_b by which the VCO is moved, or tends to be moved, from ω_c due to the bias, which has been described above, that is developed during cycle skipping.

Define

$$\Delta\omega \triangleq \overline{\omega}_{\text{out}} - \omega_{\text{in}} = -\overline{\omega}_e, \tag{8.30}$$

where $\bar{\omega}_{out}$ is the average frequency of the VCO. Then we can write the output of the phase detector in a manner that reflects our expectations that it will consist of a distorted sinusoid at a frequency $\Delta\omega$,

$$u_1(t, \Delta\omega) = \mathbf{Re}\{K'_p\, e^{j\Delta\omega t}\} + v_b + \dots. \tag{8.31}$$

The use of K'_p reflects the assumption that the distortion is small so the fundamental component has the full amplitude of the waveform; v_b is the average, or bias, voltage resulting from the distortion of the phase detector (PD) output waveform. The terms that are dropped from the end of (8.31) are harmonics of $\Delta\omega$ that are not important in the interaction under consideration.

The VCO frequency will be moved from ω_c by $u_2 K_v$, which is obtained from (8.31) as

$$\delta\omega(t, \Delta\omega) \triangleq \omega_{out}(t, \Delta\omega) - \omega_c \tag{8.32a}$$

$$= K_v u_2(t, \Delta\omega) \tag{8.32b}$$

$$= K_v K_{LF}\Big[\mathbf{Re}\{K'_p\, e^{j\Delta\omega t} F(\Delta\omega)\} + v_b\Big]. \tag{8.33}$$

Since

$$G(\Delta\omega) = K_v K_{LF} K_p F(\Delta\omega)/(-j\Delta\omega), \tag{8.34}$$

Eq. (8.33) is

$$\delta\omega(t, \Delta\omega) = \Big[-\mathbf{Re}\{j\Delta\omega G(\Delta\omega)\, e^{j\Delta\omega t}\} + v_b K_{LF} K_v\Big] \tag{8.35}$$

$$= \Delta\omega|G(\Delta\omega)|\sin[\Delta\omega t + \angle G(\Delta\omega)] + \omega_b, \tag{8.36}$$

where

$$\omega_b \triangleq v_b K_{LF} K_v = v_b K/K_p \tag{8.37}$$

is the frequency bias term that results from interaction of the first term with the reference and whose value we are attempting to compute.

The corresponding phase modulation is

$$\delta\varphi(t, \Delta\omega) = \int \delta\omega(t, \Delta\omega)\, dt = -|G(\Delta\omega)|\cos(\Delta\omega t + \angle G(\Delta\omega)) + \dots,$$

$$\tag{8.38}$$

where the bias term has been dropped. We drop it here because we are analyzing the effect of the other term, but the value of the dropped bias term is represented implicitly in $\bar{\omega}_{out}$ and therefore in $\Delta\omega$ and ω_b. Thus, while we are computing ω_b, an initial value of this variable, ω_{b0}, is implied. It might seem that we are starting with the answer, but we shall see that the results of the process are nevertheless informative.

The peak phase deviation, or modulation index, can be seen from (8.38) to be

$$m = |G(\Delta\omega)|. \tag{8.39}$$

The output from the VCO is a sinusoid with average frequency $\overline{\omega}_{out}$, which is phase modulated with the deviation given by (8.39). Assuming that the deviation is small, the VCO output voltage can be written (based on FM modulation theory; see Table 20.A.1)

$$v_{out}t = A\{\cos\overline{\omega}_{out}t - 0.5|G(\Delta\omega)|\langle\cos[(\overline{\omega}_{out} + \Delta\omega)t + \angle G(\Delta\omega) + \pi/2]$$
$$+ \cos[(\overline{\omega}_{out} - \Delta\omega)t - \angle G(\Delta\omega) - \pi/2]\rangle\}. \tag{8.40}$$

Substituting $\Delta\omega$ from (8.30) we obtain

$$v_{out} \approx A\{\cos\overline{\omega}_{out}t - 0.5|G(\Delta\omega)|\langle\cos[(2\overline{\omega}_{out} - \omega_{in})t + \angle G(\Delta\omega) + \pi/2]$$
$$+ \cos[(\omega_{in})t - \angle G(\Delta\omega) - \pi/2]\rangle\}. \tag{8.41}$$

Note that the second sideband (last term) here is at the input frequency, ω_{in}, and will thus produce a DC bias term when mixed in the phase detector with the input signal. The bias is given by Eq. (1.11) except that the magnitude is multiplied by the ratio of this sideband to A,[2]

$$v_b = 0.5|G(\Delta\omega)|K'_p \cos[\angle G(\Delta\omega) + \pi/2]$$
$$= 0.5|G(\Delta\omega)|K'_p \sin[\angle G(\Delta\omega)]. \tag{8.42}$$

This voltage at the phase detector output will be amplified by the DC gain to the output to produce an output frequency bias of

$$\omega_b = 0.5K'F(\Delta\omega \approx 0)|G(\Delta\omega)|\sin[\angle G(\Delta\omega)] \tag{8.43}$$
$$\approx 0.5K' \operatorname{Im} G(\Delta\omega). \tag{8.44}$$

This is the bias signal that causes pull-in. For example, at high modulation frequencies, a loop with a lag-lead filter will have $\angle G(\Delta\omega) \approx -\pi/2[\operatorname{sign}(\Delta\omega)]$, due to the $1/s$ term in $G(s)$, so a positive value for $\Delta\omega$ would produce $\omega_b < 0$, which would tend to cause ω_{out} to be below ω_c. If this value of ω_b were more negative than ω_{b0}, the value corresponding to

---

[2]Phase modulation passes through a mixer from either input to the output, so the values of the sidebands relative to the carrier that produces the normal PD output will be preserved from input to output of the PD.

the initially assumed ω_{out}, ω_{out} would tend to decrease toward ω_{in}, thus lowering $|\Delta\omega|$ and leading toward lock.

In the next section we will analyze the general behavior of loops, including possible anomalous behavior that can occur under certain circumstances, by determining a relative value for ω_b under various assumptions. This will tell us the direction of change and allow us to understand the general behavior to be expected (as was done in Section 8.2), including the effects of some undesired, but not necessarily unlikely, high-frequency poles. In the section after that we will solve for the transient values of v_b for a particular type of filter.

Example 8.2   Pull-in Bias   A loop with a lag-lead filter is out of lock and skipping cycles. The VCO design is such that its center frequency is 6×10^5 rad/sec higher than the current reference frequency but, at the moment, its frequency is higher than the reference by only 10^5 rad/sec. Its open-loop gain is given by

$$G(\Delta\omega) = \frac{10^7 \text{ sec}^{-1}}{j\Delta\omega} \frac{1 + j(\Delta\omega/10^3 \text{ rad/sec})}{(1 + j(\Delta\omega/1 \text{ rad/sec}))}.$$

Will the loop move toward lock or away from lock?

The bias developed, according to (8.44) is

$$\omega_b = 0.5(10^7 \text{ sec}^{-1})\mathbb{Im}\left(\frac{10^7}{j10^5} \frac{1 + j(1 \times 10^5/10^3)}{1 + j(10^5/1)} \right)$$

$$= -5 \times 10^5 \text{ rad/sec}.$$

Thus, a bias is developed that will tend to force the VCO frequency 5×10^5 rad/sec below its center frequency. But this is 10^5 rad/sec above the reference frequency, which is where we started. Thus there is equilibrium. If we change the initial frequency slightly we find that ω_b changes in the same direction but by an amount that is 4 times greater. Therefore we are at a point of unstable equilibrium, and the direction in which we will move depends on a small deviation from that point. Practically, it is uncertain because of the various approximations in developing the theory and inaccuracies in obtaining parameters for any real loop.

8.4.2   General Analysis

Suppose that, at some time t_0, there is a bias voltage on the VCO tuning line that causes its frequency to be offset from ω_c by ω_{b0}. This will result in a difference frequency $\Delta\omega(0)$ [between the VCO frequency $\omega_{out}(0)$ and the

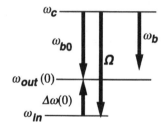

Fig. 8.13 Relationship between frequency variables.

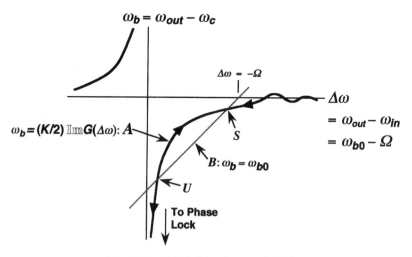

Fig. 8.14 Pull-in bias in generic PLL.

reference frequency ω_{in}] that is equal to the difference between the mistuning Ω and ω_{b0}, as shown by the following equation and illustrated in Fig. 8.13:

$$\Delta\omega(0) = \omega_{out}(0) - \omega_{in} = [\omega_{out}(0) - \omega_c] - [\omega_{in} - \omega_c] = \omega_{b0} - \Omega.$$
(8.45)

Equation (8.45) defines the line B in Fig. 8.14. Given a starting value of ω_{b0}, this line shows the corresponding value of $\Delta\omega(0)$. For that $\Delta\omega(0)$ we can compute, from Eq. (8.44), a new value of ω_b that would ultimately be produced (if $\Delta\omega$ did not change). We can tell whether ω_b, and therefore ω_{out}, will increase or decrease or remain unchanged by comparing the computed value of ω_b to the starting value ω_{b0}.

The second curve in Fig. 8.14, A, is a plot of Eq. (8.44) for a generic loop. At very low values of $\Delta\omega$, $G(\Delta\omega) \approx -jK/\Delta\omega$. Thus ω_b is very large and negative. As $\Delta\omega$ increases, $|G(\Delta\omega)|$ drops and so does ω_b. Phase shift due to the loop filter will also decrease ω_b. At very high values of $\Delta\omega$ unintended poles, for example, those found in op-amp circuitry, will cause additional phase shift and $\mathbb{Im}\, G(\Delta\omega)$ will go through zero multiple times as $|G(\Delta\omega)|$ continues to decrease.

At any value of $\Delta\omega$, the ordinate of curve B gives the corresponding value of ω_{b0} while the ordinate of curve A gives the new value. Thus, whenever curve A is above curve B, the new value of ω_b will be greater than the original value and $\Delta\omega$ will increase. Conversely, when curve A is below curve B, $\Delta\omega$ will tend to decrease. These motions are shown by the arrows on curve A. Intersections of the two curves are equilibrium points because there the initial ω_{b0} from curve B equals the resulting ω_b from curve A. Point S is one of stable equilibrium because the arrows point toward it from both sides. Conversely, U marks a point of unstable equilibrium because the arrows point away from it. If the initial value of ω_{b0} is above U (in Fig. 8.14), the operating point will move toward S and may remain there in an unlocked condition with $\omega_{out} \neq \omega_{in}$. If it starts below U, it will move downward and $\Delta\omega$ will decrease to zero and a state of true phase lock will be acquired.

It is possible that, in the vicinity of S, while each deviation of the operating point from S causes a correction toward S, each correction is so large that it overshoots S by more than the original error. Then the operating point might oscillate about S or deviate from the vicinity of S entirely. It is not obvious how the simple theory we have developed to this point can determine these dynamics,[3] but it can help us to understand the possibilities and explain observed phenomena. It is clear that, if our design includes a possible false lock point, we should be concerned. Curve A is mirrored in the graph's origin. The other part of the curve is used for $\Delta\omega < 0 \Rightarrow \omega_{out} < \omega_{in}$. This illustrates the importance of the sign of $\Delta\omega$ in Eq. (8.44).

In Fig. 8.15a, an initial offset $\Delta\omega > U$ will cause the locus to move to S at which point $\Delta\omega \approx -\Omega$, which also implies $\omega_{out} \approx \omega_c$. Thus the VCO is approximately free running. If either the oscillator center frequency is increased or the input frequency is decreased, $-\Omega = \omega_c - \omega_{in}$ will increase by the same amount, moving curve B to the right by that much and causing $\Delta\omega$ at point S to move about the same amount. Thus a change in open-loop VCO tuning will produce about the same change in frequency as would occur without the influence of the loop and, equivalently, a change in ω_{in} will have little effect upon the value of ω_{out}. For this reason, the stable operating point at S corresponds to an out-of-lock, or free running, condition.

If the center frequency of the oscillator is then moved lower, toward ω_{in}, or if ω_{in} moves higher toward ω_c, curve B moves to the left and eventually no longer intersects curve A, as shown in Fig. 8.15b. Under those conditions, curve B is always above curve A, so lock will occur from any starting point

---

[3]It is a simple matter to compute successive values of ω_b and to determine if successive deviations from **S** increase in magnitude, but this is based on a step-by-step model, as if there were a low-frequency sample-and-hold circuit in the loop. In practice, the initially assumed value will not be held until the new value is attained.

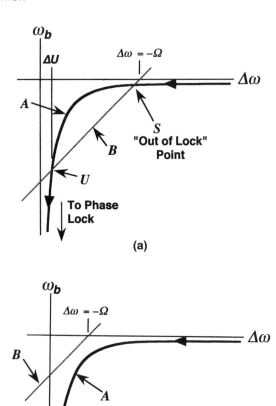

Fig. 8.15   Pull-in range: (*a*) beyond pull-in and (*b*) within pull-in.

shown on curve A. This is an illustration of the normal pull-in process that we have discussed. However, an anomaly that we have not discussed can also be illustrated.

In Fig. 8.16, curves A and B do not cross in the normal region illustrated in Fig. 8.15, but the gain is high enough that one of the arcs, which are produced by excess phase shift at high frequencies, is intersected by curve B. This can cause a stable operating point at S. If ω_c or ω_{in} is changed, curve B will move correspondingly. However, because curve A is nearly vertical at S, the change in $\Delta\omega$ will be small. If the oscillator tuning is adjusted to move ω_c (and thus curve B), the oscillator will change frequency by only a small amount compared to the change in ω_c. If ω_{in} is changed, ω_{out} will follow it.

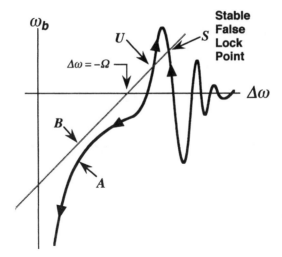

Fig. 8.16   False lock.

Thus the VCO appears to be locked to the reference, but with a frequency offset, a "false lock." One situation in which such false locks are a danger is that in which an intermediate-frequency (IF) filter is used, as shown in Fig. 8.17. Frequency-modulation sidebands may see different phase shifts in the passband of the IF filter, and that can be accentuated by multiple poles or by operating near the edge of the passband. The effect is equivalent to phase shift of the modulation waveform and therefore to phase shift in the loop.

Reducing the gain K will shrink curve B and can cause the stable point S to disappear, thus breaking the false lock. However, if S is on one of the smaller arcs, reducing the gain might cause it just to change frequency offsets. Reducing the gain might also cause curve A to shrink to the point where the false lock is replaced by an out-of-lock condition, wherein the curves intersect on a relatively horizontal part of curve A. In general, when the loop is not truly locked, a change in the input frequency will cause some

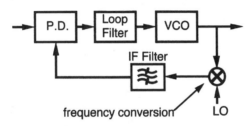

Fig. 8.17   Loop with IF filter.

smaller change in the VCO frequency. False locks and out-of-lock conditions describe two extremes of anomalous performance.

8.4.3 Pull-in Range

As our first attempt to gain quantitative information from the preceding analysis, we will compute the pull-in range under the conditions of Eq. (8.25), namely $K \gg \omega_z \gg \omega_p$.

Pull-in is certain if curves A and B (Fig. 8.14) do not intersect. The boundary of this condition occurs when they are tangent. That will occur at a point at which they have the same slope. Therefore, to find that point, we must find the value of $\Delta\omega$ at which the derivative of Eq. (8.44) with respect to $\Delta\omega$ is unity. Under the assumed conditions, Eq. (8.44) becomes

$$\omega_b \approx \frac{K}{2}\, \mathbb{Im}\left[\frac{K}{j\,\Delta\omega}\frac{\Delta\omega/\omega_z}{\Delta\omega/\omega_p}\right] = -\frac{K^2\omega_p}{2\,\omega_z}\frac{1}{\Delta\omega}. \tag{8.46}$$

The derivative is

$$\frac{d\omega_b}{d\,\Delta\omega} = \frac{K^2\omega_p}{2\,\omega_z}\frac{1}{\Delta\omega^2}, \tag{8.47}$$

which we set equal to 1, yielding

$$\Delta\omega_1 = K\sqrt{\frac{\omega_p}{2\,\omega_z}}. \tag{8.48}$$

We wish to know the value of mistuning, which is $-\Omega$, the intersection of curve B with the x axis, at which this occurs. Since curve B is a 45° line, the horizontal distance from S to $-\Omega$ is the same as the vertical distance so

$$-\Omega = \Delta\omega_1 - \omega_b(\Delta\omega_1). \tag{8.49}$$

We evaluate the last term by substituting $\Delta\omega_1$ from Eq. (8.48) into the last part of Eq. (8.46) and find that

$$-\omega_b(\Delta\omega_1) = \Delta\omega_1. \tag{8.50}$$

Thus if, in Fig. 8.15a, we are operating at the stable out-of-lock point S and we slowly decrease the mistuning, S will reach its most extreme position on curve A, on the verge of pull-in, when the beat note $\Delta\omega$ is just half of the mistuning. At that point the last two equations indicate that the absolute value of the mistuning is

$$\Omega_{PI} = |\Omega| = 2|\Delta\omega_1| = K\sqrt{\frac{2\omega_p}{\omega_z}}\,. \tag{8.51}$$

Under the assumed conditions $(K \gg \omega_z \gg \omega_p)$ we can write

$$\Omega_{PI} = \sqrt{\frac{2K^2\omega_p}{\omega_z}} = \sqrt{\frac{2K\omega_n^2}{\omega_z}} \approx 2\sqrt{K\zeta\omega_n}\,, \tag{8.52}$$

which is the same as Eq. (8.25).

8.4.4 Approximate Pull-in Time

Our next step in obtaining quantitative information from our approximate analysis is to relate the voltages at the input and output of the loop filter by Eq. (8.42), which describes how the tuning voltage causes a low-frequency term at the phase detector output, and by the filter transfer function, which describes how that term causes a tuning voltage.

By Eq. (8.42), at the input to the filter we have

$$v_b \approx 0.5K_p'\, \mathrm{Im}\, G(\Delta\omega). \tag{8.53}$$

We assume that the frequency error is large compared to ω_z so this can be written

$$v_b \approx -0.5K_p'K\omega_p/(\omega_z\,\Delta\omega). \tag{8.54}$$

The filter characteristic can be written

$$v_v(s) = K_{\mathrm{LF}}\frac{1 + s/\omega_z}{1 + s/\omega_p}v_b(s), \tag{8.55}$$

where v_v is the low-frequency tuning voltage at the output of the filter or, equivalently,

$$\left(1 + \frac{1}{\omega_p}\frac{d}{dt}\right)v_v(t) = K_{\mathrm{LF}}\left(1 + \frac{1}{\omega_z}\frac{d}{dt}\right)v_b(t). \tag{8.56}$$

Substituting v_b from Eq. (8.54) and using $K_v v_v = \omega_b = \Delta\omega + \Omega$ we obtain

$$\left(1 + \frac{1}{\omega_p}\frac{d}{dt}\right)(\Delta\omega(t) + \Omega) = -\frac{K^2 \omega_p}{2\omega_z}\left(1 + \frac{1}{\omega_z}\frac{d}{dt}\right)\frac{1}{\Delta\omega(t)} \quad (8.57)$$

$$= -\frac{\Omega_{PI}^2}{4}\left(1 + \frac{1}{\omega_z}\frac{d}{dt}\right)\frac{1}{\Delta\omega(t)}, \quad (8.58)$$

Where the last step used Eq. (8.52). Using the equality

$$\frac{d}{dt}\frac{1}{\Delta\omega(t)} = -\frac{1}{\Delta\omega(t)^2}\frac{d}{dt}\Delta\omega(t)$$

and rearranging terms, we obtain

$$\Delta\omega(t) + \Omega + \frac{\Omega_{PI}^2}{4\Delta\omega(t)} = \left(-\frac{1}{\omega_p} + \frac{\Omega_{PI}^2}{4\omega_z\,\Delta\omega^2(t)}\right)\frac{d}{dt}\Delta\omega(t), \quad (8.59)$$

$$dt = \frac{1}{\omega_p}\frac{-\Delta\omega(t)^2 + (\Omega_{PI}^2/4)(\omega_p/\omega_z)}{\Delta\omega(t)\left[\Delta\omega(t)^2 + \Omega\Delta\omega(t) + \Omega_{PI}^2/4\right]}d\Delta\omega(t). \quad (8.60)$$

We then integrate. We set the initial time to zero and, assuming that the VCO is initially at ω_c, set $\Delta\omega(0) = -\Omega$. We take acquisition to have been completed when $|\Delta\omega(T_{PI})|$ equals the seize frequency,[4] $\Omega_S = \omega_n^2/\omega_z = Kr$, where

$$r \equiv \omega_p/\omega_z, \quad (8.61)$$

$$\int_0^{T_{PI}} dt = \frac{\Omega_{PI}^2}{4\omega_z}\int_{-\Omega}^{-Kr}\frac{1}{\Delta\omega(t)\left[\Delta\omega(t)^2 + \Omega\,\Delta\omega(t) + \Omega_{PI}^2/4\right]}d\Delta\,\omega(t)$$

$$-\frac{1}{\omega_p}\int_{-\Omega}^{-Kr}\frac{\Delta\omega(t)}{\Delta\omega(t)^2 + \Omega\Delta\,\omega(t) + \Omega_{PI}^2/4}d\Delta\,\omega(t). \quad (8.62)$$

---

[4]The seize frequency is defined as a mistuning, that is, the difference between the input frequency and ω_c. Here we are assuming that the same value of $|\Delta\omega|$ will cause rapid pull-in even if the VCO is not at ω_c.

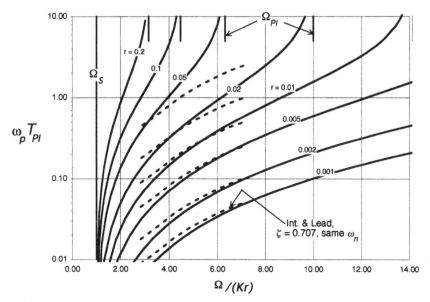

Fig. 8.18 Approximate pull-in time, $|\Delta\omega| \gg \omega_z$. Accuracy is poorest near Ω_{PI} and Ω_S.

These two integrals are standard forms, the solution of which give

$$T_{PI} = \frac{1}{\omega_p}\left\{ r\ln\left(\frac{K}{\Omega}r\right) - \frac{1+r}{2}\ln\left[2\left(r - \frac{\Omega}{K}\right) + 1\right] - \frac{(1-r)\Omega}{\sqrt{\Omega_{PI}^2 - \Omega^2}} \right.$$

$$\left. \times\left[\tan^{-1}\left(\frac{2Kr - \Omega}{\sqrt{\Omega_{PI}^2 - \Omega^2}}\right) - \tan^{-1}\left(\frac{\Omega}{\sqrt{\Omega_{PI}^2 - \Omega^2}}\right)\right]\right\}. \qquad (8.63)$$

This is plotted in Fig. 8.18.

If $\Omega \gg 2Kr$ (acquisition begins far from Ω_S),

$$T_{PI} = \frac{1}{\omega_p}\left\{ r\ln\left(\frac{K}{\Omega}r\right) - \frac{1+r}{2}\ln\left[2\left(r - \frac{\Omega}{K}\right) + 1\right] \right.$$

$$\left. + \frac{2(1-r)\Omega}{\sqrt{\Omega_{PI}^2 - \Omega^2}}\tan^{-1}\left(\frac{\Omega}{\sqrt{\Omega_{PI}^2 - \Omega^2}}\right)\right\}. \qquad (8.64)$$

When also $r \ll 1$ and $K \gg \Omega$, this equation approaches Eq. (8.27). The equation can be written in normalized form as

$$\omega_P T_{PI} = -r \ln(\Delta) - \frac{1+r}{2}\ln[1 + 2r(1 - \Delta)]$$

$$+ \frac{2(1-r)\Delta}{\sqrt{2/r - \Delta^2}}\tan^{-1}\left(\frac{\Delta}{\sqrt{2/r - \Delta^2}}\right), \qquad (8.65)$$

where

$$\Delta \equiv \Omega/(Kr). \qquad (8.66)$$

A comparison is also shown to points obtained [Sanneman and Rowbotham, 1964][5] for a loop with an integrator-and-lead filter (type 2, $r \Rightarrow 0$) but having the same values of ζ and ω_n. The dashed lines match the solid lines well for the smallest values of r, which represent loops that approach type 2. We might also say that significant deviations occur when Ω_{PI} is approached for the lag-lead filter; this is to be expected since the type 2 loops have infinite Ω_{PI}.

8.5   PHASE-PLANE ANALYSIS

In the linear range, where phase deviations are small, linear theory applies, and we have just developed expressions applicable to the opposite extreme. If we should require accurate answers for the in-between case, we must use computer simulation, experimentation, or phase-plane analysis. (See phase-plane discussions in Gardner [1979, pp. 55–58] and Stensby [1997, pp. 121–156, 168–175].)

Figure 8.19 is a phase-plane portrait. It is constructed by computing $(d\omega_e/d\varphi_e) = F(\omega_e, \varphi_e)$, so a small segment of the locus of $\omega_e = F(\varphi_e)$ can be plotted by using the computed slope. These segments can then be connected. In other words, we can go to any point in a graph of ω_e versus φ_e and extend a line a short distance from that point if we know the slope at that point. By doing this repeatedly, a set of loci can be created. Figure 8.19 is such a set. It repeats at increments of $\varphi_e = 2\pi$ and should be imagined to exist on a cylinder so that any vertical line is contiguous with a similar line 2π away. In the upper part of the figure, since $\omega_e = d\varphi/dt$ is positive, φ_e increases with time and the locus moves to the right. Each 2π increment in x, the $|y|$ can be seen to decrease. The decrease is slight for large $|y|$ and becomes larger for smaller $|y|$. Eventually the "separatrix," a division between curves that do and

---

[5]Figure 6 in Sanneman and Rowbotham is for $\zeta = 0.707$. The axes above are matched to the axes in their Fig. 6 using $\omega_p = \omega_n[1 - (1 - 2r)^{0.5}]/(2)^{0.5}$, $Kr = \omega_n r(2)^{0.5}/([1 - (1 - 2r)^{0.5}]$, which is true for $\zeta = 0.707$. Initial condition for their Fig. 6 was zero phase error.

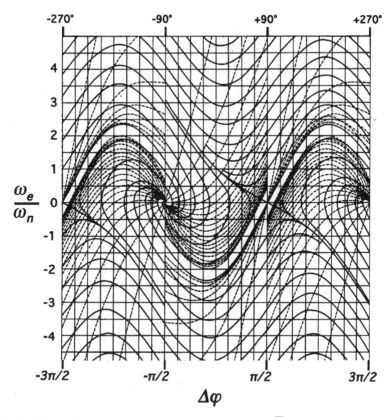

Fig. 8.19 Phase-plane plot for type 2 loop, $\zeta = 1/\sqrt{2}$. (From Sanneman and Rowbotham, 1964, Fig. 5, p. 19, ©1964 IEEE.)

do not reach the next $\pm \pi$ increment, is reached. Curves below the separatrix cross the $y = 0$ axis and turn back. From that point there is no cycle skipping. After the axis is crossed, since $y = dx/dt$ becomes negative, the direction of horizontal motion reverses, and the locus moves toward a steady-state value $(-\pi/2, 0)$. Depending upon ζ, it may move steadily to the final values or it may spiral in. A similar description applies if we begin in the lower part of the figure. Viterbi has drawn such plots for type 2 loops with ζ between 0.7 and 1.4 [Blanchard, 1976, pp. 256–259].

Figure 8.20 shows two possible loci superimposed on a one-cycle-wide region of Fig. 8.19. Assume $\Delta \varphi$ and ω_e happen to have values corresponding to point $0'$ at $t = 0$. This may be the time when the reference is connected to the loop or might just be some instant during an acquisition sequence. (Even if t does represent the time when the reference is connected, Ω cannot be read from the graph unless the initial value of $\Delta \varphi$ happens to be $\pm \pi/2$, where the phase detector output is unchanged by the connection.)

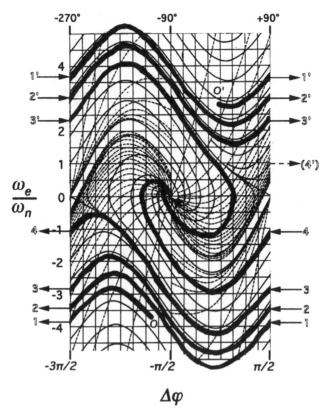

Fig. 8.20 Two loci in the phase plane. This drawing represents a cylinder connected on right and left edges. The outgoing arrow labeled n (or n') connects to the incoming arrow with the same label on the other edge. (From Sanneman and Rowbotham, 1964, Fig. 5, p. 19, ©1964 IEEE, loci added.)

Because $\omega_e > 0$ at $0'$, $\Delta\varphi$ increases with time so the locus moves to the right, guided by the curves of the graph. At $+90°$ we move to the left edge of the graph (at $1'$), although this is just a way of illustrating, on a two-dimensional graph, continuous motion around a cylinder. The locus continues left to right, moving from $0'$ to $1'$ to $2'$ to $3'$. In the process the frequency error increases above its initial value, but, at any value of $\Delta\varphi$, each locus is closer to the lock point than the previous locus. What is more, the rate at which ω_e decreases accelerates with subsequent cycles of $\Delta\varphi$ so there is a larger space between loci as time increases. After $3'$ the locus spirals into the stable lock point at $\omega_e = 0$ and $\Delta\varphi = -90°$. A very slightly higher initial value of ω_e would have resulted in the locus after $3'$ being above the separatrix, and it would have gone through $4'$ before spiraling into the lock point. The spiraling nature of the approach to the lock point represents overshoot in ω_e.

A similar process occurs if we begin at point 0, but now we move to the left because the negative value of ω_e represents decreasing $\Delta\varphi$. In this plot, segment 3–4 is a little below the separatrix so another cycle is skipped.

We can develop the equations needed to plot a phase-plane portrait for the lag-lead filter by beginning with one similar to Eq. (8.56), except it is written for the total filter input and output voltages rather than the low-frequency component. Relating these voltages to the corresponding frequency and phase errors, we obtain

$$\left(1 + \frac{1}{\omega_p}\frac{d}{dt}\right)\left(\frac{\Omega - \omega_e(t)}{K_v}\right) = K_{LF}\left(1 + \frac{1}{\omega_z}\frac{d}{dt}\right)K_p\cos\Delta\varphi(t), \quad (8.67)$$

$$\frac{d\omega_e(t)}{dt} + \omega_p\left(1 - \frac{K}{\omega_z}\sin\Delta\varphi(t)\right)\omega_e(t) + K\omega_p\cos\Delta\varphi(t) = \omega_p\Omega. \quad (8.68)$$

Using

$$\frac{d\omega_e(t)}{dt} = \frac{d\omega_e}{d\Delta\varphi(t)}\frac{d\Delta\varphi(t)}{dt} = \frac{d\omega_e}{d\Delta\varphi(t)}\omega_e, \quad (8.69)$$

and defining the coordinates of the plot at

$$y \equiv \frac{\omega_e}{\omega_n} \quad \text{and} \quad x = \Delta\varphi, \quad (8.70)$$

Eq. (8.68) can be written as

$$\frac{dy}{dx} = \frac{\omega_n}{\omega_z}\sin x - \frac{\omega_n}{K} + \frac{\Omega/K - \cos x}{y}. \quad (8.71)$$

If we let $K \Rightarrow \infty$, we obtain the equivalent expression for a type 2 loop,

$$\frac{dy}{dx} = 2\zeta\sin x - \frac{\cos x}{y}, \quad (8.72)$$

which is plotted in Fig. 8.19 for $\zeta = 1/\sqrt{2}$.

This figure also includes isochrones (dashed lines), which mark fixed time increments along the loci. Since $y \equiv (1/\omega_n)\,d\Delta\varphi/dt$, a small time increment Δt corresponds to a phase change $\Delta(\Delta\varphi) = y\omega_n\,\Delta t$. The isochrones in Fig. 8.19 are drawn for $\Delta t = 1/(4\omega_n)$ so, moving along a locus, the change in x between isochrones will be $\Delta(\Delta\varphi) = y/4$. The time to get from one phase and frequency to another can be determined by counting the number of isochrone increments and dividing by $4\omega_n$. It is by this method that the dashed curves in Fig. 8.18 were computed from Fig. 8.19.

8.6 PULL-OUT

In Chapter 6 we described the phase error that results from a frequency step (Fig. 6.9) but that was for a linear system. Of particular interest is the maximum frequency step ω_{po} that will not push $\Delta\varphi$ to the point where a cycle is skipped. This is inherently a nonlinear problem. Beginning at the lock point in Fig. 8.19, if the frequency is stepped (vertically, at constant $\Delta\varphi$) beyond the separatrix at $(-\pi/2, 3.2)$, a cycle will be skipped as the operating point returns to the lock point. For this type 2 loop with $\zeta = 1/\sqrt{2}$, the pull-out frequency is therefore $\omega_{po} = 3.2\,\omega_n$. By studying such plots, Gardner has determined, for type 2 loops, that [Gardner, 1979, p. 57]

$$\omega_{po} = 1.8\omega_n(1 + \zeta) \quad \text{if} \quad 0.5 < \zeta < 1.4. \tag{8.73}$$

8.7 EFFECT OF OFFSETS

Here we consider how to maximize pull-in range in the presence of a phase detector or op-amp offset. Assume the general lag-lead filter and let the offset be equivalent to a phase $\sin^{-1}\varphi_{os}$, approximately φ_{os} for small values. This will produce an output offset $\omega_{os} = K'\varphi_{os}$. Presumably φ_{os} is not controllable so its sign is unknown and the loop may have to overcome both it and the mistuning that would otherwise exist. Therefore the pull-in range is effectively reduced by $|\omega_{os}|$ and the effective pull-in range is

$$\Omega_{PI\,os} = \Omega_{PI} - \omega_{os} = K\!\left(\sqrt{4x} - |\varphi_{os}|\right), \tag{8.74}$$

where x is given by (8.24).

If K is to be limited, perhaps to prevent components from saturating, then the largest value of $\Omega_{PI\,os}$ will occur at the highest value of x, which is 0.5 and which occurs for a first-order loop. However, the restriction of (8.25) would be violated for $x = 0.5$ and Ω_{PI} would not exceed $K(1 - |\varphi_{os}|)$.

If $\Omega_{PI\,os}$ is to be maximized in the presence of a fixed loop bandwidth ω_L, then we rewrite (8.74) in terms of ω_L, which is $K\omega_p/\omega_z = 2Kx$, assuming the restrictions of (8.25) apply.

$$\frac{\Omega_{PI\,os}}{\omega_L} = \sqrt{\frac{1}{x} - \frac{|\varphi_{os}|}{2x}}. \tag{8.75}$$

To maximize this ratio, we take the derivative of the right side with respect to x and set it equal to zero. This gives

$$x = |\varphi_{os}|^2. \tag{8.76}$$

Substituting this into (8.75), we obtain the maximum pull-in range as

$$\Omega_{PI\,os} \leq \frac{\omega_L}{2|\varphi_{os}|}. \tag{8.77}$$

8.8 EFFECT OF COMPONENT SATURATION

Suppose some component has a limited range that can be effectively modeled as clipping of the tuning voltage. That is, the tuning voltage, and consequently the VCO frequency, stops at some minimum or maximum value and can be moved no further. Assume that there are no other consequences of the component's saturation (e.g., strange or sluggish operation). What effect will this clipping have on pull-in?

Obviously, the reference frequency must be within the clip limits or no lock can occur. If, during pull-in, the frequency overshoots to a limit, there will be some change in the response. Suppose though that the lock frequency is within the clip limits but the VCO center frequency ω_c is not. Then the initial VCO frequency is not what would occur in a linear system with zero output from the phase detector, but corresponds rather to a clip limit. If the mistuning (in the absence of clipping) is within the seize range, the loop should lock because the high-frequency gain is adequate for the phase detector to tune the VCO to the input frequency. However, further out in the pull-in range, if the range of frequencies that would be produced during the initial cycle of ω_e is beyond the clipping limit, the VCO will not move from its initial frequency. Then there will be no feedback to develop a pull-in voltage and the loop will not lock.

8.9 HANGUP

If a loop is at the steady-state frequency but the phase is at a point of unstable equilibrium (e.g., π radians from steady-state for a type 2 loop), there will be no error signal to cause the loop to move to the stable lock point. Any small deviation from the unstable-equilibrium point, perhaps due to noise, can initiate acquisition of the correct steady-state phase, but the error signal will initially be small so the first part of the transient will be slow. This phenomenon is called hangup [Gardner, 1977]. An operating point on the separatrix of the phase plane leads to an unstable equilibrium point, for example, the saddle point at $(\pi/2, 0)$ in Fig. 8.19, but any slight deviation from the separatrix leads to phase lock. However, note how close are the isochrones in that region, indicating slow motion. Everlasting hangup is a zero-probability event, but finite-duration hangup has finite probability and can sometimes be a problem. One might be inclined to use a sawtooth PD, which theoretically has essentially no unstable equilibrium point of finite extent, but we will find, in Chapter 13, that noise can cause the sawtooth to lose its sharp edge.

8.A   APPENDIX: SUMMARY OF ACQUISITION FORMULAS FOR SECOND-ORDER LOOPS

Table 8.A.1 provides a summary of acquisition formulas for second-order loops.

8.M   APPENDIX: NONLINEAR SIMULATION

The MATLAB script NLPhP simulates the nonlinear acquisition behavior of a phase-locked loop (PLL). With this script we can reproduce the loci in Fig. 8.20 and obtain other enlightening results. The output is an approximation whose degree of accuracy improves as we allow more computer operations for a given problem.[6] It is more complex than previous programs but employs the same basic techniques. The program is given in Section 8.M.11.

8.M.1   Sampling and Simulation

Rather than performing computations on the closed-loop response, we will use MATLAB to compute the open-loop response, and we will close the loop explicitly on a sampled basis. That will permit us to introduce the nonlinear characteristic of the PD, one sample at a time. The process is represented by Fig. 8.M.1.

The output phase φ_{out} is subtracted from the input phase φ_{in} and the error φ_e is processed by the nonlinear PD characteristic. This modified phase error is sampled and held at a regular sampling rate. The HOLD output that depends on the value of φ_{out} at the end of sampling period j provides the input for the open-loop transfer function $G(s)$ during sample period $j + 1$. The methods described in Section 6.10 are used to compute the output from $G(s)$ at the end of each period. The difference between this simulation and the true case is that $G(s)$ is excited by a stair-step waveform rather than a continuous waveform but, as the sampling period shrinks, so does the inaccuracy. A good approximation of the effect of the added sampler and HOLD, for frequencies that are low compared to the sampling frequency, is that it introduces a phase shift that is linear with frequency and reaches $-\pi$ radians at the sampling frequency [Egan, 1981, pp. 123–127]. NLPhP computes this phase shift at ω_L and displays its value.

---

[6]Using MATLAB v.3.5 on a MacII, 15.7-MHz clock, circa 1988, each plot shown in this section requires less than 3 mins. Using MATLAB v.4a on a 90-MHz Mac 7200, 12 sec are required.

TABLE 8.A.1. Acquisition Formulas

Parameter (ranges are $\pm$ given values)	PD Type	Value of Parameter[a]	Conditions
Hold-in, Ω_H	Sine	K	
	Triangle	$\dfrac{\pi}{2}K$	
	Sawtooth	πK	
	PFD	$2\pi K$	
Seize (lock-in) Ω_S approximately	Sine[b]	ω_L	$\omega_z \ll \omega_L$
	Triangle[c]	$\dfrac{\pi}{2}\omega_L$	
	Sawtooth[d]	$\pi\omega_L$	
	PFD	Unlimited by PD	
Pull-in (acquisition) Ω_{PI} approximately	Sine	$2K\sqrt{x-x^2}\,;\ x = \dfrac{\omega_p}{2\omega_z}$	Lag-lead filter[e] $K \gg \omega_z$
		$3\zeta K\sqrt{\sqrt{0.423 + 1.19\zeta^4} - 1.092\zeta^2}$	Low-pass filter[f] $\Omega_{PI} < \Omega_H$
	Triangle	$\pi\sqrt{\zeta K\omega_n}$	Lag-lead filter[g] $\omega_n \ll K;\ \zeta < 1$
		Ω_H	Low-pass filter[h] $\zeta > 1$
	Sawtooth	$2\pi K\sqrt{\dfrac{\omega_p}{3\omega_z}} \approx 2\pi\sqrt{\dfrac{2}{3}K\zeta\omega_n}$	Lag-lead filter[d] $K \gg \omega_z > 2\omega_p$
		$\pi K \tanh\left\{\dfrac{\pi}{2\sqrt{\dfrac{1}{\zeta^2}-1}}\right\}$	Low-pass filter[i] $\zeta \leq 1$
		πK	Low-pass filter[i] $\zeta > 1$
	PFD	Unlimited by PD	

[a] See Egan (1981, pp. 211–220) for values under other conditions.
[b] See Gardner (1979, p. 70) and Kroupa (1973, p. 177).
[c] Egan (1981, p. 214).
[d] Byrne (1962, p. 588).
[e] See Greenstein (1974) and Rey (1960).
[f] Richman (1954a).
[g] Cahn (1962).
[h] Protonotarios (1969).
[i] Goldstein (1962).

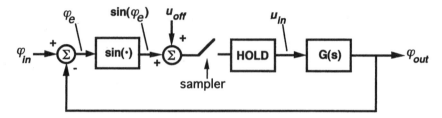

Fig. 8.M.1   Computation process.

For reasons of accuracy, a high sampling rate must be maintained, but MATLAB limits the number of points in the array that is plotted. The solution is to not plot every point (e.g., every fifteenth computed point was plotted for the figures in this section).

Continuity of phase and frequency out of $G(s)$ requires that their values at the beginning of each period equal their values at the end of the previous period. The theory developed in Section 6.10.2 is used for this purpose.

8.M.2   Comparing Phase-Plane Plots

Figure 8.M.2 is output from NLPhP under the same conditions that apply to the phase-plane plot of Fig. 8.19. Note that $\Delta\varphi = \varphi_e - \pi/2$ rad so the stable points in Fig. 8.19 are at $-\pi/2$ radians whereas they are at 0 in Fig. 8.M.2. Frequency in Fig. 8.M.2 is normalized to ω_n, as it is in Fig. 8.19, since we set $W_n \triangleq \omega_n$ sec/rad = 1. Here we have chosen to plot $\omega_{out}(\varphi_{out})$ rather than $\omega_e(\varphi_e)$, but the trajectories are the same, as can be seen from the symmetry of Fig. 8.19 or by changing variables in Eqs. (6.50), (6.51), and (6.48) with $\varphi_{in} = 0$.

In both plots, the locus that peaks near 4 rad/sec passes through final phase [i.e., $\varphi_e = 0, \Delta\varphi = -(\pi/2)$ rad/sec] with a frequency of about 2.8 rad/sec, goes through zero frequency at a phase of slightly less than 2 rad above the final phase, undershoots by about 1.2 rad/sec, then rapidly moves to the final phase and frequency. However, in both cases, the locus at final phase and 3.5 rad/sec (i.e., the starting point in Fig. 8.M.2) is above the separatrix and so skips a cycle, peaking at about 4 rad/sec in the process. (The existing curve in Fig. 8.19 is slightly above 3.5 rad/sec at final phase. It is apparent that a curve drawn through 3.5 rad/sec would separate further from that curve while moving to the right so it would come closer to peaking at 4 rad/sec than does the existing curve.) In other words, the program output closely matches the phase-plane plot of Fig. 8.19.

8.M.3   Truncating Phase

We have the choice of truncating phase so it is restricted to a $\pm\pi$ rad range (i.e., throwing away phase changes in whole cycle increments) or of showing multiple cycles of phase. Figures 8.M.2 and 8.M.3 represent the same

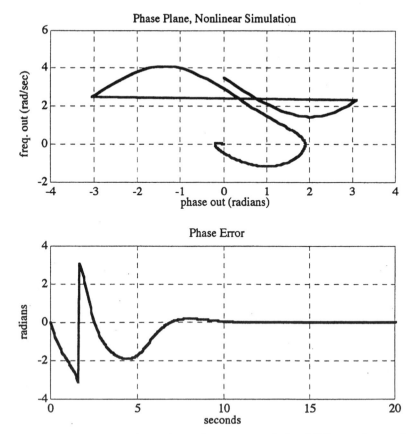

Fig. 8.M.2 Phase plane and transient, phase truncated. $\zeta = 0.707$, $\alpha = 1$, $\omega_n = 1$, 15 samples per plotted point, point each 0.05 sec.

responses (from 3.5 rad/sec initial frequency, zero initial phase, $\zeta = 0.707$, $\alpha = 1$, zero phase input), but phase in Fig. 8.M.2 is restricted to the range $\pm \pi$ rad whereas in Fig. 8.M.3 it is not. This is only a matter of how we plot the responses; both pairs of figures represent the same transient.

8.M.4 Effect of α

Our experimentation with linear loops showed that, if the excitation and initial conditions were the same for two loops that differed only in α, the responses were identical. Figure 8.M.4 is the response of the same loop whose response is represented by Fig. 8.M.2 except that α differs between the two. Apparently, in the nonlinear case, α does make a difference. In Fig. 8.M.4, with $\alpha = 0$, the phase goes just past the peak of the PD characteristic at $90° = 1.57$ rad and then turns back. With $\alpha = 1$, the

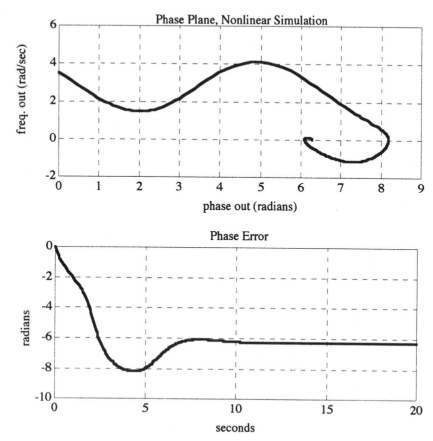

Fig. 8.M.3   Transient and phase plane, phase not truncated. Same responses as in Fig. 8.M.2.

initial frequency error must be reduced from 3.5 to 2.8 rad/sec before the initial overshoot is the same as in Fig. 8.M.4. However, if we reduce the initial frequency offset from 3.5 to 0.35 rad/sec, so the response becomes almost linear, it looks very much like that in Fig. 8.M.4 scaled down by a factor of 10, regardless of α.

8.M.5   Observing Pull-in

Figure 8.M.5 shows various kinds of output that are available from NLPhP to help us understand the pull-in process and confirm what we have learned in this chapter. The parameters here are the same as for Figs. 8.M.2 and 8.M.3 except the initial phase error has been increased from 3.5 to 4.5 rads. Note also that a different set of phase and frequency units has been selected.

In Figs. 8.M.5a and b the phase plane and the time transient of the phase error are shown without truncation. The same are shown in Figs. 8.M.5c and

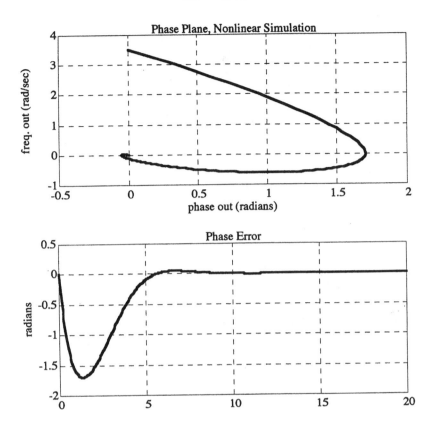

Fig. 8.M.4 Response for same conditions as for Figs. 8.M.2 and 8.M.3. But $\alpha = 0$.

d with truncation.[7] The frequency and PD outputs are shown as a function of time in Figs. 8.M.5e and f.

Note how the beat frequency (Fig. 8.M.5f) slowly decreases at first and how the distortion of the beat note develops an average value in the loop filter that causes the output frequency to decrease toward lock (Fig. 8.M.5e). Observe what happens 10 sec into the transient. The plot in Fig. 8.M.5d shows that the phase just passes a point of minimum PD output at -0.25 cycle when it turns around and moves toward the lock value. This can be seen in the PD output (Fig. 8.M.5f), which takes a slight inverted dip before going again to a minimum on the way to its final value. Note how the frequency error (Fig. 8.M.5e) goes to zero at 10 sec, corresponding to the zero slope of the phase (Fig. 8.M.5d) and of the PD output (Fig. 8.M.5f). We can also

[7]These plots have been modified slightly for print. In particular, the program will not deemphasize the retrace in the phase-plane plot as has been done here. The somewhat irregular starting and ending points for the retrace in the phase plane and the extremes in the phase-versus-time plot are due to the finite time steps in the program.

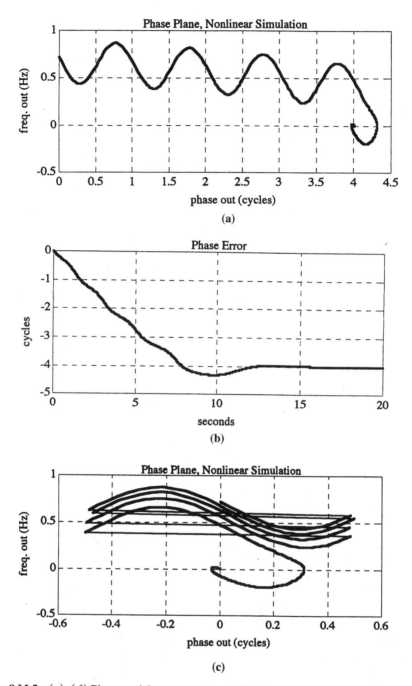

Fig. 8.M.5 (a)–(f) Phase and frequency plots. Initial frequency error increased from 3.5 to 4.5.

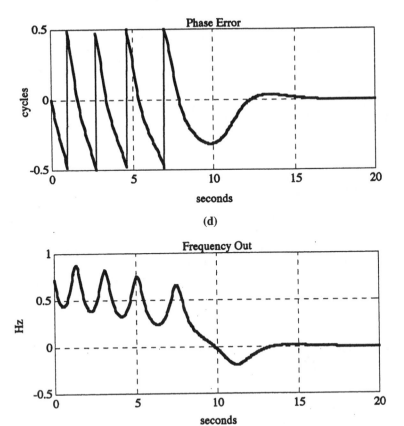

(d)

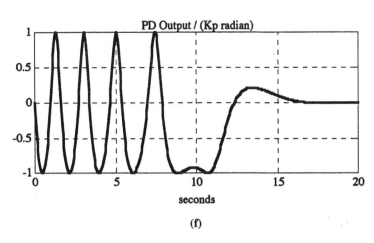

(e)

(f)

Fig. 8.M.5 (*Continued*)

correlate this minimum phase error (Fig. 8.M.5d) at 10 sec with the loop-back at freq = 0 and phase = 0.31 cycles (Fig. 8.M.5c).

Note the similarity between Figs. 8.M.5e and f and Fig. 8.3. Understandably a difference can be seen, due to the effect of the low-pass filter associated with Fig. 8.3 and the integrator-and-lead filter with Fig. 8.M.5c, in the ripple magnitude as the beat frequency decreases.

These plots have been for one set of loop parameters and initial conditions. An infinite number of other combinations can be obtained using NLPhP.

8.M.6   Introducing a Phase Offset

NLPhP permits us to introduce an offset at the PD output so we can see how acquisition proceeds when the final PD output is not zero. This differs from an input phase step. For example, with $\alpha = 1$ (integrator and lead), the input to $G(s)$ must settle to zero. With an offset, the corresponding final phase error will be established somewhere on the sinusoidal characteristic where the slope (gain) is less than maximum whereas, with a phase step input (and no offset), the final value from the PD would be at the maximum gain point.

8.M.7   Introducing a Frequency Step

A phase step is provided by giving φ_{in} a steady value during the simulation, and a frequency step is provided by increasing φ_{in} at a constant rate. These functions can be present when initial conditions are specified, but, when a step response (SR) is selected,[8] the initial conditions are overridden so the loop can begin at steady state. Other driving functions can be provided at the input by following a similar procedure.

A phase step has no discernible effect on the final state of a loop (unless it causes the loop to break lock and remain out of lock). The output phase changes by the same amounts as the input phase. But, with a type 1 loop, a frequency step changes the final phase error. The response can be affected by both the initial phase error and the final phase error, since both affect the region of operation in the nonlinearity. We can choose a combination of phase offset and frequency step to simulate any initial and final state for a frequency step.

8.M.8   Customizing the Nonlinearity

The nonlinearity has been explicitly stated (i.e., broken out) at two places in NLPhP to make it easy to find so we can replace it with other nonlinearities. For example, we might create a sawtooth PD characteristic by replacing the

---

[8]Other requirements, in order to obtain this type of simulation: Initial phase and frequency zero and a nonzero input phase or frequency.

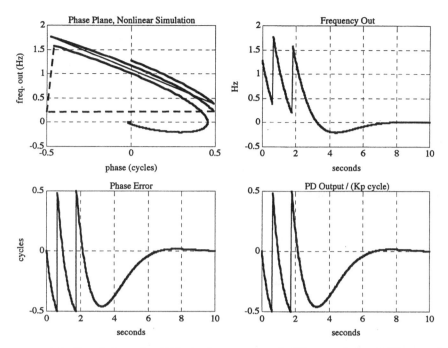

Fig. 8.M.6 Simulation of a PLL having a sawtooth PD. $\alpha = 1$, $\zeta = 0.707$, $\omega_n = 1$ rad/sec. Dashed lines in phase plane illustrate plotting anomaly associated with truncation.

sine nonlinearity by its argument and using truncation, like NLPhP employs on the variable e (see the region marked by "`>>>TRUNCATION`").

Figure 8.M.6 shows the results of changing to a sawtooth PD characteristic NLsaw.[9] The initial frequency error has been increased to 8 rad/sec because of the greater seize range with the sawtooth PD. The sampling period and period between outputs has also been halved because that happens to allow us to see an anomaly, which is marked by the dashed line in the phase-plane plot. It results from an unfortunate lack of synchronization between the occurrence of the sharp nonlinearity in the sawtooth PD characteristic and the occurrence of truncation in the plot (the anomaly would not be apparent in untruncated plots). The plot truncation occurs as soon as 0.5 cycle is exceeded, but the frequency does not respond until the end of the next sample period, during which the value from the HOLD circuit reflects the severe change in PD output. Moreover, the display of the frequency change is

---

[9] The PD output display (lower right) has also been changed to be more suitable to the new PD characteristic, but it is the same as the phase error plot when phase is displayed in cycles. We can see that the displayed time has been halved. The period between samples has also been halved.

delayed 14 samples (for this simulation) until the next displayed point. Usually this anomaly occurs at one of the 14 undisplayed points between displayed points, in which case it is not seen. In this case, however, the phase plane shows one occurrence of a step in phase due to truncation followed by a step in frequency due to the loop response, rather than showing both at the same time. This anomalous display could be prevented at the cost of program complexity.

8.M.9 Verifying Acquisition Equations

NLPhP gives us a tool to verify the acquisition formulas in Section 8.3 and their limits of applicability. After modifying the nonlinearity, we can do the same for loops employing other PD characteristics [Egan, 1981, pp. 211–220].

8.M.10 Some Experiments

Here are some suggested experiments using NLPhP. The spreadsheets in Appendix 8.S can be helpful in computing theoretical acquisition parameters for comparison to simulations.

8.M.10.1 Type 1 Loop with Low-Pass Filter, Pull-in. Use SR = Winit = Ap = alpha = Offset = 0, Wn = 1, z = 0.2, OutInc = 0.05/Wn, ending = 40/Wn, and SmpPerOut = 15. Winit = 0 means that the output frequency is initially zero, the steady-state frequency at the center of the PD characteristic. Ap = 0 means that the input phase is not stepped. We will set certain values for Phinit, the initial output phase, and certain values for Af, implying a reference frequency other than zero, that is, mistuning.

If we had chosen Af = 0 and Winit ≠ 0, we would be moving from some mistuning to zero mistuning. The responses would be different because our locus on the PD characteristic and the corresponding K_p would change with time in a different manner in the two cases.

Ap = 0 and Phinit = k gives the same results as Ap = $-k$ and Phinit = 0. A step response (SR = 1) with Ap = $-k$ would also give the same results when alpha is zero since the initial output will not be influenced by a step when the filter is low-pass.

What is the theoretical pull-in frequency? Try a simulation with Af = 1.15 and Phinit = 3.14. Do the results agree with the theoretical prediction? Change Af to 1.1. Try various values for Phinit. Is the predicted value as accurate as you had expected?

8.M.10.2 Type 1 Loop with Low-Pass Filter, Hangup. Use SR = Winit = Ap = alpha = Offset = 0, Wn = 1, z = 0.2, Phinit = 3.14, OutInc = 0.1/Wn, ending = 25/Wn, and SmpPerOut = 5. Observe the very slow start of the transient. What would happen if Phinit were set closer to π?

8.M.10.3 *Type 2 Loop, Integrator-and-Lead Filter, Seize and Speed.*
Use SR = Ap = Af = Offset = 0, alpha = 1, z = 2, OutInc = 0.05/Wn, ending = 40/Wn, and SmpPerOut = 15, Winit = − 10.

Since the DC gain is infinite, Winit = k produces the same results as Af = $-k$. In either case the steady-state phase error at start and end of transient is zero.

We computed pull-in frequency for the previous loop. For this type 2 loop we will compute and measure seize frequency and pull-in time. (Recall that a loop with a low-pass filter has no seize frequency and a type 2 loop has theoretically infinite pull-in frequency.)

We will measure seize frequency by looking for the highest initial frequency error from which the loop locks without skipping a cycle at any initial phase. We will be simulating the case in which the PD output is initially zero because there is no reference, and then a reference of a given frequency and phase suddenly appears.

We can also estimate seize frequency from the phase-plane plot. When that plot goes through a phase of zero or π radians, the frequency error is the same as is the mistuning when a reference is suddenly applied, if the phase of that reference is such as to produce zero PD output.[10] The seize frequency, which is for arbitrary phase, occurs between the frequency at the next-to-last crossing of zero phase and the frequency at the subsequent point where the phase is π, so it will be somewhat uncertain. (If the resolution of the plotted points were fine enough, the frequency at π radians could be read where the retrace crosses zero phase. When the step size of the plotted phase is significant, that point must be estimated. The estimate could be helped by choosing not to truncate the phase [truncatePh $\Rightarrow$ 0] so a smooth curve exists between points on either side of π radians.) By observing

---

[10] Justification for the seize frequency being between the next-to-last zero phase crossings and the subsequent π radians point in the phase plane of a type 2 loop: In the phase plane, the mistuning can only be observed at zero or π radians phase error in a type 2 loop. No current flows in the filter at zero PD output so the state is determined entirely by phase and frequency or, equivalently, filter capacitor charge. If the reference were removed, the PD output would not change, nor would the output frequency; thus the frequency error under this condition is the mistuning, the value of frequency error before a connection is made.

We identify the last time when the phase error is zero before lock as T_{-1} and the next previous such time as T_{-2}. All phases occur between T_{-2} and T_{-1}. Moreover, since stable equilibrium occurs at zero phase error, as the locus leaves such a point ($T_{-2} +$), the filter capacitor charges in such a direction as to reduce the error. When π error is reached (at T_π), it will begin charging in the opposite direction and continue to do so until T_{-1}, but the minimum error will have occurred at T_π. Therefore, between T_{-2} and T_{-1} the capacitor will always have a charge between its values at T_{-2} and T_π so the corresponding mistuning is between the mistunings at T_{-2} and at T_π. Thus, the locus between T_{-2} and T_{-1} contains all possible phases, and each point corresponds to a mistuning between the mistuning at T_{-2} and T_π, both of which can be read from the phase-plane plot. Therefore, the mistuning that will allow lock without cycle skipping regardless of phase lies between the mistunings at T_{-2} and at T_π, which are observable in the phase plane.

this range from the phase plane produced under various initial conditions, we could estimate the seize frequency.

Start with `Phinit = 3.14` (rad). At this phase, determine the boundary between a value of `Winit` that will allow lock without cycle skipping ("seizing") and one that will not. `Winit = 5.2` (rad/sec) is suggested as a starting point. With Winit set to that highest value, try several other phases to show that the loop will lock without skipping at the highest frequency at which it did so with `Phinit = 3.14`. Does this frequency compare well with the predicted seize frequency? Are the conditions of validity for Eq. (8.23) met?

Set `Winit = -10` and estimate the seize frequency from the phase plane. Start with `Phinit = 1.57`. Observe also the pull-in time, estimating the time at which the frequency error equals the seize frequency for the end of the pull-in period. Now try `Phinit = -1.57`. Notice how much the time to seize can vary. The longest time at any phase is the required value. You can try some other initial phases. How does the pull-in time obtained here compare with theory. How does the seize frequency compare between the two methods and with theory. Are the conditions of Eqs. (8.26) and (8.27) well met?

8.M.10.4 *Type 1 Loop, Lag-Lead Filter, Seize.* Use `Winit = Phinit = Offset = 0`, `SR = 1`, `Af = 9`, `Wn = 1.0005`, `alpha = 0.99701`, `z = 1.6725`, `OutInc = 0.05/Wn`, `ending = 50/Wn`, and `SmpPerOut = 30`.

Set `Ap` to 3.14 and various other phases. Show that the seize frequency is between 4.45 and 4.5 rad/sec. To do this, show that `Af = 4.45` (rad/sec) seizes at all phases and 4.5 fails at some initial phase. What is the theoretical value and how well are the conditions met?

Observe the phase planes with `Ap = 0` and various values of `Phinit`. How would an estimate of seize frequency based on the phase plane, as for $\alpha = 1$, compare to the value obtained by trying various frequency steps. The theory that supports the equivalence for $\alpha = 1$ does not do so for $\alpha < 1$.[11] Nevertheless, if we take the pull-in time to end when the average frequency over a cycle equals the seize frequency, the frequency at zero phase is a good estimate of that average, at least in this case. What is the estimated pull-in time from the plots? How do these values compare to theory and how well are conditions met?

8.M.10.5 *Type 1 Loop, Lag-Lead Filter, Offset.* For an offset of 0.15 rad, Eq. (8.76) gives an optimum value of $x = 0.0225$. By Eq. (8.24), this implies a ratio of pole to zero frequency of 0.045. Choose $\omega_p = 4.5$ rad/sec, $\omega_z = 100$ rad/sec, and $\omega_L = 1000$ rad/sec. Using `Acqpz`, we obtain, from these

---

[11] Zero phase does not imply steady state. The frequency does not uniquely identify the state. Even with zero phase error, the filter capacitor could be discharging.

parameters, $K = 22{,}112 \text{ sec}^{-1}$, $\zeta = 1.584351299$, $\alpha = 0.995497962$, $\omega_n = 315.444$ rad/sec and $\Omega_{PI} = 6559$ rad/sec. Enter alpha, z, and Wn in NLPhP.

The development of the equations of Section 8.7 is based on the idea that an offset amounts to a mistuning and thus reduces the (additional) mistuning allowed for lock to be assured. Unfortunately, a simulation to show pull-in range can be very time consuming because the pull-in time grows as the pull-in limits is reached. Instead, we will observe that the offset is effectively added to the initial mistuning. We will do this by obtaining the same pull-in time for a given mistuning with zero offset and for a mistuning that is reduced by the effective mistuning when an offset exists. Initially, set Offset = 0, Ap = -1.57, and Af equal to 80% of the pull-in frequency, 5247 rad/sec. Also set SR = 1, Phinit = Winit = 0, OutInc = 0.05/Wn, ending = 200/Wn, and SmpPerOut = 7. Determine how long the loop takes to seize. By Eq. (8.75), ω_{PI} will be reduced by $\omega_L \times 0.15 \text{ rad}/(2|\varphi_{os}|) = 3333$ rad/sec in the presence of the offset. Reduce the initial frequency by 3333 rad/sec to give Af = 1914 (rad/sec). Rerun NLPhP with Offset = -0.15 and the reduced value of Af. How close is the time to seize to what was obtained before.

8.M.11   Simulation Using Approximate Method

The MATLAB program NLPhx performs the same function as NLPhP but uses the approximate algorithm described in Section 6.11 for the repeated computations. In some tests, outputs appeared to be the same as obtained with NLPhP and the simulations were faster, but only by about 9%. Probably its main value would be for use when the matrix manipulation programs, such as are contained in MATLAB, are not available. (Because they are available, MATLAB's matrix mathematics are used where appropriate in NLPhx.)

NLPhP

```
%TIME RESPONSE & PHASE PLANE PLOT
% SINE NONLINEARITY
clear;

% MODIFY PARAMETERS BELOW & GO
%*****************************
SR = 0; % 1 for step response (if also Phinit=Winit=0 and Ap or
                        Af not 0).,
      % Otherwise start at initial conditions. For step response,
      % initial conditions correspond to steady-state conditions
      % with zero PD output.
Phinit = 0; % radian initial phase
Winit = 3.5; % radian/sec initial frequency
Ap = 0;    % radian phase input step
```

```
Af = 0;    % rad/sec input frequency step
Wn = 1;    % rad/sec natural frequency
z = .707;  % damping factor
alpha = 1; % Eq. (6.5)
OutInc = .05/Wn;% period between outputs
ending = 20/Wn; % Student MATLAB limits ending/inc to1024 in
                          v.3.5, 8192 in v.4, 16384 in v.5
SmpPerOut = 15; % samples per output period
Offset = 0;    % offset at PD output (same sign as steady-state
                          output)
truncatePh = 1; % 1:truncate phase; 0:do not
Punits = 0;    % 0:radians 1:cycles 2:degrees
Funits = 0;    % 0:radians/sec  1:Hz
% Choose Output
% op=1: Phase Plane = Freq. vs. Phase
% op=2: Phase Plane + Phase Error vs. Time
% op=3: Frequency + PD Output vs. Time
% op=4: All of above = 4 plots
op = 4;
%******************************
fprintf('\nLoop with SINUSOIDAL Phase Detector Characteristic')
fprintf('\nWn = %g rad/sec; zeta = %g; alpha =
                          %g\n',Wn,z,alpha)
% Set step response or not and output that choice
% and other parameters chosen above
if (SR == 1)&(Phinit == 0)&(Winit == 0)&((Ap ~= 0)|(Af ~= 0)),
   stp = 1;
   fprintf('Phase In = %g rad; Freq In = %g rad/sec;',Ap,Af);
   fprintf(' Offset = %g rad\n',Offset);
   disp('STEP RESPONSE, zero initial phase and frequency');
else
   stp = 0;
   fprintf('Phase In = %g rad; Freq In = %g rad/sec;',Ap,Af);
   fprintf(' Offset = %g rad\n',Offset);
   disp('STARTING AT INITIAL CONDITIONS');
   fprintf('Init Phase = %g rad; Init Freq. = %g
                          rad/sec\n',Phinit,Winit);
end % if

last = 1 + ending/OutInc; % index of last output
t = linspace(0,ending,last); % linear sequence of times for
                          plots
dt = (t(2) - t(1))/SmpPerOut; % sample time increment
clg; % clear previous graphs; for v.4, clf ########

%Polynomials for open loop gain:
   N = [0    2*z*Wn*alpha  Wn^2]; % numerator
   D = [1    2*z*Wn*(1-alpha)  0 ]; % denominator
[a,b,c,d] = tf2ss(N,D); % elements of dynamic matrix equation
y = Phinit; % initial phase at output
```

```
% To help judge inaccuracy of ignoring sampling,
%  compute extra phase shift produced at unity gain
factor = N(2)^2-D(2)^2;
WL = sqrt((factor+sqrt(factor^2+4*N(3)^2))/2); % open-loop gain
                          = 1
SampPhShift = pi*WL*OutInc/SmpPerOut; % radians lag from
                          sampling at WL
%disp('Added phase due to sampling (radians):')
%disp(-SampPhShift)
fprintf('Added phase due to sampling = %g rad = %g deg.\n\n',-
                          SampPhShift, -SampPhShift*180/pi)


%  BEGIN SIMULATION

NL = sin(Ap-Phinit); % <<<<<<<<<<<< THE NON-LINEARITY
if stp % initial state established with no reference input
   if Offset
      Winit = Offset*Wn/(2*z*(1-alpha));
         % steady state output frequency due to Offset
   else
      Winit = 0;
   end
   x0(1) = (Winit-N(2)*Offset)/(N(3)-D(2)*N(2));
   x0(2) = -N(2)*x0(1)/N(3);
   e = Ap; % initial phase error
   freq(1) = NL*2*z*Wn*alpha + Winit;
         % step from initial value theorem + prior steady state
else % state variables to give specified initial conditions
      % X2 and X3 respectively in Eq. (6.50) and (6.51) with
                          D(3) = 0:
   x0(1) = (Winit-N(2)*(NL+Offset))/(N(3)-D(2)*N(2));
   x0(2) = (Phinit-N(2)*x0(1))/N(3);
   e = Ap-Phinit; % initial phase error
   freq(1) = Winit; % initial frequency at output
end % if
if truncatePh, % truncate e range to ±pi %
                          <<<<<<<<<<<<TRUNCATION
   if e>0
      e = rem(e+pi,2*pi) - pi;
   else
      e = rem(e-pi,2*pi) + pi;
   end
end % if truncatePh
er(1) = e;
[Phi, Gamma] = C2D(a,b,dt); % values for use in solution
                          equation [see manual]
```

```
for i = 1:last-1, % compute loop responses
    for j = 1:SmpPerOut, % loop through unrecorded samples
        PhIn = Ap + Af*dt*(j-.5+(i-1)*SmpPerOut); % phase step
                                plus phase ramp
        NL = sin(PhIn - y); % <<<<<<<<<<<<<< THE NON-LINEARITY
        u1(1) = NL + Offset; % initial input to sampler
        u1(2) = u1(1); % constant Hold output -> same at
                                beginning and end of dt
        x = ltitr(Phi,Gamma,u1',x0); % solutions at period start
                                and end [see manual]
        yy = c*x'; % corresponding outputs (d is null for
                                polynomials above)
        y = yy(2); % output at end of period is needed
        x0 = x(2,:)'; % final condition gives initial condition
                                for next step
        % For step response, let initial freq equal first
                                computed value
        % to prevent infinitesimal time at zero from appearing to
                                be finite
        if (stp & i==1 & j==1),
            freq(1) = N(2)*u1(2) + x0(1)*(N(3)-N(2)*D(2));
        end % if
    end % for j
    e = Ap + Af*dt*SmpPerOut*i - y; % untruncated error
    if truncatePh, % truncate e range to ±pi %
                                <<<<<<<<<<<<TRUNCATION
        if e>0
            e = rem(e+pi,2*pi) - pi;
        else
            e = rem(e-pi,2*pi) + pi;
        end
    end % if truncatePh
    er(i+1) = e;
    % corresponding frequency out from Eq. (6.43) with d3 = 0
    freq(i+1) = N(2)*u1(2) + x0(1)*(N(3)-N(2)*D(2));
end % for i

% END SIMULATION. The rest is plotting.

if Punits == 1,
    Pscale = 1/(2*pi);
elseif Punits == 2,
    Pscale = 180/pi;
else,
    Pscale = 1;
end % if Punits
```

```
if Funits == 1,
   Fscale = 1/(2*pi);
else
   Fscale = 1;
end % if Funits

if op ~= 3, % selecting phase-plane plot
   if op==1,
      subplot(111),
   elseif op==2,
      subplot(211),
   elseif op==4,
      subplot(221),
   end % if op==1
   plot(-Pscale*er,Fscale*freq,'r')
   title('Phase Plane, Nonlinear Simulation')
   grid
   if Punits == 1, % Set display of phase units
      xlabel('phase out (cycles)');
   elseif Punits == 2,
      xlabel('phase out (degrees)');
   else,
      xlabel('phase out (radians)')
   end % if Punits
   if Funits == 1, % Set display of frequency units
      ylabel('freq. out (Hz)');
   else,
      ylabel('freq. out (rad/sec)');
   end % if Funits
end % if op~=3
if (op == 2)|(op == 4), % selecting phase error vs. time
   if op == 2,
      subplot (212),
   else,
      subplot (223),
   end % if op==2
   plot(t,Pscale*er,'g')
   title('Phase Error')
   grid
   xlabel('seconds')
   if Punits == 1, % Set display of phase units
      ylabel('cycles');
   elseif Punits == 2,
      ylabel('degrees');
   else,
      ylabel('radians');
   end % if Punits
end % if ()|()
```

```
if (op == 3)|(op == 4),
   if op == 3, % selecting frequency vs. time
      subplot(211);
   else,
      subplot(222),
   end % if op==3
   plot(t,Fscale*freq,'b')
   title('Frequency Out')
   grid
   xlabel('seconds');
   if Funits == 1, % Set display of frequency units
      ylabel('Hz');
   else,
    ylabel('radians / second');
   end % if Funits
   if op == 3, % selecting PD output vs. time
      subplot(212);
   else,
      subplot(224),
   end % if op==3
   plot(t,sin(er))
   title('PD Output / (Kp radian)')
   grid
   xlabel('seconds')
end % if ()|()
```

8.S APPENDIX: ACQUISITION SPREADSHEET

The Acq2 spreadsheet provides the acquisition parameters given in Section 8.3 from the values of α, ζ, ω_n, and mistuning Ω. It also gives K, ω_p, and ω_z in the process.

The spreadsheet Acqpz provides acquisition information based on given values of ω_p, ω_z, ω_L, and mistuning. It also gives K, ω_n, ζ, and α.

PROBLEMS

8.1 A loop has a balanced mixer phase detector with a (maximum) gain of $K'_p = 0.1$ V/cycle. The tuning characteristic of the oscillator has a 40-MHz/V slope. The loop filter is shown in Fig. P8.1. The VCO is 90° out of phase with an input signal to which it is locked when the frequency is 30 MHz.

    (a) If the input frequency drifts, how high can it go before the loop will lose lock? Give frequencies in both radians/second and hertz.

    (b) After lock is broken, the input frequency is lowered again. At what frequency can lock be reacquired? What are the restrictions on the formula that you used and how well are they met (give numerical values)?

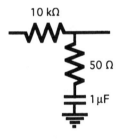

Fig. P8.1

(c) At what difference between input frequency and VCO center frequency ω_c will cycle skipping stop? What are the restrictions on the formula that you used and how well are they met (give numerical values)?

(d) How long will it take to stop cycle skipping if the input signal is 14 kHz above f_c? What are the restrictions on the formula that you used and how well are they met (give numerical values)? Compare the answer you get from a formula to the value given by Fig. 8.18.

(e) Repeat (d) for an initial 6-kHz mistuning.

8.2 The filter in the loop described in Problem 8.1 is changed to that shown in Fig. P8.2, but the phase detector and VCO gain constants remain the same.

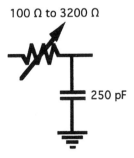

Fig. P8.2

(a) What is the total range over which the input frequency can be acquired (difference between minimum and maximum frequencies) when the variable resistor is set at its highest value?

(b) What is the total range when the resistor goes to minimum value?

8.3 A loop false locks 100 kHz above the input frequency. At that offset, a narrow IF filter has caused 120° of effective phase lag and a 3-dB gain drop in the loop. The loop has an lag-lead filter with the pole at 2 kHz and the zero at 20 kHz. What is K under the following conditions:

(a) The VCO center frequency f_c equals the input frequency f_{in}?

(b) The VCO center frequency f_c is 50 kHz higher than the input frequency f_{in}?

8.4 A loop must have 100-kHz bandwidth. It has a lag-lead filter with a 10-kHz zero frequency. Pull-in range is to be ± 500 kHz.

(a) What is the largest uncompensated phase detector offset that can be tolerated?

(b) What should the filter pole frequency be?

CHAPTER 9

ACQUISITION AIDS

Sometimes the pull-in range is inadequate or acquisition is too slow. An obvious solution is to increase the loop bandwidth, but other considerations, such as a filtering action that may be desired from the loop, can make that impractical. An acquisition aid often provides the solution. It can permit the loop to lock when its mistuning falls between the hold-in range and the pull-in range ($\Omega_{PI} \leq |\Omega| < \Omega_H$) or it can speed up a pull-in that might otherwise be delayed while many cycles are skipped. We will consider several acquisition aids in this chapter.

9.1 COHERENT DETECTION—LOCK INDICATOR

We can use the coherent detector as part of an acquisition aid or to warn when the loop has become unlocked. Detection refers to extraction of a signal proportional to the amplitude (or power[1]) of the detected signal. Coherent detection is a method of detection that employs a second, reference, signal of identical frequency and proper phase. Coherent detection has an advantage of sensitivity over other methods, but the necessity to obtain the reference signal adds complexity.

The coherent detector can be a balanced mixer, structurally identical to the phase detector (in which case $K'_{pd} = K'_p$. The pertinent term in Eq. (3.5) is $\cos[\varphi_1(t) - \varphi_2(t)]$, the same term that produces phase detection. However, for coherent detection, $\varphi_1(t) - \varphi_2(t)$ is approximately 0°, corresponding to

---

[1] This is called square-law detection.

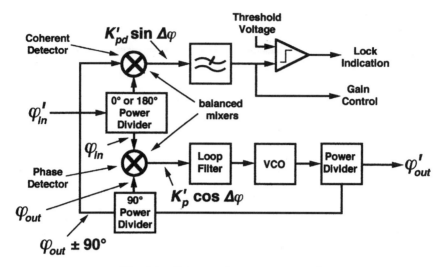

Fig. 9.1 Coherent detector with PLL.[2]

maximum output, rather than the $-90°$ required for phase detection, which corresponds to maximum slope. Because of this relationship the coherent detector in Fig. 9.1 achieves its maximum output when the phase error in the loop is zero ($\Delta\varphi = -\pi/2$). A signal to which the loop is not locked will cause an approximately sinusoidal output from the coherent detector. That can be attenuated by a low-pass filter so that only locked signals will cause sufficient output to break threshold and indicate lock.

The output of the coherent detector will be[3] [Eq. (1.7)]

$$\nu_{CD} = K'_{pd} \cos(\Delta\varphi + \pi/2) = -K'_{pd} \sin \Delta\varphi. \qquad (9.1)$$

9.1.1 During Acquisition

The output from the coherent detector is only approximately sinusoidal, just as the output from phase detector is only approximately sinusoidal, and for the same reason. The little feedback that does exist when the loop is out of lock perturbs the sinusoid. In Chapter 8, we developed Eq. (8.42) to give the bias component from the phase detector that results from this process. We can use the same equation to give us the bias component from the coherent detector by adding the 90° phase shift that is appropriate to the latter, thus

---

[2]Power dividers split a signal into several signals, usually providing the same impedance at all ports. Different types of power dividers provide different phase relationships between signals, usually multiples of 90°.

[3]The output could be positive or negative, depending on the sign of the relative phase shifts in the power dividers.

obtaining

$$|v_{bC}| \approx 0.5|G(\Delta\omega)|K'_{\text{pd}}\cos[\angle G(\Delta\omega)] \le 0.5|G(\Delta\omega)|K'_{\text{pd}},   (9.2)$$

where $\Delta\omega$ is again the error frequency during pull-in. The threshold voltage must be set high enough so v_{bC} does not exceed it when the loop is not locked. Otherwise the coherent detector might indicate phase lock during a false lock. Ideally, a loop with an integrator-and-lead or lag-lead filter will have $\angle G \approx -90°$ at high frequencies so $|v_{bC}| \approx 0$ when $\Delta\omega$ is high compared to ω_z. We must control excess phase shift that leads to false lock until $|\Delta\omega| \gg \omega_L$ so v_{bC} will be small at any false lock. [We have already assumed $|G(\Delta\omega)| \ll 1$ in developing Eq. (8.42), and therefore (9.2).[4] Under those circumstances, $|v_{bC}| \ll 0.5 \, K_{\text{pd}}$ holds whereas $|v_{bC}| \approx K_{\text{pd}}$ is true during a normal lock if the phase error is kept low. However, there are circumstances, particularly in the presence of noise, when it is difficult or impossible to determine the presence of a false lock with the coherent detector [Gardner, 1979, p. 155].

9.1.2 During Sweep, Locked Loop

We have shown that the coherent detector can develop a voltage when the loop is not locked. We now show that its voltage during lock can be diminished by frequency sweeping. These two effects set limits for the threshold setting. By Eq. (6.18), the phase error during a linear frequency sweep settles to

$$\varphi_e(t') = \{\dot\omega_{\text{in}} - \dot\omega_c\}\left\{\frac{1 - 4\zeta^2(1 - \alpha)}{\omega_n^2} + 2\zeta\frac{1 - \alpha}{\omega_n}t'\right\}.   (9.3)$$

If the input frequency is sweeping while the voltage-controlled oscillator (VCO) center frequency (not its actual instantaneous frequency) is stationary, $\dot\omega_c = 0$; if the input frequency is steady while the VCO center frequency is swept, perhaps by a ramp of voltage added to its tuning input, $\dot\omega_{\text{in}} = 0$; and, if the VCO center frequency is swept in synchronism with the input frequency, $\dot\omega_e = \{\dot\omega_{\text{in}} - \dot\omega_c\} = 0$.

With an integrator-and-lead filter ($\alpha = 1$), the phase error is constant at $\dot\omega_e/\omega_n^2$, but with a lag-lead or lag filter it increases with time. This is the error relative to steady state, when there is no sweep. The question arises, from what event are we measuring t? When is $t = 0$? If we look at the lag-lead shown in Fig. 3.23f, we see that it differs from an integrator-and-lead (Fig. 3.23c) by the shunt resistor R_p, which passes some of the current that would otherwise go into C. Thus a changing phase error is required to

---

[4]Otherwise 0.5 $|G(\Delta\omega)|$ cannot be used as an approximation for J_1 ($|G(\Delta\omega)|$).

provide this changing current as the voltage across R_p changes. However, at the moment when there is no voltage across R_p, the phase error will be the same as if R_p were absent. If we call this moment $t = 0$, then we can write Eq. (9.3) as

$$\varphi_e = \{\dot{\omega}_{\text{in}} - \dot{\omega}_c\}\left\{\frac{1}{\omega_n^2} + 2\zeta\frac{1 - \alpha}{\omega_n}t\right\} = \{\dot{\omega}_{\text{in}} - \dot{\omega}_c\}\left(\frac{1}{\omega_n^2} + \frac{t}{K}\right). \quad (9.4)$$

Here t equals t' plus a constant, and the constant is chosen such that, at $t = 0$, when the voltage across R_p is zero, the equation is the same as it would be for an integrator-and-lead filter.

If, during steady-state lock (no sweep), $\Delta\varphi = -\pi/2$, then, during sweep, it will be $\Delta\varphi = \varphi_e - \pi/2$, so Eq. (9.1) will become $\nu_{\text{CD}} = K'_{\text{pd}}\cos(\varphi_e)$. This is approximately equal to K'_{pd} for small φ_e. For larger errors Eq. (9.4) must be replaced by a form that accounts for the sinusoidal characteristic of the phase detector,[5]

$$\sin(\varphi_e) = \{\dot{\omega}_{\text{in}} - \dot{\omega}_c\}\left\{\frac{1}{\omega_n^2} + \frac{t}{K'}\right\}, \quad (9.5)$$

and Eq. (9.1) then becomes

$$\nu_{\text{CD}} = K'_{\text{pd}}\cos\varphi_e = K'_{\text{pd}}\sqrt{1 - \sin^2\varphi_e} = K'_{\text{pd}}\sqrt{1 - \left\{[\dot{\omega}_{\text{in}} - \dot{\omega}_c]\left[\frac{1}{\omega_n^2} + \frac{t}{K'}\right]\right\}^2}. \quad (9.6)$$

Thus the output of the coherent detector decreases when ω_{in} or ω_c is swept while the loop maintains phase lock. This must be taken into account in setting the threshold voltage.

9.2 CHANGING LOOP PARAMETERS TEMPORARILY

9.2.1 Coherent Automatic Gain Control

Loop parameters can be changed to speed up acquisition. One method is to control the gain of an amplifier that precedes the circuit in Fig. 9.1 by the output voltage from the coherent detector. This arrangement is shown in Fig. 9.2. It must be done in such a manner that, when the loop is not locked, the gain is maximum. Assuming that the phase detector acts like a multiplier (or a balanced mixer in which the input signal is weak compared to the VCO

[5]This assumes that zero input to the filter occurs when $\Delta\varphi = -90°$.

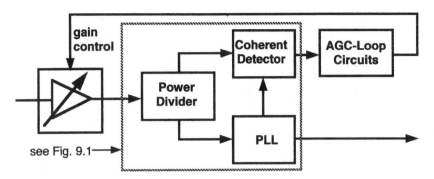

Fig. 9.2 Loop gain control by coherent AGC.

output), the loop gain will then be high because K_p' is proportional to the amplitude of the input signal. When the loop pulls into lock, the output from the coherent detector contains a strong DC component that will pass through the low-pass filter and decrease the preceding amplifier's gain; K_p' will then decrease, resulting in a lower loop bandwidth. This coherently detected voltage can then be used to maintain a constant input level and thus constant phase-locked loop (PLL) parameters. We must ensure that the PLL is stable for all the possible gains as the acquisition process is taking place. Note that the total system now consists of two interdependent loops, the PLL and the automatic gain control (AGC) loop, both of which must be designed for proper performance.

If the AGC loop has high DC gain, it will hold the output of the coherent detector approximately constant, and it will no longer be useful for differentiating between true and false locks. However, the signal that controls the variable-gain amplifier will indicate the strength of the signal at the amplifier's input, and it will thus be useful for false lock detection.

Whether or not coherent AGC is employed, proper false lock detection requires that the strength of the input signal be well enough known. Otherwise we cannot differentiate lock to a weak signal from a DC voltage generated by a strong signal when the loop is not locked. To make this so, it may be necessary to precede the circuit of Fig. 9.1 or 9.2 with a separate, noncoherent, AGC to establish a constant signal level. However, once again, excessive noise can defeat the system by causing the gain to be controlled by the noise rather than the signal.

9.2.2 Filter Modification

We can also modify the loop filter time constants while leaving the low-frequency gain unchanged [Rey, 1960]. During acquisition the filter can be effectively eliminated by shorting the series resistor, possibly turning the loop into a first-order loop. In Fig. 9.3a, R_{1b} might represent the series resistance

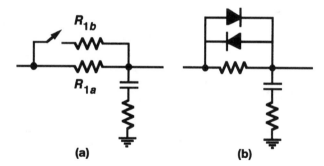

Fig. 9.3 Modification of filter for acquisition.

of the switch, which will usually be a semiconductor device; it could be negligible or it might be purposely added to control gain in the acquisition mode, perhaps to avoid instability. Once acquisition has occurred rapidly in the wide-bandwidth loop, the switch can be opened[6] to restore the desired loop parameters. This process must be controlled from external circuitry, perhaps a timer that is initiated when the loop center frequency is tuned to a new value or the output from a lock indicator such as the coherent detector.

Figure 9.3b shows a circuit that automatically widens the bandwidth for large changes in u_1. Once the loop has locked, the changes will be small enough (in the absence of excessive noise) that the diodes will have a very high impedance but, during pull-in, they will conduct and increase the filter pole frequency. The circuit must be carefully planned so this transition occurs at the best point in the pull-in process. There is also a danger that transients from the phase detector (especially from certain digital types) will be large enough to pass through the diodes, thus severely reducing the effectiveness of the filter.

9.2.3 Comparison of Two Types of Parameter Modifications

Figure 9.4 compares the two methods described above by showing the Bode plots of the unmodified loop and of both modified loops.

9.3 AUTOMATIC TUNING OF ω_c, FREQUENCY DISCRIMINATOR

A frequency discriminator can be used to produce a voltage that is proportional to the received frequency and that voltage can be added to u_2 to tune ω_c closer to the input frequency. The quadricorrelator [Richman, 1954b,

---

[6]We must take care that the switch control signal does not couple excessively into the filter and cause a transient that could undo the process.

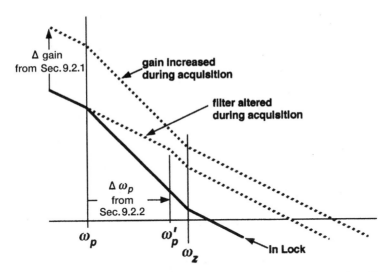

Fig. 9.4 Bode plots corresponding to loop during acquisition. These plots illustrate the modification of loop parameters based on a linearized loop although that is not an accurate representation during much of the pull-in process.

Egan, 1981, p. 243] is a particularly applicable type of discriminator. It produces a voltage that is proportional to the difference between two input frequencies. (The output of an ordinary discriminator is proportional to the frequency offset from some constant value. Thus it is not suitable for indicating the difference between two frequencies that might both be variable. It is also subject to inaccuracies due to component drifts that might change its center frequency or that of the VCO or reference.) This can be used to tune the VCO as shown in Fig. 9.5. At low-frequency errors, its output diminishes so it has little effect on the loop once lock has occurred.

The loop input and the VCO output are compared in two multipliers. One of the signals is shifted by 90° before entering one of the multipliers. As a result, the outputs from the two multipliers, which are at the difference frequency $\Delta \omega$, differ in phase by 90°. One of these is differentiated, giving it a 90° phase shift and a frequency-dependent gain. The two signals that enter the final multiplier are in phase (coherent detection), and one of them has an amplitude that is proportional to $\Delta \omega$. The low-frequency output of the final multiplier is proportional to $-\Delta \omega = \omega_{in} - \omega_{out}$ so it can be added to u_2 in the loop to decrease $\Delta \omega$. Once again, the response of the PLL during pull-in depends on the combined action of two loops. This time a frequency control loop has been added.

Figure 9.6 shows a more practical model of the quadricorrelator. It has amplifiers and the multipliers have been replaced by balanced mixers. The differentiator has been replaced by a high-pass filter, which functions as a

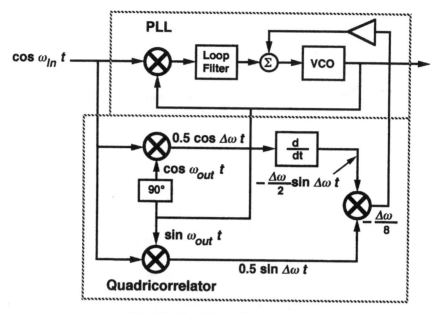

Fig. 9.5   Quadricorrelator concept.

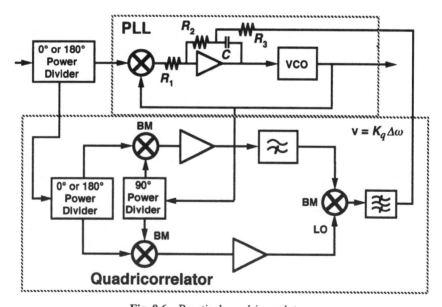

Fig. 9.6   Practical quadricorrelator.

differentiator at frequencies that are well below its cutoff frequency ω_{co}:

$$F(\omega) = \frac{j\omega}{\omega_{co} + j\omega} \approx j\frac{\omega}{\omega_{co}} \qquad \text{for } \omega \ll \omega_{co}. \qquad (9.7)$$

Some options have been shown. A bandpass filter has been placed at the output. We generally assume low-pass function after balanced mixers and multipliers—it eliminates the sum frequency component—but the high-pass function can be added to decrease further the influence of the quadricorrelator on the loop during normal in-lock operation, if desired.

The quadricorrelator output is now shown as being added to the PLL at its loop filter [Gardner, 1979, p. 85]. The loop filter here becomes a filter, an integrator, for the quadricorrelator loop. To see this, assume no DC output from the phase detector (PD), either because the PLL is out of lock or by superposition. Then no current flows through R_1. Since the input to the op-amp is a virtual ground into which no current can flow, no current will then flow through R_2 either, so the junction of C and R_3 is effectively connected to the op-amp input. Since the quadricorrelator otherwise has no pole at the origin (there is no integration to convert frequency to phase as in the PLL), the integrator causes it to change from a zero-order to a first-order loop (to the degree that this integrator is a true integrator).

9.4 ACQUISITION AIDING LOGIC

Several schemes are available that cause the phase detector to aid the acquisition process by becoming effectively a frequency discriminator under conditions where it would otherwise skip cycles (the loop being out of lock) [Egan, 1981, pp. 247–252]. The best known of these is the phase-frequency detector. It is most often used when the signals are, or have been made, digital and is not generally employed in locking to noisy signals.

Compare Fig. 9.7 with Fig. 3.1. The output at C is the same as before, and as before, it goes to 1 (positive) when triggered by the waveform at A. The difference is that, here, B resets the upper flip-flop only indirectly, although the results appear to be the same; B sets the lower flip-flop, which in turn provides a second 1 input to the AND gate (the first being C). This causes a 1 at the AND output, which then resets both flip-flops. The results are the same as before for the case shown. However, if τ should go to zero and then negative, B then preceding A, the results are considerably different. Now the output pulse appears at D and is proportional to $-\tau$. The output at D will eventually be inverted relative to that at C, possibly by entering the opposite input to the loop filter op-amp, so the pulse width at D represents negative phase relative to that at C. This phase detector has a linearly

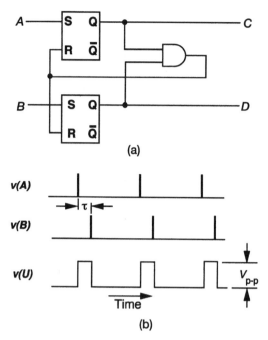

Fig. 9.7 Representation of the phase-frequency detector.

changing output (average pulse width) from τ equal to -1 reference period to $+1$ period, a range of $\pm 360°$. (These linear characteristics have been described in Section 3.1.3.)

Even more important than the extended range is what happens when the loop is out of lock. If B is slightly lower in frequency than A, τ will increase until it equals a reference period. Then two A pulses will occur between B pulses and the output pulse width will drop back toward zero. Thus the characteristic will look like that in Fig. 3.1. There is a significant difference, however. The center of the lock range in Fig. 3.1 is in the middle of the sawtooth waveform, at $180°$ phase difference, whereas it is at the bottom of the waveform, at $0°$, here. Thus, rather than generating an average value in the center of the range when out of lock, this phase detector generates an average value that corresponds to $180°$ above the normal lock point. In a high-gain loop the resulting voltage will drive the VCO toward the reference frequency.

If the frequency of B is higher than that of A, the same process will occur at D. Thus, when the loop is out of lock, a sawtooth will occur at C or at D, depending on which frequency is higher, and the average value produced will drive the loop toward lock.

The $180°$-equivalent output occurs as the ratio of the two frequencies approaches unity and if the effect of feedback can be neglected. For higher frequency ratios, the offset becomes even greater, approaching the equivalent of 360° at large ratios. If one input should be lost, the output would become a steady value equivalent to 360° phase. In all cases the polarity will be correct to bring the frequency error toward zero.

9.5 SWEEPING ω_c, TYPE 2 LOOP

Various methods can be employed to cause ω_c to be swept past the input frequency so $\Omega \Rightarrow 0$ and lock will occur. The sweep may be discontinued when lock occurs or it may continue to be applied but be effectively overcome by the phase lock.

9.5.1 Maximum Sweep Speed, Closed-Loop Sweeping

The maximum output from the phase detector is K'_p rad. This will cause a current of K'_p rad$/R_1$ into C in the filter of Fig. 3.23c. This in turn will cause a voltage ramp of slope K'_p rad$/(R_1 C)$ and that will produce a frequency change of slope

$$d\Omega/dt = K'_p K_v/(R_1 C) = \omega_n^2/\text{rad}. \tag{9.8}$$

The frequency cannot ramp faster. It can move faster for a short time due to the zero in the filter—a sudden change in u_1 will be multiplied by R_2/R_1 and immediately appear at the filter output. However, the sustained ramp cannot move faster than given by Eq. (9.8). A low-pass pole, created by placing R across C, will only reduce the current into the C and therefore the slope. Even if the loop is locked it cannot follow a frequency slope greater than that given by Eq. (9.8).

By analyzing phase-plane plots, corresponding to equations such as Eq. (8.72), with $\zeta = 1/\sqrt{2}$, but with a term representing a linearly changing Ω, Viterbi [1966, p. 58] confirmed Eq. (9.8). He also found that, as the sweep rate is lowered slightly, lock *may* occur if Ω is swept *through* zero. Whether it does depends on initial values (phase and frequency). If Ω is sweeping away from zero, the loop will not lock unless it does so immediately, without skipping a cycle (i.e., if it is almost locked to start with). But, when the sweep rate falls to

$$d\Omega/dt < \left(\omega_n^2/2 \text{ rad} \right), \tag{9.9}$$

the loop will always lock if Ω is swept through zero. The effect of much lower sweep rates is to establish a pull-in range (noninfinite) for values of Ω

TABLE 9.1. Lock Probability vs. Sweep Speed, $\zeta = 1/\sqrt{2}$, I & L Filter, Closed-Loop Sweeping

$R = \left\| \dfrac{(d\Omega/dt)\text{rad}}{\omega_n^2} \right\|$	Sweeping Away $d\|\Omega\|/dt > 0$	Sweeping Toward $d\|\Omega\|/dt < 0$
$1 < R$	No lock; cannot maintain lock	
$0.5 < R < 1$	No lock[b]	Transition between 0 and 100% probability 50% probability at $R \approx 0.92$[a]
$R = 0.5$	"	100% probability of lock
$R \ll 0.5$	$\Omega_{PI} \approx \omega_n^2 \omega_L/(2\text{ rad } d\Omega/dt)$	"

[a]Gardner (1979, Fig. 5.8).
[b]Except if initial values are such that no cycle is skipped.

that correspond to sweeping away from lock,

$$\Omega_{PI} = \frac{\omega_n^2 \omega_L}{2\dot{\Omega}\text{ rad}}, \qquad |\Delta\omega| \text{ increasing}, \qquad \dot{\Omega} \ll \frac{\omega_n^2}{2\text{ rad}}. \qquad (9.10)$$

Results for a high-gain (approaching integrator and lead) loop with $\zeta = 1/\sqrt{2}$ are summarized in Table 9.1. While this only applies precisely for $\zeta = 1/\sqrt{2}$, values of ζ that are close to $1/\sqrt{2}$ tend to be desirable for other reasons, so the information for this damping ratio is quite useful [Gardner, 1979, p. 80].

9.5.2 Open-Loop Sweeping

One way to acquire lock is to sweep ω_c with the loop open, detect when $\Omega = 0$, stop the sweep at that point, and close the loop. Detection is performed by mixing the loop input and output signals, passing the mixer output through a low-pass "decision" filter, and detecting the filter output using a rectifier. See Fig. 9.8. The detected signal is then compared to a fixed threshold and the sweep is stopped when the threshold is surpassed. The bandwidth of the low-pass decision filter determines how large the frequency error can be when threshold is broken.

The detector might be a half-wave rectifier as shown in Fig. 9.9a or a full-wave rectifier as in Fig. 9.9b. When the loop is being swept at a high rate, it may be that only a part of a cycle of ω_e will be seen, so the response depends on the phase that happens to occur as the signal passes through the low-pass filter [Blanchard, 1976, pp. 281–287]. Since the half-wave rectifier will have a maximum response only when the waveform is at a positive peak, whereas the full-wave rectifier has a maximum response to either a positive or a negative peak, the latter will have a greater probability of a strong response. To further increase the likelihood of a strong response, the

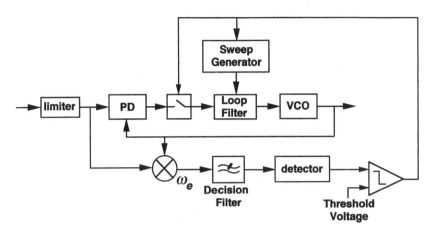

Fig. 9.8 Open-loop sweeping. When the error frequency is small enough to pass through the low-pass decision filter, the threshold is exceeded. This causes the sweep generator to stop and the loop to close.

detection circuitry can be repeated, except with one of the mixer inputs shifted 90°, and the output of the second threshold circuit can be ORed with the first. Then a maximum response will exist through one of the four paths every 90° of phase, and the response will be close to maximum regardless of the phase.[7]

The narrower the decision filter, the slower must be the sweep. The time during which the difference frequency is within the pass band of the decision filter is proportional to the filter bandwidth B and inversely proportional to the sweep speed.

$$T_p \sim B/\dot{\Omega}. \tag{9.11}$$

The rate at which the envelope of a pulse can rise in a filter is also proportional to the filter bandwidth, so, to reach a given relative threshold $r_T = V_T/K'_p$, the bandwidth must be proportional to the inverse of the pulse width,

$$B \sim 1/T_p \sim \dot{\Omega}/B. \tag{9.12}$$

Therefore the maximum sweep rate that will permit the detected output to rise to within r_T of its maximum possible value can be related to the filter

---

[7]If the video filtering is not required, the detectors can be eliminated, and the threshold circuit(s) can sense the waveforms directly. The analog to the full-wave rectifier is a threshold circuit set to detect both positive and negative excursions.

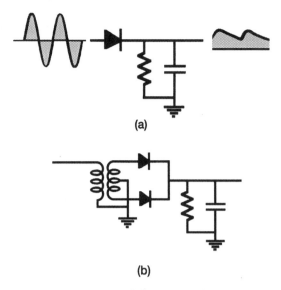

Fig. 9.9   Detectors: (*a*) half and (*b*) full wave.

bandwidth by a function of r_T,

$$\dot{\Omega}_M/B^2 = c(r_T).$$ (9.13)

Blanchard presents[8] experimental curves giving $c(r_T)$ for various probabilities of successful detection (stopping of the sweep) between 50 and 99%. For $0.2 < r_T < 0.8$, his curve for the ratio of maximum sweep rate to decision filter bandwidth at 99%[9] success can be expressed approximately as

$$\dot{\Omega}_M/B^2 \approx \frac{(167)^{(1-r_T)}}{2\pi \text{ rad}}.$$ (9.14)

On the basis of comparison with other data given by Blanchard,[10] the detector used in his experiment is probably a full-wave rectifier or a pair of quadrature rectifiers.

---

[8]See Blanchard (1976), Fig. 11.6; from unpublished document by S. Vialle, May 1969. Presumably the loop is a high-gain loop with a damping factor on the order of $1/\sqrt{2}$.

[9]When probability of lock changes from 99 to 50%, with $\alpha = 0.5$, the value $\dot{\Omega}_M/B^2$ almost doubles.

[10]Blanchard (1976), Fig. 11.5, gives probability of stopping versus sweep rate for a particular set of conditions that are also covered by Fig. 11.6. Figure 11.5 gives three curves, one for each of the configurations discussed above. The data from Fig. 11.5 falls between that for a full-wave rectifier and for two quadrature full-wave rectifiers.

The question naturally arises: What is the advantage of open-loop sweeping over closed-loop sweeping, if any? We can compare them by recognizing that the bandwidth of the loop that is to be locked after the sweep stops influences both the allowed parameters in Eq. (9.13) and the speed at which it can be swept in closed loop. The wider the decision filter the greater will be the maximum frequency error when the sweep stops and the wider must be the loop if it is to acquire rapidly at that time. But a wide loop can also be swept rapidly when it is closed, and that sweep rate can be compared to the open-loop rate. An approximate analysis based on such considerations is given in Appendix 9.A and results are shown in Fig. 9.10, which gives the ratio of maximum open-loop sweep rate to maximum closed-loop rate for various values of threshold r_T. From the 99% probability curve we see that there can be an advantage in open-loop sweeping at low threshold levels. Here the high sensitivity of the frequency error sensing mechanism allows the loop to lock at faster sweep rates than it could in closed loop. However, the practical limitation on low thresholds is increased susceptibility to noise. (Our analysis has not included the effects of noise.)

Example 9.1   Sweeping to Acquire   The VCO center frequency in a PLL is to be swept to aid in acquiring lock. The natural frequency is 2 kHz. Due to noise considerations, we can set detection threshold no lower than 40% of maximum. How fast can we sweep for 99% probability of lock?

Figure 9.10 indicates that we can sweep faster at $r_T = 0.4$ with the loop open than with it closed. We can solve (9.14) for $\dot{\Omega}_M / B^2$ and substitute this

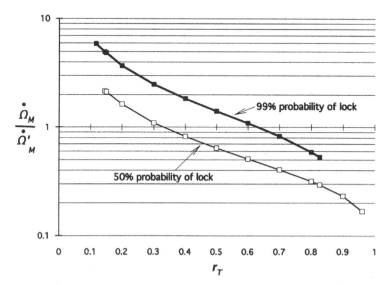

Fig. 9.10   Approximate ratio of maximum open-loop sweep rate $\dot{\Omega}_M$ to maximum closed-loop rate $\dot{\Omega}'_M$ versus relative threshold.

into (9.20) to relate f_n to B. Equation (9.14) gives $\dot{\Omega}_M/B^2$ as 3.43. Equation (9.20) then becomes

$$f_n = 1.95B \quad \text{so} \quad \underline{B = 6470 \text{ rad/sec} = 1030 \text{ Hz}.}$$

Since we have obtained $\dot{\Omega}_M/B^2$ and B, we can obtain

$$\dot{\Omega}_M = 3.43(6470/\text{sec})^2 = 1.44 \times 10^8 \text{ rad/sec}^2 \Rightarrow 22.9 \text{ MHz/sec.}$$

If we had used closed-loop sweeping, the rate for 100% lock probability (and approximately the same as the 99% probability rate) would be, from (9.9),

$$\dot{\Omega}'_M = \frac{\omega_n^2}{2 \text{ rad}} \Rightarrow 12.6 \text{ MHz/sec.}$$

This is 1.8 times slower, consistent with Fig. 9.10.

9.5.3 Combined Techniques

It is also possible to use a decision circuit to terminate the sweep, as in open-loop sweeping, but while allowing the loop to remain closed. This may provide advantages from both techniques, permitting the feedback to slow the sweep as zero error is approached but still stopping it based on the detector output.

9.6 SWEEP CIRCUITS

Now that we have discussed some theory concerning the affect of sweep rate on locking, let us consider how we might generate the sweep.

9.6.1 Switched Current Source

The VCO can be caused to sweep when it is out of lock by injection of a current into the filter feedback capacitor as in Fig. 9.11. It is important that the current not pass through either of the resistors, as it would if it were

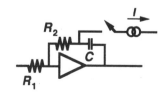

Fig. 9.11 Switched sweep current.

injected at either end of R_1. Otherwise there will be a step of output voltage when the current stops and the lock, once achieved, could be broken as a result. When the loop locks, even though the current continues, the loop will cancel the injected current and sweeping will stop (in this type 2 loop). There will be a steady phase error that causes I to pass through R_1 and R_2 and thus not enter C. The value of this phase offset can be easily computed from

$$\cos \Delta \varphi = IR_1/K'_p. \qquad (9.15)$$

We can substitute this into the expression for the sweep rate before acquisition, obtaining

$$-d\Omega/dt = K_v I/C = K'_p \cos \Delta \varphi \, K_v/R_1 C = \omega_n^2 \cos \Delta \varphi, \qquad (9.16)$$

which again shows that $|d\Omega/dt|$ cannot exceed ω_n^2. The phase offset can be eliminated by stopping the current I, perhaps based on the output from a lock indicator.

9.6.2   Automatic Sweep Circuit—Sinusoidal

The circuit of Fig. 9.12 can be designed to cause frequency sweeping when the PLL is out of lock. The loop filter is part of two loops, the PLL and the sweep loop. The latter is designed to be an oscillator at a frequency that is low enough that $d\Omega/dt$ will permit acquisition. When the PLL locks, it tries to cancel the input from the sweep loop. The PLL now changes the transfer function across the loop filter as seen by the sweep loop. This reduces the gain in the sweep loop so that oscillation ceases. The effect of each loop on the other should be taken into account by replacing the transfer function across the loop filter by the appropriate (from filter input to filter output) closed-loop transfer function of the other loop, assuming the other loop is locked. When the PLL is not locked, there can still be feedback [Egan, 1981,

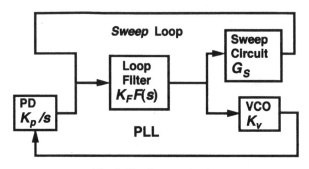

Fig. 9.12   Sweep circuit.

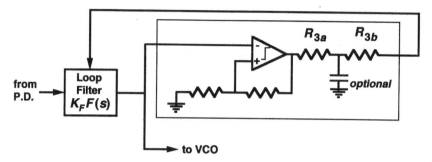

Fig. 9.13 Sweep generator with hysteresis. Hysteresis is created by the voltage divider, which gives positive feedback.

p. 246] due to the false lock phenomenon previously discussed, and the sweep loop must be designed to continue sweeping under these circumstances.

9.6.3 Automatic Sweep Circuit—Nonsinusoidal

Figure 9.13 shows a modification of the sweep circuit that causes the sweep to be more linear and eliminates the influence of the sweep loop on the PLL during lock. This is done by including a high-gain amplifier or comparator in the sweep loop. When the tuning voltage reaches an upper limit, it causes the output polarity of the comparator to reverse. Positive feedback provides hysteresis, preventing reversal from taking place until the lower limit is reached. The sweep may be produced by injection of the comparator output into the loop filter through R_3, as in Fig. 9.6, for example, or the loop filter may act primarily as an amplifier for the output of an optional low-pass filter. The low-pass then acts approximately as an integrator, converting the level change from the comparator into an approximate ramp. Of course, the circuit components must be chosen to ensure sweeping between the desired limits when the loop is out of lock and to ensure that the loop is not pulled out of its hold-in range when sweep stops [Egan, 1981, pp. 244–246]. This could happen if the steady-state sweep input to the filter were too large for the phase detector to cancel.

9.A APPENDIX: MAXIMUM SWEEP RATE, OPEN-LOOP VS. CLOSED-LOOP

Suppose we want ω_L to be as great as the value of $|\Delta \omega|$ at the moment when the sweep stops so the loop will then lock rapidly. There will be a frequency error due to the time T from $\Omega = 0$ until the sweep stops. Blanchard (1976) has also observed a dispersion of $\pm B$ in the stop frequency. Thus the total error at the maximum sweep rate could be

$$\Delta\omega \leq \dot{\Omega}_M T + B. \tag{9.17}$$

We can set this equal to ω_L, which, for $\zeta = 1/\sqrt{2}$, equals $\sqrt{2}\,\omega_n$. We now make the rough approximation that the filter output is like the response to a step occurring at $\Omega = 0$. The time to rise to a relative threshold level of r_T through a single-pole low-pass filter is obtained from

$$r_T = 1 - e^{-TB}. \tag{9.18}$$

Combining these we obtain

$$T = \frac{\sqrt{2}\,\omega_n - B}{\dot{\Omega}_M} = \frac{\ln[1/(1-r_T)]}{B}, \tag{9.19}$$

and we can solve this for the natural frequency of the loop that is just wide enough to acquire rapidly.

$$\omega_n = \frac{1}{\sqrt{2}}\left[\frac{\dot{\Omega}_M}{B}\ln\left(\frac{1}{1-r_T}\right) + B\right]. \tag{9.20}$$

The maximum sweep rate for closed-loop sweeping is

$$\dot{\Omega}'_M = \gamma\omega_n^2, \tag{9.21}$$

where $\gamma \approx 0.5$ for 100% probability of lock and 0.92 for 50% [Gardner, 1979, Fig. 5.8]. Combining the last two equations, we obtain

$$\dot{\Omega}'_M = \frac{\gamma}{2}\left[\frac{\dot{\Omega}_M}{B}\ln\left(\frac{1}{1-\nu_T}\right) + B\right]^2, \tag{9.22}$$

$$\frac{\dot{\Omega}'_M}{B^2} = \frac{\gamma}{2}\left[\frac{\dot{\Omega}_M}{B^2}\ln\left(\frac{1}{1-\nu_T}\right) + 1\right]^2. \tag{9.23}$$

The ratio of maximum open-loop sweeping rate to maximum closed-loop sweeping rate for the same ω_n is thus

$$r \equiv \frac{\dot{\Omega}_M}{\dot{\Omega}'_M} \approx \frac{2}{\gamma}\frac{\dot{\Omega}_M/B^2}{\left\{(\dot{\Omega}_M/B^2)\ln[1/(1-\nu_T)] + 1\right\}^2}. \tag{9.24}$$

This ratio is plotted in Fig. 9.10 based on Blanchard's (1976) curves for Eq. (9.13).

PROBLEMS

9.1 A loop has $\zeta = 0.7$, $\omega_n = 500$ rad/sec, and $K_p' = 0.2$ V/rad. The loop filter is integrator and lead.

(a) How fast can the VCO's center frequency be swept past the reference frequency with certainty that the loop will lock?

(b) What steady-state phase error will it acquire after lock at this sweep speed?

(c) If the same phase detector is used as a coherent detector (assuming they act as ideal multipliers), how high can the threshold at its output be set to ensure that lock will be detected at this maximum sweep rate?

(d) After the sweep input is removed, the loop remaining in lock, what will become the steady-state output from the coherent detector?

9.2 A loop has $\omega_n = 500$ rad/sec, and $K_p' = 0.2$ V/rad. The loop filter is lag-lead. The reference frequency is swept past the VCO center frequency at 1 kHz/sec. After the loop locks to the sweeping signal, the phase detector output ramps at a rate of 1.8V/sec. What is K?

9.3 The reference input is sweeping at 10 kHz/sec when it is connected to the loop. The loop is a "high-gain loop" with a natural frequency of 140 Hz and a filter zero at 100 Hz. How far can the reference have swept past the VCO center frequency when the connection is made if lock is to be certain?

9.4 What is the maximum open-loop sweep rate for 99% probability of detection if the decision filter has a bandwidth of 1 kHz and the threshold is set at 60% of the maximum output from the coherent detector?

9.5 Refer to Figure P9.5.

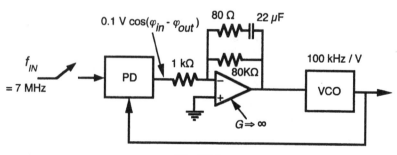

Fig. P9.5

(a) After the switch is closed, over what range of center frequencies f_c (which occurs when the phase detector output is zero) will the loop lock without skipping cycles?

(b) Over what range of f_c will the loop eventually lock?

(c) If the center frequency is 7.00500 MHz, how long will acquisition take?

(d) Once lock is achieved, how high can the VCO center frequency drift without loss of lock?

(e) What is the shortest time required to sweep (closed loop) the VCO center frequency by 2% (from $0.99 f_{IN}$ to $1.01 f_{IN}$) to ensure acquisition in the absence of noise?

CHAPTER 10

APPLICATIONS AND EXTENSIONS

We can apply the basic theory that we have studied to a broad range of circuits that may appear at first to be significantly different from those for which the theory was developed. The challenge is to recognize the equivalence. In this chapter we briefly discuss a number of such applications. The references provide a guide to further information.

10.1 GENERALIZED VOLTAGE-CONTROLLED OSCILLATOR

By identifying certain collections of components as being equivalent to the voltage-controlled oscillator (VCO) in the loops that we have studied, we can use the theory that has been developed to treat more complex circuits. Thus the "VCO" in Fig. 10.1 can be a representation of a group of functional blocks.

10.1.1 Frequency Synthesis, Frequency Division

A phase-locked frequency synthesizer can be represented by substituting the circuit of Fig. 10.2 for the "VCO" in the generic loop. The frequency divider ($\div N$) divides the oscillator frequency (and phase) by the divide ratio N. Loop responses can be determined for the loop of Fig. 10.1, writing the tuning sensitivity of the "VCO" block as

$$K_v = K'_v/N, \tag{10.1}$$

241

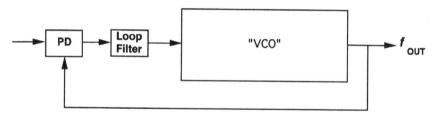

Fig. 10.1   Generalized VCO in loop.

where K'_v is the actual VCO tuning sensitivity. The usable output is taken from the output of the actual VCO since the object is to vary that frequency by changing N. We know that, in steady state, $f_{out} = f_{in}$ for the phase-locked loop (PLL) so

$$f_v = Nf_{out} = Nf_{in}, \tag{10.2}$$

where f_{in} is the loop's reference frequency, of course, but the term may also be applied to a higher frequency that has been divided by a constant to provide f_{in}. The synthesizer produces, at the VCO output, frequencies that are, during steady state, multiples of f_{in}. Thus f_{in} is the minimum step size, or resolution, of the synthesizer. Since it is usually a precise frequency derived from a crystal oscillator or other frequency standard, the synthesized output has the same relative accuracy but can produce many frequencies rather than the single frequency available from the reference. From Eq. (10.2) can be seen that a given relative change (e.g., expressed in percentage or parts per million) in f_{in} causes the same relative change in f_v.

The loop filter usually includes a low-pass filter. In our previous studies, we have assumed a low-pass filter following the phase detector to eliminate all but the desired, difference frequency, output component. We have also assumed that its cutoff frequency is so high that it has little effect on loop response. In a synthesizer this latter assumption is not necessarily valid and the low-pass filter that eliminates undesired components at the phase detector output often must be considered part of the loop filter. We have studied the second-order loop with a low-pass filter ($\alpha = 0$) but, if a type -2 loop is

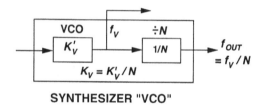

SYNTHESIZER "VCO"

Fig. 10.2   Frequency divider in a "VCO."

employed, the additional low-pass filter implies a third-order loop. While we have also studied such higher order loops, using state-space methods and computer analysis, a detailed analysis of a class of third-order loops that is particularly useful for frequency synthesizers is contained in Appendix 10.A.

Phase-locked loop theory, such as we have studied, is used to determine the synthesizer's loop stability, its transient response, and its response to various noise sources.

10.1.1.1 Stability.
Stability is evaluated using the methods in Chapter 5 with K_v given by Eq. (10.1). A complication that is sometimes significant in stability analysis, as well as in the determination of loop response in general, is due to the bandwidths of synthesizer loops often being extended so close to the reference frequency (e.g., $\omega_L = 0.1\omega_{in}$) that we can no longer consider only the low-frequency component of the phase detector's output. To permit such wide bandwidths, without excessive modulation of the output by the undesired components from the phase detector, at f_{in} and harmonics thereof, special phase detectors are employed. These minimize undesired outputs. However, the fact that information is only updated once per cycle of f_{in} affects the loop response [Egan, 1991]. While this is essentially true in all PLLs, it is not significant when the loops are narrow enough relative to the input frequency.

If the phase detector is of the sample hold type, an additional phase shift of

$$\varphi_s = -\pi f_m/f_{in} \tag{10.3}$$

(corresponding to a delay of half a sample period) can be included in $G(\omega)$ to improve the accuracy of the stability analysis. This can also be used as a conservative approximation with other types of phase detectors. Such additional phase shift also causes a reduction in phase margin, leading to greater peaking in transient and frequency responses.

10.1.1.2 Transient Response.
The primary transient experienced occurs when N is changed in order to change f_v. Because this is equivalent to a change in f_{in}, it can be analyzed by the methods studied in Chapter 6. If N is changed from N_0 to N_1, the final value of the VCO frequency will be $f_{v1} = N_1 f_{in}$. Just after N is changed, however, f_v will still be the previously synthesized frequency, $f_{v0} = N_0 f_{in}$.[1] This is also the frequency that would be produced with the final value of $N = N_1$ by a reference frequency of

$$f_{in,0} = f_{v0}/N_1 = N_0 f_{in}/N_1 \tag{10.4}$$

---

[1]More universally, this is true the instant before the change is made, but practical synthesizers have low-pass filters that do not permit a phase change to be transmitted instantly to the output.

so the equivalent input step is

$$\Delta f_{\text{in}} = f_{\text{in}}[1 - N_0/N_1].$$

(10.5)

A step that may equivalently be introduced after f_v is

$$\Delta f_v \equiv f_{v1} - f_{v0} = N_1 \Delta f_{\text{in}} = N_1 f_{\text{in}}[1 - N_0/N_1] = f_{\text{in}}[N_1 - N_0],$$

(10.6)

and to which f_v will respond in the typical low-pass fashion of $H(s)$.

10.1.1.3 *Response to Noise.* Noise considerations are usually very important in synthesizer design since high-purity signals are sought. The loop imposes low-pass filtering (H) upon phase noise that enters with the reference and high-pass filtering $(1 - H)$ on the open-loop noise of the VCO. Noise introduced through various loop components, such as the op-amp, is also transferred to the output. The processes are the same as we have studied in Chapter 7.

While there are many additional considerations peculiar to frequency synthesis, [Egan, 1981], the theory that we have studied is basic to synthesizer design.

10.1.2 Heterodyning (Frequency Mixing)

The VCO plus a frequency mixer can also be represented as an equivalent VCO. In Fig. 10.3, the frequency before the mixer is

$$f_v = f_c + K_v u_2.$$

(10.7)

After the mixer the frequency is

$$f_v' = f_v \pm f_{\text{MIX}} = f_c' + K_v u_2,$$

(10.8)

where

$$f_c' = f_c \pm f_{\text{MIX}}.$$

(10.9)

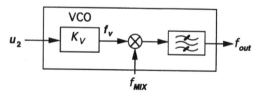

Fig. 10.3   Equivalent VCO with frequency mixing.

Thus the mixing process modifies the center frequency of the "VCO." If Eq. (10.9) produces a negative sign, there is also an additional inversion, or 180° phase change, to be accounted for. This is easily done by changing the sign of one of the blocks in the loop. Most computations, such as response to modulation or step response, involve changes from steady state anyway, so the center frequency is not of concern. It is, of course, important from a practical standpoint.

The desired output is often f_v with the mixing process used to facilitate realization of the PLL.

If the filter following the mixer is not very wide, it must be accounted for in determining loop performance. How it affects modulation phase is of importance to loop response. For small phase deviation and with a symmetrical filter, an equivalent low-pass filter can be introduced into the loop, one that affects the modulation in a manner that produces the same effect on the modulation sidebands as would the actual bandpass filter.

10.2   LONG LOOP

A configuration called a long loop is shown in Fig. 10.4. The output frequency is multiplied by N and mixed with the input. (Recall that phase undergoes summation, multiplication, or division simultaneously with frequency.) The resulting difference frequency is phase detected against the output signal. The phases of the various signals are shown in Fig. 10.4.

An equivalent single-loop model is shown in Fig. 10.5. Note that the output of the phase detector is the same in each model. We again create an equivalent VCO (Fig. 10.6) to substitute into our generic loop so we can see clearly how to apply our mathematical tools. Thus the long loop is analyzable by techniques that we have developed, but what are we getting for the added complexity? What we gain is the relatively narrow bandpass filter that enables us to limit the band of signals that enters the phase detector, thus rejecting unwanted energy at the input and, especially, noise. We could, of

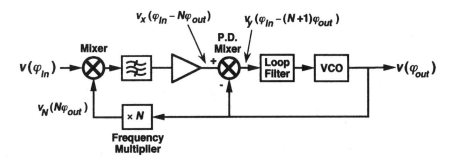

Fig. 10.4   Long loop.

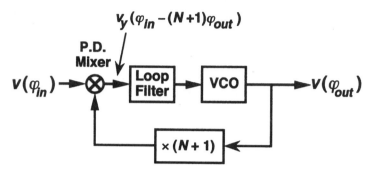

Fig. 10.5 Long loop first equivalent.

course, pass the input to any loop through a bandpass filter, but this filter can be narrower. If the input signal frequencies vary over ΔF, the passband of a filter preceding an ordinary loop would have to be ΔF wide, but the passband in a long loop can be much narrower.

From Fig. 10.4, the passband width of the filter must be

$$\Delta F \geq \Delta f_{\text{in}} - N\Delta f_{\text{out}}. \tag{10.10}$$

From Fig. 10.5, the input frequency is $N + 1$ times the output frequency, so Eq. (10.10) gives

$$\Delta F \geq \Delta f_{\text{in}} - [N/(N + 1)]\Delta f_{\text{in}} = \Delta f_{\text{in}}/(N + 1). \tag{10.11}$$

Thus the necessary width of the input filter is reduced by a factor $N + 1$ in the long loop. However, the VCO must be tuned to the correct frequency to start since the loop cannot see the input signal unless it passes through the narrow input pass band. Again, if the bandpass filter is not wide compared to the loop, the response of the modulation to the equivalent low-pass filtering must be taken into account.

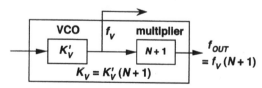

Fig. 10.6 Long loop equivalent VCO. Substitutes for the VCO and $\times(N + 1)$ in Fig. 10.5 and forms a standard loop with the PD mixer and loop filter of Fig. 10.4.

10.3   CARRIER RECOVERY

One prominent use of PLLs is in carrier recovery of phase-coded signals.

10.3.1   Biphase Costas Loop

The simplest of such signals is the biphase coded (BPSK for binary phase shift keyed) signal. Information is carried in the phase, which has two possible values, 180° apart. To decode the received signal, an unmodulated reference, operating at the same frequency, to which the phase can be compared, is needed. The problem of recovering the reference from the signal is difficult because a randomly modulated BPSK signal has no discrete energy line at the carrier frequency. The energy spectrum has the same shape as that of the individual (constant-phase) pulses of which the signal is composed with no discrete line at the carrier, or premodulation, frequency. We can imagine trying to lock a PLL to such a signal. The output from the phase detector (PD) would rapidly change polarity, as the phase changed in 180° increments with equal rapidity, producing zero average output. Some process is needed that is able to ignore the phase transitions.

Figure 10.7 shows a carrier-recovery and demodulation loop for biphase signals (BPSK), called a Costas loop. The VCO and input signals are multiplied in two places, with one of the signals shifted 90° in one place. As a result, two quadrature outputs are obtained that, when multiplied together, produce a $\sin(2\Delta\varphi)$ function. That has a typical shape for a phase detector output—the slope near zero output can be used to develop u_1—but the phase is doubled. That increases the slope and, what is more important, makes the function insensitive to 180° changes:

$$\sin[2(\varphi + \Theta)] = \sin[2(\varphi + \Theta + \pi)]. \qquad (10.12)$$

The usual process can be used for linear analysis, with K_p taken from the shape of the $\sin(2\varphi)$ curve.

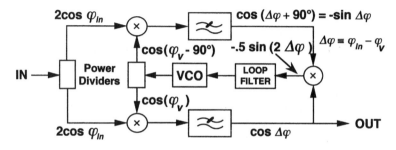

Fig. 10.7   Costas loop.

Quadrature PD Out

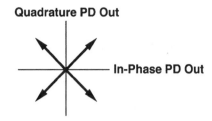

In-Phase PD Out

Fig. 10.8 QPSK.

Another advantage of this Costas loop is that it provides detection simultaneously with carrier recovery. The lower phase detector in Fig. 10.7 has in-phase inputs so it is a coherent detector, producing a maximum-amplitude positive or negative output.

10.3.2 *N*-Phase Costas Loop

In addition to two-phase signals, four- and eight-phase signals are also common. A four-phase (QPSK for quadrature phase shift keyed) signal is illustrated in Fig. 10.8. More complex versions of the Costas loop [Lindsey and Simon, 1973, pp. 74–75; Ziemer, 1985; Gardner, 1979, Chapter 11] employing four or eight multiplications of the VCO by the input, rather than the two shown above, can be used to derive a carrier from these. One of these circuits is described in Appendix 10.B.

10.3.3 Multiply and Divide

Another method for carrier recovery from *M*-phase signals involves the multiplication of the signal by *M*, as illustrated in Fig. 10.9. The allowed phases of the coded signal are

$$\varphi_{in} = \varphi + i2\pi/M, \tag{10.13}$$

where $0 \le i \le M - 1$. After multiplication by *M*, the phases are

$$\varphi_M = M\varphi_{in} = M\varphi + i2\pi \Rightarrow M\varphi, \tag{10.14}$$

regardless of which of the allowed states (*i*) exists. This must be filtered because, in practice, the signals lose amplitude, sometimes going through

Fig. 10.9 Carrier recovery by multiplication with a PLL providing filtering.

zero, as they transition from one state to another. A narrow filter is needed to filter out this amplitude modulation, and possibly accompanying noise, and the PLL is often used for this purpose.

10.4 DATA SYNCHRONIZATION

In addition to the need for carrier recovery, recovery of the clock that synchronizes the data is also important. This process too involves various circuits to which basic PLL principles can be applied. [Gardner, 1979, Chapter 11; Spilker, 1977]. We will discuss some of them briefly.

10.4.1 Early-Late Gate Bit Synchronizer

An early-late gate bit synchronizer is shown in Fig. 10.10. Waveforms are shown in Fig. 10.11.

The VCO output is split, and one path is delayed by one quarter of the bit period while the other is advanced by that much. The time-shifted pulses are used to control integrate-and-dump circuits. Integration of the input pulse train begins at the beginning of the delayed or advanced VCO pulse and stops at the end of that pulse. The integrator output is then sampled and held, after which the integrator is dumped, remaining in that state until the start of the next time-shifted pulse. The sample-and-hold (S & H) outputs are converted to their absolute values and then subtracted from each other. The values of the individual voltages, v_a and v_b, are shown in Fig. 10.12 as a function of the delay of the incoming pulse relative the position shown in Fig. 10.11. The resulting average voltage into the loop filter equals $v_a - v_b$, which is plotted in Fig. 10.12. Either a positive or a negative pulse will produce an output of AT_b when shifted from the position in which they are

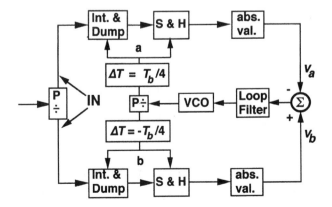

Fig. 10.10   Early-late gate bit synchronizer.

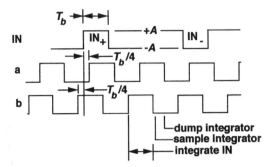

Fig. 10.11 Bit synchronizer waveforms.

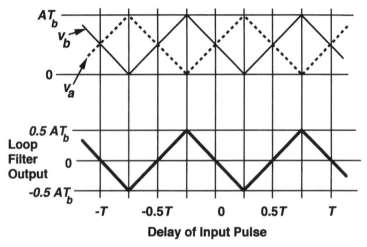

Fig. 10.12 PD characteristic and its components.

shown in Fig. 10.11 by $T_b/4$. However, when there is no transition (no change in bit value), $v_a - v_b = 0$. Since there is a 50% probability of a transition, the average voltage into the loop filter is half of the value produced after a transition. Thus the loop filter output goes from $-0.5AT_b$ to $+0.5AT_b$ with a triangular characteristic, giving $K_p = 2AT_b$/cycle.

The PLL theory we have studied, especially as it pertains to triangular PD characteristic, applies to this loop. The quarter bit period delay and advance could be obtained by decoding four-stage shift frequency divider driven by the VCO operating at four times the data rate.

10.4.2 Synchronizing to a Pseudorandom Bit Sequence

A pseudorandom bit sequence of length M has an autocorrelation function such as shown in Fig. 10.13. If the incoming signal is composed of repetitions

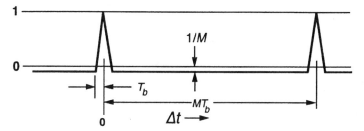

Fig. 10.13 Autocorrelation function of pseudorandom sequence.

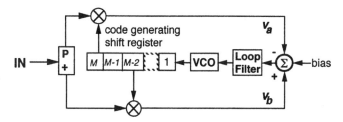

Fig. 10.14 Code synchronizing loop.

of this sequence, we can synchronize to it using a method that has similarities to that which was employed for the simple code in the previous section. We do this by multiplying the incoming code sequence by the identical sequence synchronized by the VCO and by the same sequence shifted by two bits and then subtracting the results. This is illustrated in Fig. 10.14.

In this "delay lock loop," [Spilker, 1963; Spilker and Magill, 1961] the bit streams entering the multiplier have zero average value (e.g., ±1). The circuit could also be realized using ExNOR gates and binary (0 and 1) signals. The average value of v_a varies with the time offset between the incoming code and the code generator with the same characteristic as the autocorrelation function. So does the average of v_b, but its characteristic is offset from the characteristic of v_a by two bit periods (smaller offsets are also used). These two characteristics are superimposed in Fig. 10.15, where v_b has also been given the minus sign it acquires at the summer. The resulting phase detector characteristic is shown in Fig. 10.16.

These diagrams have been normalized so 1 represents the voltage that appears at the loop filter output due to fully correlated (in phase) codes in one of the two correlators. The resulting phase detector gain (still so normalized) is

$$K_p = \frac{(M + 1/M)}{T_b} \frac{MT_b}{\text{cycle}} = \frac{M + 1}{\text{cycle}}. \qquad (10.15)$$

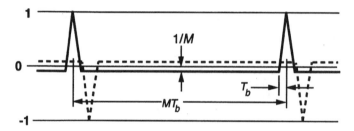

Fig. 10.15 Characteristics of v_a and $-v_b$ (normalized).

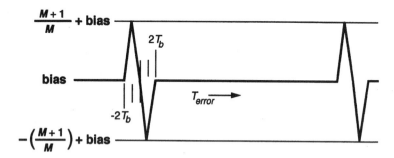

Fig. 10.16 Phase discriminator characteristic (normalized).

Except in the narrow region between $\pm 2T_b$, the phase detector has zero output and there is no feedback. We then depend upon the difference between the incoming bit rate and the VCO frequency to move the operating point toward the region where the PD has gain. The bias voltage might be used as part of a strategy to ensure acquisition of lock in an acceptable time.

10.4.3 Delay-and-Multiply Synchronizer

Another method manipulates the data stream to create a spectral line at the data rate and then uses the PLL as a narrow tracking filter. This method is discussed in Appendix 10.C.

10.5 ALL-DIGITAL PHASE-LOCKED LOOP (ADPLL)

Another loop that is analyzable by the theory that we have developed is the all-digital PLL (ADPLL). Sometimes an ordinary PLL that uses a phase detector that is driven by, and provides, logic-level signals is called a digital PLL. However, we refer here to a loop that is truly digital. In fact, it has no continuously variable signal. State variables that were voltages or currents in analog loops are numbers in the ADPLL. Of course, the numbers are represented by voltages in a binary form, but the variables of interest are the represented numbers. Many different ADPLLs are discussed in the literature [Lindsey and Chie, 1981]. We will discuss two in order to show how the general principles of PLLs can be applied to ADPLLs. The many possible implementations of ADPLLs and various special considerations applicable to them provide opportunities for additional study that is beyond the scope of this book.

10.5.1 Basic Digital Implementation

10.5.1.1 The Loop. First consider the block diagram of a simple ADPLL shown in Fig. 10.17. It is labeled "too simple" for reasons we will discover, but it will serve as an introduction to the equivalencies that can be drawn between the digital blocks and those which we have studied as well as to some new potential problems.

All of the signals are numbers, represented electronically. The input is a binary clock of frequency f_{in}, and the output is to be a number that, in the manner of phase-locked loops, has an average repetition rate of f_{in} but follows the input with a closeness that depends on the loop parameters. The VCO has been replaced by an output accumulator (OA). Its output is a number that changes each clock cycle by an amount equal to its input N_2. Each time the OA reaches its capacity N_v, it recycles to 0 ($N_v - 1$ being the highest number that can be attained). Thus one output cycle is represented by N_v and the output phase of the OA is

$$\varphi_{out} = (N_{out}/N_v) \text{ cycles.} \tag{10.16}$$

The output frequency is

$$f_{out} = \Delta\varphi_{out}/\Delta t = (N_2/N_v)f_{clock} \tag{10.17}$$

since the output is incremented by N_2 each cycle ($T = 1$ c/f_{clock}) of the OA clock. If we consider N_2 to be the tuning signal, we can obtain the OA gain constant from (10.17) as

$$K_v = df_{out}/dN_2 = f_{clock}/N_v. \tag{10.18}$$

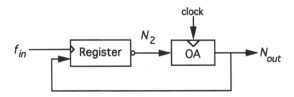

Fig. 10.17 Too simple ADPLL.

The register stores the value of N_{out}, representing the loop's output phase, each cycle of the input signal. Thus the stored number represents the output phase at the last transition of f_{in}, and this can be considered the phase difference between the output and input signals. The register thus functions as a phase detector (and zero-order hold), but we need a phase inversion. One simple way to do this is to make N_2 the complement of the sampled value of N_{out} (represented by the small circle at the output of the register). During the nth input cycle (following the nth input clock transition), we have

$$N_2 = -N_{out}(\varphi_{in} = 2\pi n). \tag{10.19}$$

The phase detector gain constant is

$$K_p = -\frac{\partial N_2}{\partial \varphi_{out}} = -\frac{\partial N_2}{\partial N_{out}} \frac{\partial N_{out}}{\partial \varphi_{out}} = -(-1)\frac{N_v}{1 \text{ cycle}} = N_v \text{ cycle}^{-1}. \tag{10.20}$$

The forward gain constant is

$$K = K_p K_v = f_{clock} \text{ cycle}^{-1}. \tag{10.21}$$

We have the essential parts of a PLL, an OA whose output phase advances at a rate proportional to the control signal that is the output from a phase detector/register. The mathematical block diagram is given by Fig. 2.4 with $K_{LF} = 1$ and φ_{out}, K_p and K_v defined above. For this first-order loop, $u_1 = u_2 = N_2$.

Let us consider what the output looks like as the frequency changes; it will be affected by its discrete or sampled nature. For small values of N_2, the output frequency will advance with increasing N_2, beginning with $f_{out} = f_{clock}/N_v$ when $N_2 = 1$. For small N_2 it will be a relatively smoothly increasing sawtooth value. As N_2 increases, the number of output changes between overflows will eventually become small so the output will no longer resemble a smoothly increasing phase. At $N_2 = N_v/2$, the output will change by half of the capacity of the OA each input period and will appear to be a square wave. At higher values of N_2, the output frequency will be negative, as demonstrated by the slope of the phase change with time, and decrease until, at the maximum value of $N_2 = N_v - 1$, it will appear to be again at a frequency of f_{clock}/N_v, but with a negative sign. These appearances are due to the fact that the advancing value of N_{out} is sampled at a frequency of f_{clock}.

10.5.1.2 Sampling and Stability.
There are two sampling processes occurring in the simple loop, one in the OA at f_{clock} and one in the register at f_{in}. To make the OA operate like a VCO that has approximately a continuous phase change at its output, it is necessary that

$$f_{clock} \gg f_{in}, \tag{10.22}$$

since f_{in} is the frequency of the OA's output during lock. If this inequality is not met, the output will appear to be a few steps rather than a ramp of phase. Suppose f_{in} should have a discrete phase change of $\Delta\varphi$ between two of the transitions that clock the register. At the first input sample after this step, a frequency change will occur at N_{out} equal to

$$\Delta f_{out} = K\,\Delta\varphi = f_{clock}\,\Delta\varphi\,\text{cycle}^{-1}. \tag{10.23}$$

This results in a phase change at the next input sample, which occurs 1 cycle$/f_{in}$ later, of

$$\Delta\phi = -\Delta f_{out}\,\text{cycle}/f_{in} = -\Delta\varphi f_{clock}/f_{in}. \tag{10.24}$$

So the phase error grows by $-f_{clock}/f_{in}$ each input sample and, according to (10.22), this is a substantial increase each time. This "first-order loop" is unstable due to the sampling process. It repeatedly overshoots the final value by an amount much larger than the pervious error, causing a growing oscillation. To solve this dilemma, let us divide N_2 by 2^q by shifting bits between the register and the OA, connecting 2^i from the register to 2^{i-q} at the OA input. Now we have reduced the gain by 2^q, but we have also reduced the maximum value of N_2 by the same amount so now

$$N_2 \le (N_v - 1)2^{-q} \tag{10.25}$$

so the highest frequency we can accommodate will be

$$f_{in} = f_{out} \le f_{clock}\frac{N_v - 1}{N_v}2^{-q} \approx 2^{-q}f_{clock}. \tag{10.26}$$

The ratio of phase change magnitudes between adjacent samples that could be obtained from (10.24) is now, due to our gain change of 2^{-q},

$$-\frac{\Delta\phi}{\Delta\varphi} = 2^{-q}\frac{f_{clock}}{f_{in}} > 1, \tag{10.27}$$

where unity is approached at maximum f_{in}. Unity is desired in Eq. (10.27), since the output phase will then follow the input step in one input sample period. Instability occurs when $-\Delta\phi/\Delta\varphi = 2$, since the magnitude of the error in $\Delta\varphi$ will then increase each input sample. Thus we are constrained to operate in a narrow range of frequencies, depending on the value of f_{clock} and q. We could alleviate this restriction by obtaining N_2 by adding a constant to the PD output so the value of N_2 would not be tied so directly to f_{in}. Then Eqs. (10.25) and (10.26) would not hold, and we could lower the gain by operating at higher input frequencies than allowed by (10.26). With all of these changes, the block diagram would look like Fig. 10.18.

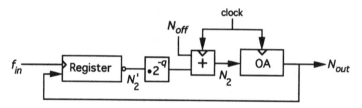

Fig. 10.18 Workable simple ADPLL.

We can see these stability problems more succinctly from the z-transform representation of the loop. Again we will assume Eq. (10.22) so we can simplify analysis by ignoring the faster sampling in the OA. The mathematical block diagram with the sampler shown is Fig. 10.19, where the blocks are represented in terms of Laplace transforms. Now, however, Eq. (10.20) has changed to

$$K_p = N_v 2^{-q} \text{ cycle}^{-1} \tag{10.28}$$

so the forward transfer function is

$$G(s) = \frac{1 - e^{-Ts}}{s^2} f_{\text{clock}} 2^{-q} \text{ cycle}^{-1}, \tag{10.29}$$

in Laplace terms, and the corresponding z-transform is

$$G(z) = 2^{-q} \frac{f_{\text{clock}}}{\text{cycle}} \frac{(1 - z^{-1})Tz^{-1}}{(1 - z^{-1})^2} = 2^{-q} \frac{f_{\text{clock}}}{f_{\text{in}}} \frac{1}{z - 1}. \tag{10.30}$$

Given the position of the sampler, the z-transform of the closed-loop response is

$$H(z) = \frac{G(z)}{1 + G(z)} = \frac{C}{z - 1 + C}, \tag{10.31}$$

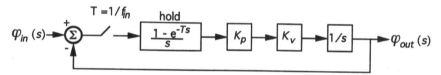

Fig. 10.19 Mathematical diagram of ADPLL in Fig. 10.18.

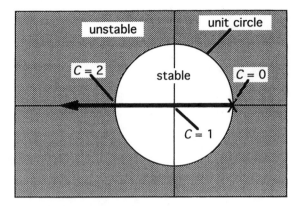

Fig. 10.20 Z-plane representation of the loop.

where

$$C = 2^{-q} \frac{f_{\text{clock}}}{f_{\text{in}}}. \tag{10.32}$$

The open- and closed-loop poles for this loop are shown on the z-plane in Fig. 10.20. The closed-loop pole locus begins at the open-loop pole when $C = 0$ and moves along the real axis as C increases. The best response will be in the center of the unit circle, where $C = 1$ and oscillation occurs when the locus reaches the unit circle and enters the region of instability outside of it, at $C = 2$. This corresponds to what we ascertained using the simple analysis above. Without an offset added to the PD output, we were restricted to operating where $C > 1$, in the left half plane only.

10.5.1.3   *Choice of Values*

Example 10.1   First-Order ADPLL, A Particular Implementation   Given: $N_v = 2^{12} = 4096$; $f_{\text{clock}} = 8.192$ MHz; $f_{\text{in}} = 400$ kHz. Find: (a) a good value for q; (b) N_2; (c) N_{off} if required; (d) ω_L; (e) OA output range.

a. We would like to have $C = 1$. Since $f_{\text{clock}}/f_{\text{in}} \approx 21$, we consider $q = 4$ and 5 for $2^{-q} = \frac{1}{16}$ and $\frac{1}{32}$, respectively. If we use the former, $C = 1.31$, which is not far from optimum.

b. From Eq. (10.17),

$$f_{\text{in}} = f_{\text{out}} = 4 \times 10^5 \text{ Hz} = (N_2/4096)8.192 \times 10^6 \text{ Hz} = N_2 \times 2 \text{ kHz},$$

$$N_2 = 200,$$

$$N_2' = 16N_2 = 3200.$$

c. No offset is required since $N_2' = 3200 < N_v$.

d. For this first-order loop,

$$\omega_L = K = 2^{-4} f_{\text{clock}} \text{ cycle}^{-1} = 5.12 \times 10^5 \text{ sec}^{-1} \Rightarrow f_L = 81.49 \text{ kHz}.$$

e. Since N_2 can vary from 0 to $\text{INT}(4095/16) = 255$, Eq. (10.17) also shows that the possible range of f_{out} is from 0 to

$$f_{\text{out, max}} = (255/4096)8.192 \text{ MHz} = 510 \text{ kHz}.$$

Also note in the example that N_2 can be constant if f_{in} is a power-of-two multiple of the step size, 2 kHz, but, for other frequencies, it will cycle between two numbers such that the average output frequency, which cycles between two adjacent multiples of 2 kHz, equals the input frequency. This is an example of jitter or ripple, an effect that is due to quantization, the fact that the signals are quantized rather than continuously variable. False locks, wherein the input and output frequencies are related by a ratio of small numbers, are another unwelcome possibility in sampled systems. [Egan, 1981, pp. 228–230]. The program described in Appendix 10.M.2 should be helpful for exploring some of these effects.

10.5.1.4 *Higher Order Loops.* We have shown a simple loop with no loop filter, but loop filters can be implemented digitally. Figure 10.21 shows a simple integrator-and-lead filter. The $1/s$ integration block can be realized with an accumulator. The values of K_1 and K_2 can be adjusted by how the digits are shifted in the two legs. We can write the resulting transfer function as

$$\frac{y}{x}(s) = K_1 + \frac{K_2}{s} \tag{10.33}$$

$$= \frac{K_2(1 + sK_1/K_2)}{s}. \tag{10.34}$$

This is the same general form as Eq. (3.38) with $\omega_z = K_2/K_1$.

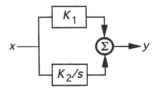

Fig. 10.21 Digital integrator- and -lead filter.

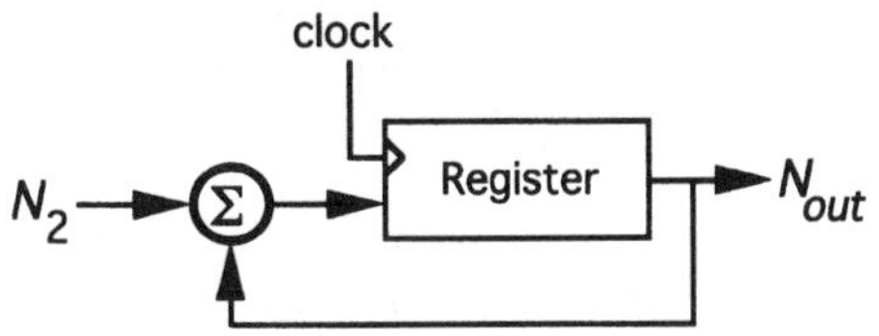

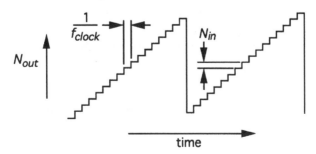

Fig. 10.22 Output accumulator.

With this filter we should ensure that a number in the middle of the output range of the PD is interpreted as zero, lower numbers being considered negative. This will allow steady state, where the accumulator output is fixed, to occur in the middle of the PD range. Such DC shifts do not affect mathematical blocks but can be necessary or advisable in various implementations.

10.5.2 OA, NCO, DDS

The OA is shown in more detail in Fig. 10.22. The control input, N_2, is added (numerically) to N_{out}. Each clock cycle the sum is entered into the register. (We assume that delays or some other mechanism prevents the register output from changing before the new input has been completely entered into the register. This is the kind of practical detail that will not be addressed here.) Figure 10.23 shows how N_{out} advances each clock cycle until it recycles when the register's capacity has been exceeded. The OA is an accumulator, a combination of memory and adder, that we have designated OA to differentiate it from accumulators that may be used elsewhere in the ADPLL.

Fig. 10.23 OA output.

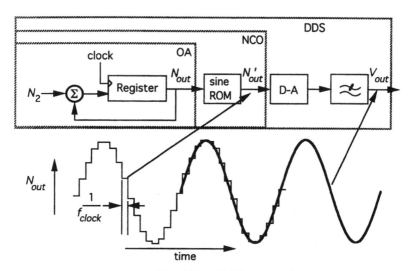

Fig. 10.24 OA, NCO, DDS, and waveforms.

Adding a read-only memory (ROM) to the OA, as in Fig. 10.24, produces a numerically controlled oscillator (NCO).[2] The ROM produces the sine (or cosine) of the phase represented by the output of the OA, which provides the address to the ROM. The output N'_{out}, is now a numerical representation of a sinusoid.

Adding a digital-to-analog (DA) converter and a low-pass filter produces a direct digital synthesizer (DDS). The DA converts the numerical sine wave to a stepped voltage representation of the sine wave, and the low-pass filter rejects all but the fundamental component, so we finally arrive at a sine voltage V_{out} such as would be produced by a VCO.

The ADPLL with the OA output N_{out} fed back to the PD performs like an analog loop with a sawtooth phase detector whereas a ADPLL in which the NCO output N'_{out} is fed back performs more like an analog loop with a sinusoidal phase detector. Their linear small signal representations are the same, but their large signal performances differ. In some cases the sampling type of PD may not be appropriate (e.g., when the input is not a simple square wave but a sinusoid, to which we must lock, plus noise or other signals). We can achieve phase detection that is similar to analog multiplication (see Section 3.1.4.3) by multiplying N'_{out} by a digital representation of an analog input signal and passing the result through a digital low-pass filter. We also could use V_{out} with an analog phase detector and convert the phase detector's output to a digital signal using an AD, thus producing a hybrid digital/analog PLL.

---

[2]NCO is a common term whereas OA is not.

We might use the transition of the most significant bit (MSB) of N_{out} or N'_{out} as a clock edge. (This suggests a different set of PLLs in which phase detection is accomplished by sampling an appropriate input waveform at a time determined by the output waveform.) However, it has an inherent jitter, with a total range of one clock cycle. To understand this peculiarity of the sampled nature of the digital representation, let us compare the transitions of the most significant bit to a clock edge that is obtained by processing V_{out} through a limiter to produce a rectangular clock waveform, $V'_{out} = \text{sign}(V_{out})$.

If $N_2 = 2^i$, where i is a whole number, each cycle of N_{out} and N'_{out} will be the same as each other cycle, and there will be no jitter. If we add a small number ε to N_2 for one clock cycle only, it will change all subsequent values of N_{out} by ε, but it will not necessarily affect the transition time of the MSB. If it does, it will shift it by $(1/f_{clock}) \gg 1/\varepsilon$. Transition of the MSB indicates that N_v has been reached sometime in the previous sample period, but the information as to just when is contained in other bits. However, while the MSB transition moves by zero or $1/f_{clock}$, the transition time of V'_{out} will change by ε/N_v cycles, as is appropriate, because V_{out} is affected by all of the bits in N'_{out}.

If $N_2 \neq 2^i$, all cycles will not be identical. Some cycles of the most significant bits of N_{out} and N'_{out} will be one sample period longer than others (this is jitter or "ripple"). The cycles of V_{out} and V'_{out}, however, will all be of the same duration.

10.5.3   Implementing an ADPLL by Pulse Addition and Removal

A second ADPLL implementation is shown in Fig. 10.25. This diagram corresponds to an available integrated circuit (IC).[3] We will study it to improve our ability to recognize the correspondence between the fundamentals we have studied and structures that look much different at first observation. Additional description and analysis, including for a second-order loop, is in Appendix 10.D. More information can be obtained from the IC data sheet and from an application note [Troha, 1994].

The VCO function is provided by removing pulses from, or adding pulses to, a fixed-frequency clock and dividing the result by a large number so the jitter is relatively small. (Adjacent cycles are approximately of the same duration because a small number of pulses, relative to the total, have been removed.)

We begin our examination at the switch, which is open for purposes of discussion only. The signal to the right of the switch we call I. This is a series of pulses with value 1 or -1 and an average net frequency of Δf (number of

---

[3]All the blocks in Fig. 10.25 are available in the TI SN54LS297 IC excepting the $\div N$, which must be supplied externally. A flip-flop phase detector is also included. See the data sheet but notice some difference in symbols from what we use here.

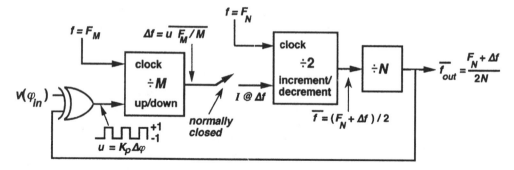

Fig. 10.25 All digital PLL using pulse removal.

positive pulses less number of negative pulses divided by time). The subsequent $\div 2$ circuit divides the clock at frequency F_N by 2 but, when it receives an increment $(+1)$ or decrement (-1) command at I, it adds one clock edge or subtracts one, respectively. Thus the average frequency from the $\div 2$ circuit is

$$\bar{f} = \frac{F_N + \Delta f}{2}. \tag{10.35}$$

The output frequency is this value divided by N:

$$\bar{f}_{out} = \frac{F_N + \Delta f}{2N} \tag{10.36}$$

10.5.3.1 Transfer Function.

The phase detector we are using is an Exclusive-OR gate (as in Section 3.1.2); we will call its output u and define it as true $= 1$ or false $= -1$. When its inputs are exactly in phase, its output is -1. As the relative phase of its inputs shifts, its average output becomes more positive until it equals $+1$ when the inputs are exactly out of phase. During the half-cycle change in the relative phase of its inputs, its average output changes proportionately from -1 to $+1$ and could be represented by the triangle characteristic in Fig. 3.2 with those limits. Thus the phase detector gain constant is

$$K_p = 2/(0.5 \text{ cycle}) = 4/\text{cycle}. \tag{10.37}$$

When the PD output is $+1$ the $\div M$ circuit counts up; when it is -1 the circuit counts down. The output frequency from the $\div M$ therefore has an average value of

$$\Delta f = uF_M/M. \tag{10.38}$$

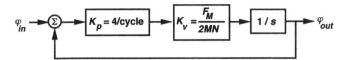

Fig. 10.26   Digital PLL, mathematical block diagram.

The gain of the VCO block can be obtained from Eq. (10.36) and (10.38) as

$$K_v = \frac{\overline{df_{out}}}{du} = \frac{F_M}{2MN}. \tag{10.39}$$

A block diagram using these constants is shown in Fig. 10.26. The $1/s$ block, as usual, converts frequency to phase.

Figure 10.26 represents a first-order loop with unity-gain bandwidth of

$$\omega_L = K = [2F_M/(MN)]\,\text{rad/cycle}. \tag{10.40}$$

Transfer functions are given by Eqs. (2.25b) and (2.26b). Transient responses are as given in Table 2.1.

10.5.3.2 Tuning Range. The hold-in range can be seen from Fig. 10.25 and Eqs. (10.36) and (10.38) to be

$$F_H = \pm F_M/(2MN), \tag{10.41}$$

the limits corresponding to $+1$ or -1 from the phase detector.

From Eq. (10.36) the center frequency can be seen to be

$$f_c = F_N/(2N). \tag{10.42}$$

Thus both the center frequency and the bandwidth can be controlled independently by means of external clocks (F_M and F_N) or by setting divider ratios (M and N).

10.6 SUMMARY

There are many other circuits and software functions to which these principles can be applied. Some types of PLLs require further considerations (e.g., control of jitter) that are peculiar to those types, but this chapter has illustrated how the theory that we have studied can be used to obtain a basic understanding of diverse circuits that employ the common properties of phase-locked loops.

10.A APPENDIX: EXACT ANALYSIS OF A SPECIAL-CASE THIRD-ORDER LOOP

In this appendix we will analyze a special case of a third-order loop that is particularly valuable when a type 2 loop is desired, but an additional significant high-frequency pole is also needed. As suggested in Section 10.1.1, this is often the case for frequency synthesizers. Two possible loop filters for this third-order loop are shown in Fig. 10.A.1.

The transfer function for either filter is

$$-K_{LF} F(s) = \frac{1}{R_1 C_1} \frac{1 + \dfrac{s}{\omega_z}}{s\left(1 + \dfrac{s}{\omega_p}\right)}. \tag{10.A.1}$$

For Fig. 10.A.1a, the zero and pole are at

$$\omega_z = \frac{1}{R_2 C_1} \tag{10.A.2}$$

and

$$\omega_p = \frac{1}{R_{p2} C_{p2}} \tag{10.A.3}$$

while for Fig. 10.A.1b, they are at

$$\omega_z = \frac{1}{R_2(C_1 + C_2)} \tag{10.A.4}$$

and

$$\omega_p = \frac{1}{R_2 C_2}. \tag{10.A.5}$$

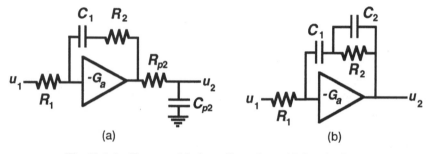

Fig. 10.A.1   Two possible loop filters for a third-order loop.

A loop in which unity open-loop gain occurs midway between the filter zero and the second (finite) pole on a log plot is of special significance. As will be shown, it gives maximum phase margin for any given ratio of pole to zero frequency. Of course, we could increase the phase margin by moving the pole frequency higher or the zero frequency lower. However, there are good reasons to not do either. Increasing the pole frequency decreases the rejection of undesired outputs from the phase detector; that is particularly important in frequency synthesizers. Decreasing the zero frequency can have several undesired effects.

To lower the zero frequency, a larger capacitor is required in the loop filter and that can present a problem if space is limited. In addition, lowering the zero frequency, while holding ω_L constant, decreases the low-frequency gain of the loop and thus the suppression of VCO noise at those frequencies (see Section 12.2). We also can expect slower response at lower zero frequencies. We can see this trend in the type-2 loop that results from ignoring the pole at ω_p. For example, from Fig. 6.6, the $\zeta = 4$ curve reaches a relative error of $10^{-3.4}$ at $t = 30/\omega_n$ whereas the $\zeta = 2$ curve reaches the same level at $t = 20.2/\omega_n$. Using Eqs. (6.14) and (4.16), we find that, for approximately the same ω_L, an ω_z decrease of 4 has produced a time increase of 3. Thus we would often like to have the zero occur as high in occur as high in frequency as possible. Therefore a good compromise can often be obtained by placing unity open-loop gain at the point of maximum phase margin, midway (on a log scale) between the zero frequency and the second pole frequency, as shown in Fig. 10.A.2.

We define r to be the ratio

$$r \triangleq \frac{\omega_p}{\omega_L} = \frac{\omega_L}{\omega_z}. \tag{10.A.6}$$

Figure 10.A.3 shows phase margin versus r. Zero phase margin occurs at $r = 1$, where the zero and pole are coincident, since they then cancel leaving

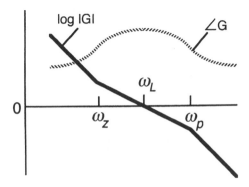

Fig. 10.A.2   Gain and phase of a type 2, third-order loop.

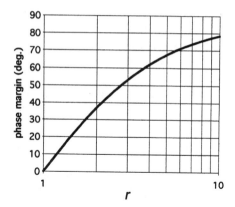

Fig. 10.A.3   Phase margin vs. r.

a pure integrator. Fortunately, we can obtain the transient response for this loop with reasonable ease.

10.A.1   Loop Response

The open-loop transfer function

$$G(s) = \frac{\omega_L^2}{r} \frac{1 + \dfrac{s}{\omega_L/r}}{s^2[1 + s/(\omega_L r)]} = \omega_L^2 r \frac{s + \omega_L/r}{s^2(s + \omega_L r)} \qquad (10.A.7)$$

has two poles at zero frequency and a pole and a zero equally spaced, on a log scale, about the unity-gain frequency ω_L, as desired. The corresponding closed-loop transfer function is

$$H(s) = \frac{G(s)}{1 + G(s)} = \omega_L^2 r \frac{s + \omega_L/r}{s^3 + s^2\omega_L r + s\omega_L^2 r + \omega_L^3} \qquad (10.A.8)$$

We can factor $(s + \omega_L)$ out of the denominator,[4] giving

$$\frac{G(s)}{1 + G(s)} = \omega_L^2 r \frac{s + \omega_L/r}{(s + \omega_L)[s^2 + \omega_L(r - 1)s + \omega_L^2]}. \qquad (10.A.9)$$

This can be written as a partial-fraction expansion [Truxal, 1955, p. 14],

$$\frac{G(s)}{1 + G(s)} = \frac{a}{s + \omega_L} + f(s), \qquad (10.A.10)$$

---

[4]This is shown by Encinas (1993), pp. 30–41.

where a and $f(s)$ are to be determined. We can solve for a using a common method wherein we multiply both sides by $(s + \omega_L)$ and then set s equal to $-\omega_L$, giving

$$a = \omega_L^2 r \frac{-\omega_L + \omega_L/r}{\omega_L^2 + \omega_L^2(1 - r) + \omega_L^2} = \omega_L \frac{r - 1}{r - 3}. \qquad (10.A.11)$$

Therefore, (10.A.10) can be written

$$\frac{G(s)}{1 + G(s)} = \omega_L \frac{r - 1}{r - 3} \frac{1}{s + \omega_L} + f(s). \qquad (10.A.12)$$

By equating (10.A.12) to (10.A.9) we can obtain

$$f(s) = \frac{\omega_L^2 r(s + \omega_L/r) - \omega(r - 1/r - 3)\left[s^2 + \omega_L(r - 1)s + \omega_L^2\right]}{(s + \omega_L)\left[s^2 + \omega_L(r - 1)s + \omega_L^2\right]}$$

$$(10.A.13)$$

$$= -\frac{\omega_L}{r - 3} \frac{(r - 1)s^2 + (r + 1)\omega_L s + 2\omega_L^2}{(s + \omega_L)\left[s^2 + \omega_L(r - 1)s + \omega_L^2\right]} \qquad (10.A.14)$$

$$= -\frac{\omega_L}{r - 3} \frac{(r - 1)s + 2\omega_L}{s^2 + \omega_L(r - 1)s + \omega_L^2}. \qquad (10.A.15)$$

This part of the response is second order with standard parameters

$$\omega_n = \omega_L \triangleq \omega_{n2} \qquad (10.A.16)$$

and

$$\zeta = (r - 1)/2 \triangleq \zeta_2. \qquad (10.A.17)$$

We add the subscript 2 to emphasize that these parameters apply only to the part of the response represented by $f(s)$. The overall response [as in Eq. (10.A.8)] is third-order and therefore such second-order parameters do not apply to it. After inserting (10.A.15) into (10.A.12), the closed-loop transfer function can be written in terms of these parameters as

$$H(s) = \frac{G(s)}{1 + G(s)} = \frac{\omega_{n2}}{\zeta_2 - 1}\left[\frac{\zeta_2}{s + \omega_{n2}} - \frac{\zeta_2 s + \omega_{n2}}{s^2 + \omega_{n2}(2\zeta_2)s + \omega_{n2}^2}\right].$$

$$(10.A.18)$$

We write this as the sum of two responses,

$$H(s) = \frac{\zeta_2}{\zeta_2 - 1} H_1(s) - \frac{1}{\zeta_2 - 1} H_2(s), \qquad (10.A.19)$$

where

$$H_1(s) = \frac{\omega_{n2}}{s + \omega_n 2} \qquad (10.A.20)$$

is the response of a first-order loop (or of a low-pass filter—see Section 2.2.1 or Table 2.1) and

$$H_2(s) = \omega_{n2} \frac{\zeta_2 s + \omega_{n2}}{s^2 + \omega_{n2}(2\zeta_2)s + \omega_{n2}^2} \qquad (10.A.21)$$

is the response of a second-order loop with $\alpha = 0.5$ [see Eq. (6.4)]. The responses, for various types of inputs (e.g., step, ramp), to this third-order loop can thus be obtained by combining the responses for the two parts, which are available in Table 2.1 and in Chapter 6.

The corresponding error response is

$$1 - H(s) = \frac{1}{1 + G(s)} = \frac{\zeta_2}{\zeta_2 - 1}[1 - H_1(s)] - \frac{1}{\zeta_2 - 1}[1 - H_2(s)],$$

$$(10.A.22)$$

where

$$1 - H_1(s) = \frac{s}{s + \omega_{n2}} \qquad (10.A.23)$$

and

$$1 - H_2(s) = \frac{s(s + \omega_{n2}\zeta_2)}{s^2 + \omega_{n2}(2\zeta_2)s + \omega_{n2}^2}. \qquad (10.A.24)$$

This is the error response of a first-order loop [or a high-pass filter—see Eq. (2.12a)] plus the error response of a second-order loop with $\alpha = 0.5$. Again, this information is available in Table 2.1 and in Chapter 6. Some of the third-order responses that are more likely to be of value in synthesizer design are provided below.

10.A.2   Final Values

Combining Eq. (10.A.22) to (10.A.24) we can obtain

$$1 - H(s) = \frac{s^2[s + \omega_{n2}(2\zeta_2 + 1)]}{(s + \omega_{n2})(s^2 + 2\zeta_2\omega_{n2}s + \omega_{n2}^2)} \qquad (10.A.25)$$

with a limit for small s of

$$\lim_{s \to 0}[1 - H(s)] = \frac{2\zeta_2 + 1}{\omega_{n2}^2}s^2 = \frac{r}{\omega_{n2}^2}s^2. \qquad (10.A.26)$$

By the final value theorem, when the loop is driven by $1/s^n$, the error response approaches a final value of

$$\lim_{s \to 0} \frac{s[1 - H(s)]}{s^n} = \frac{r}{\omega_{n2}^2} s^{3-n}. \tag{10.A.27}$$

This indicates that the phase error resulting from a phase step ($n = 1$) or a frequency step ($n = 2$) will settle to zero, and the phase error will not reach steady state for $n > 3$. A frequency ramp ($n = 3$) will produce a steady-state phase error of r/ω_L^2 which is also $1/(\omega_z \omega_L)$, the same phase error that occurs in a second-order type 2 loop with $\omega_z \leq \omega_L$ (as is true here). This is not surprising since one way to realize the third-order loop is to follow the integrator-and-lead of Fig. 3.23c with a passive low-pass filter, as in Fig. 10.A.1. The output of a low-pass filter driven by a ramp is another ramp with the same slope, offset from the input by a constant voltage.

10.A.3   Triple Roots

For $\zeta_2 = 1$ ($r = 3$), Eq. (10.A.25) becomes

$$[1 - H(s)]_{\zeta=1} = \frac{s^2(s + 3\omega_n)}{(s + \omega_n)^3}. \tag{10.A.28}$$

The step response can be obtained by partial fraction expansion:

$$\frac{1}{s}[1 - H(s)]_{\zeta=1} = \frac{s(s + 3\omega_n)}{(s + \omega_n)^3} = \frac{1}{(s + \omega_n)} + \frac{\omega_n}{(s + \omega_n)^2} - \frac{2\omega_n^2}{(s + \omega_n)^3}. \tag{10.A.29}$$

From this the time response to a unit step can be written in terms of normalized time,

$$x \equiv \omega_n t, \tag{10.A.30}$$

as

$$Y_{u, \text{error}}(x) = e^{-x}(1 + x - x^2). \tag{10.A.31}$$

10.A.4   Step Response

The step response corresponding to Eq. (10.A.24) is given by Eq. (6.9) or (6.11) with $\alpha = 0.5$. [Eq. (6.10) corresponds to the triple-root case that is covered by Eq. (10.A.31)]. With this we combine, as in Eq. (10.A.22), the high-pass response corresponding to (10.A.23), to obtain

$$Y_{u,\,error}(t) = \frac{\zeta_2}{\zeta_2 - 1}e^{-\omega_L t} - \frac{1}{\zeta_2 - 1}e^{-\zeta_2 \omega_L t}\left[\cosh\left(\omega_L t\sqrt{\zeta_2^2 - 1}\right)\right]$$

(10.A.32a)

$$= \frac{r - 1}{r - 3}e^{-\omega_L t} - \frac{2}{r - 3}e^{-[(r-1)/2]\omega_L t}\left[\cosh\left(\omega_L t\sqrt{\left(\frac{r - 1}{2}\right)^2 - 1}\right)\right]$$

(10.A.32b)

for $\zeta_2 > 1 (r > 3)$, or

$$Y_{u,\,error}(t) = \frac{\zeta_2}{\zeta_2 - 1}e^{-\omega_L t} - \frac{1}{\zeta_2 - 1}e^{-\zeta_2 \omega_L t}\left[\cos\left(\omega_L t\sqrt{1 - \zeta_2^2}\right)\right] \quad (10.A.33a)$$

$$= \frac{r - 1}{r - 3}e^{-\omega_L t} - \frac{2}{r - 3}e^{-[(r-1)/2]\omega_L t}\left[\cos\left(\omega_L t\sqrt{1 - \left(\frac{r - 1}{2}\right)^2}\right)\right]$$

(10.A.33b)

for $\zeta_2 < 1 (r < 3)$.

These step responses are shown in Fig. 10.A.4. The envelope for large times is given in Fig. 10.A.5. Figure 10.A.6 gives the error response to a ramp, which can be used to compute phase error during a phase ramp, that is, a frequency step. The ramp response was obtained by numerically integrating the step response.

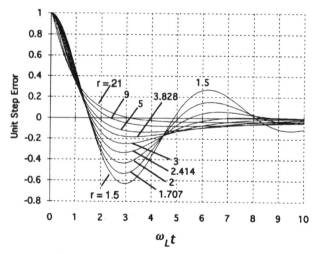

Fig. 10.A.4 Error response to unit step, third-order loop with ω_z, ω_p symmetrical about ω_L.

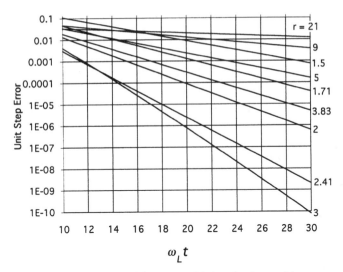

Fig. 10.A.5 Log of the envelope of the error, third-order loop with ω_z, ω_p symmetrical about ω_L.

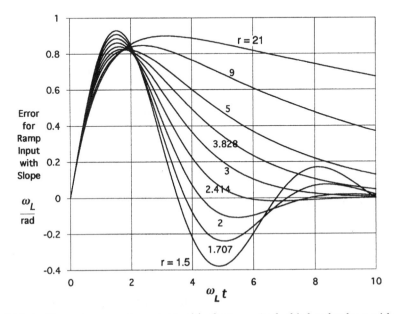

Fig. 10.A.6 Error response to a ramp with slope ω_L/rad, third-order loop with ω_z, ω_p symmetrical about ω_L.

Example 10.A.1

Problem: What is the frequency error 40 μsec after a commanded frequency change of 2 MHz if the loop has unity gain at 10 kHz and a loop filter that is an integrator plus a zero at 4142 Hz and a pole at 24.14 kHz? What is the error at 400 μsec? Use the method of combining first- and second-order responses and compare results to those given by Fig. 10.A.4 or 10.A.5. (This will verify the equivalence and demonstrate how other third-order responses, for example, to parabolic inputs, could be obtained.)

Solution: The ratio of f_p to f_L is $r = 2.414$, which is also the ratio of f_L to f_z (which is, of course, the way we designed the loop). From Eq. (10.A.16), $f_{n2} = f_L = 10$ kHz. From (10.A.17), $\zeta_2 = 0.707$. The error response from Eq. (10.A.22) is

$$[1 - H(s)] = 2 \text{ MHz}\{-2.413 [1 - H_1(s)] + 3.413 [1 - H_2(s)]\},$$

so the related time functions are

$$f_{\text{error}}(t) = 2 \text{ MHz}\{-2.413 f_{\text{error},1}(t) + 3.413 f_{\text{error},2}(t)\}.$$

$1 - H_1(s)$ from Eq. (10.A.23) is the impulse response of a high-pass filter. At

$$\omega_n t = 2\pi 10^4(40 \times 10^{-6}) = 2.51$$

this first-order part of the response to a 2-MHz step is

$$f_{\text{error},1} = 2 \text{ MHz}\{-2.413 \exp(-2.51)\} = -0.395 \text{ MHz}.$$

The second-order part of the response can be found from the upper and lower curves in Fig. 6.3 at $\omega_n t = 2.51$ (but we will use corresponding equations for better accuracy). The error response to a unit step for $\alpha = 1$ and $\zeta = 0.707$ is 0.0132 and for $\alpha = 0$ is -0.200 so the error response, for $\alpha = 0.5$, is the average of these or -0.034. The second part of the response thus equals

$$f_{\text{error},2} = 2 \text{ MHz}\{3.413(-0.034)\} = -0.235 \text{ MHz}$$

at 40 μsec. The sum of the two parts is

$$f_{\text{error}} = -0.395 \text{ MHz} - 0.235 \text{ MHz} = \underline{-0.63 \text{ MHz}}.$$

Figure 10.A.4 gives an error for a unit step of -0.314 at $\omega_{n2} t = 2.51$. Multiplying by the 2-MHz excitation we again get -0.63 MHz.

If we repeat the process at 400 μsec, the first-order part of the error after a unit step is -3×10^{-11}. From Fig. 6.5, the envelope of the error response to a unit step at $\omega_n t = 25.1$, when $\zeta = 0.707$ and $\alpha = 0.5$, is $10^{-7.7} = 1.96 \times 10^{-8}$. Multiplying by 3.413 we obtain 6.7×10^{-8}, which is obviously much larger than the first-order part and essentially equals the total error envelop. This is also the answer obtained from Fig. 10.A.5. Multiplying by 2 MHz, we obtain an error

$$f_{\text{error}} (400 \ \mu\text{sec}) \leq 0.13 \ \text{Hz}.$$

Example 10.A.2

Problem: What is the peak phase error in a frequency synthesizer that steps from 120 to 150 MHz if $r = 2.4$, unity open-loop gain occurs at 5 kHz, and the reference frequency is 100 kHz?

Solution: Refer to Figs. 10.1 and 10.2. During the transient, $N = 150$ MHz/100 kHz $= 1500$ so the frequency step at the output of the "VCO" is $\Delta f = d\varphi/dt = (150 \ \text{MHz} - 120 \ \text{MHz})/1500 = 20 \ \text{kHz} = 2 \times 10^4$ cycles/sec.
By Fig. 10.A.6, a ramp with slope ω_L/rad will produce a peak phase error of 0.84 for $r = 2.4$. For this loop, $\omega_L = 5 \times 10^3$ Hz (2π rad/cycle) $= 31,400$ rad/sec. Thus a phase ramp of $\omega_L/\text{rad} = 31,400/\text{sec}$ produces a peak phase error of 0.84. We have a step of 2×10^4 cycles/sec, so it will produce a peak phase error of

$$\varphi_{\text{peak}} = 0.84 \ (2 \times 10^4 \ \text{cycles/sec})/(31,400/\text{sec}) = 0.54 \ \text{cycle}.$$

This is larger than the linear range of a sinusoidal PD but is within the range of some types of PDs that are typically used in frequency synthesizers.

10.A.5 Modulation Response

As in Chapter 7, we will write the frequency response in terms of a normalized variable. Here we will use

$$\Omega_L \triangleq \omega/\omega_L \tag{10.A.34}$$

and the parameter for the individual curves will be r rather than ζ. Equation (10.A.7) can be written in these terms as

$$G(\Omega_L) = -\frac{j\Omega_L r + 1}{\Omega_L^2 (j\Omega_L + r)}, \tag{10.A.35}$$

leading to expressions for gain and phase:

$$|G(\Omega_L)|^2 = \frac{\Omega_L^2 r^2 + 1}{\Omega_L^4 (\Omega_L^2 + r^2)}, \tag{10.A.36}$$

$$\angle G(\Omega_L) = \tan^{-1}(\Omega_L r) - \tan^{-1}(\Omega_L / r) - 180°. \tag{10.A.37}$$

These are shown in Fig. 10.A.7, where the optimum nature of this function, in terms of phase margin, is apparent.

Similarly, the closed-loop responses can be developed from Eq. (10.A.8):

$$H(\Omega_L) = \frac{j\Omega_L r + 1}{-j\Omega_L^3 - \Omega_L^2 r + j\Omega_L r + 1}, \tag{10.A.38}$$

$$|H(\Omega_L)|^2 = \frac{(\Omega_L r)^2 + 1}{\Omega_L^2 (\Omega_L^2 - r)^2 + (\Omega_L^2 r - 1)^2}, \tag{10.A.39}$$

$$\angle H(\Omega_L) = \tan^{-1}(\Omega_L r) - \tan^{-1}\left[\frac{\Omega_L(\Omega_L^2 - r)}{\Omega_L^2 r - 1}\right]. \tag{10.A.40}$$

These relationships are shown in Fig. 10.A.8.

The magnitude and phase of the error response can be obtained from Eq. (10.A.38):

$$1 - H(\Omega_L) = \frac{j\Omega_L^3 + \Omega_L^2 r}{+j\Omega_L^3 + \Omega_L^2 r - j\Omega_L r - 1}, \tag{10.A.41}$$

$$|1 - H(\Omega_L)|^2 = \frac{\Omega_L^4 (\Omega_L^2 + r^2)}{\Omega_L^2 (\Omega_L^2 - r)^2 + (\Omega_L^2 r - 1)^2}, \tag{10.A.42}$$

$$\angle[1 - H(\Omega_L)] = \tan^{-1}\left(\frac{\Omega_L}{r}\right) - \tan^{-1}\left[\frac{\Omega_L(\Omega_L^2 - r)}{\Omega_L^2 r - 1}\right]. \tag{10.A.43}$$

In the second term in Eqs. (10.A.40) and (10.A.43), the ambiguity in the quadrant of the $\tan^{-1}$ must be solved so the value of the $\tan^{-1}$ is always increasing.

We can easily show, by comparing Eqs. (10.A.39) and (10.A.42) that

$$|1 - H(\Omega_L)|^2 = \left|H\left(\frac{1}{\Omega_L}\right)\right|^2. \tag{10.A.44}$$

Thus we can use Fig. 10.A.8 to give the error also by taking the value from a

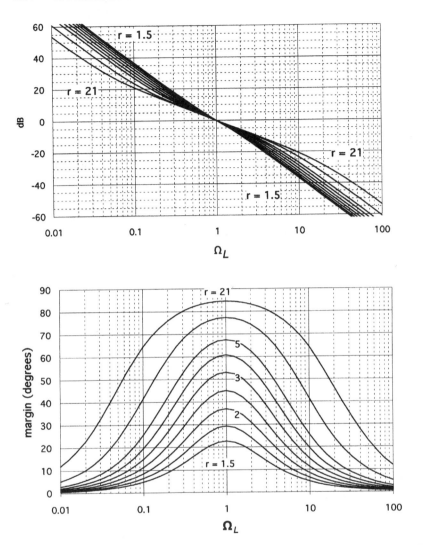

Fig. 10.A.7 Open-loop gain and phase for a third-order loop with ω_z, ω_p symmetrical about ω_L. $r \triangleq \omega_p/\omega_L = 1.5, 1.707, 2, 2.414, 3, 3.828, 5, 9,$ and 21.

frequency equal to the reciprocal of the frequency of interest. (For example, for $r = 2$, $\Omega_L = 0.5$, Fig. 10.A.8 shows $|H| \approx 2.7$ dB. This will also be the value of $|1 - H|$ at $\Omega_L = 2$.)

Similarly, we obtain

$$\angle H(\Omega_L) = -\angle[1 - H(1/\Omega_L)]. \tag{10.A.45}$$

Thus we can obtain the phase of $1 - H(\Omega_L)$ by reading the phase of $H(1/\Omega_L)$ from Fig. 10.A.8.

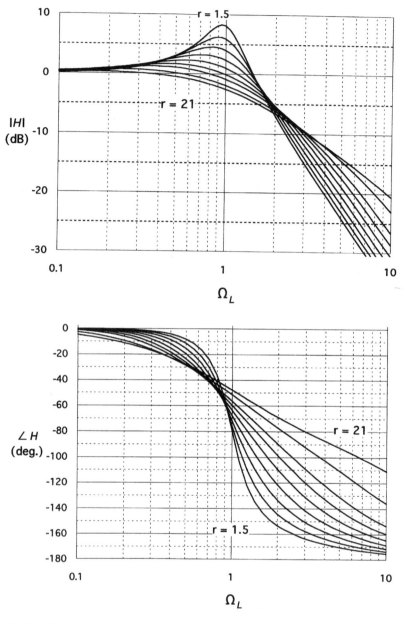

Fig. 10.A.8 Magnitude and phase of $H(\Omega)$ for a third-order loop with ω_z, ω_p symmetrical about ω_L. $r \triangleq \omega_p/\omega_L = 1.5$, 1.707, 2, 2.414, 3, 3.828, 5, 9, and 21.

10.B APPENDIX: COSTAS LOOP FOR *N* PHASES

A general representation of a loop for carrier recovery with an n-phase signal is shown in Fig. 10.B.1. The biphase Costas loop of Fig. 10.7 is a special case of this loop.

The number of multiplications of the incoming signal and the VCO output is n, the number of equally spaced phases (usually 2, 4, or 8) in the code, and the phase shifts given to the n VCO outputs are multiples of

$$\theta = \pi/n. \tag{10.B.1}$$

After low passing, all of the multiplier outputs are multiplied together to produce v_x, as shown. The derivative of v_x with respect to the phase difference is

$$\frac{dv_x}{d\Delta\varphi} = -\left[\prod_{i=0}^{n-1} \cos(\Delta\varphi + i\theta)\right] \sum_{i=0}^{n-1} \frac{\sin(\Delta\varphi + i\theta)}{\cos(\Delta\varphi + i\theta)}, \tag{10.B.2}$$

where, as before,

$$\Delta\varphi \equiv \varphi_{in} - \varphi_v. \tag{10.B.3}$$

In order to concentrate on regions near the zero crossings of v_x, we define α as the phase change from such a point, where $i = k$,

$$\alpha = \Delta\varphi + k\theta - \frac{\pi}{2}. \tag{10.B.4}$$

Then the derivative near such a point is

$$\frac{dv_x}{d\Delta\varphi}\bigg|_{\alpha\approx 0} \approx \left[\prod_{i=0}^{n-1} \cos\left(\alpha - k\theta + \frac{\pi}{2} + i\theta\right)\right] \frac{\cos\alpha}{\sin\alpha} \tag{10.B.5}$$

$$\approx (-1)^n \left[\prod_{j=-k}^{n-1-k} \sin(\alpha + j\theta)\right] \frac{1}{\sin\alpha}. \tag{10.B.6}$$

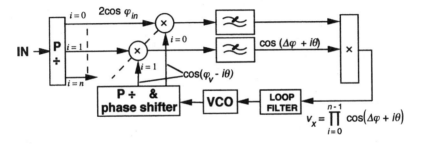

Fig. 10.B.1 *N*-phase Costas loop.

The approximation of Eq. (10.B.5) is valid because, near $\alpha = 0$, one of the terms in the summation of (10.B.2) approaches infinity; that is, the term for which the denominator is $\sin \alpha$. Even though this term is much larger than the other terms in the summation, v_x does not approach infinity because the denominator term is identical to a term within the preceding product of terms. Normally n is some power of 2 $(2, 4, 8, 16)$, so the first factor in (10.B.6) is then equal to 1. Therefore, we will drop $(-1)^n$ from Eq. (10.B.6) at this point.

Note that, if $k = 0$, the first term in the product in Eq. (10.B.6) cancels the denominator on the right and the expression can then be written

$$\left.\frac{dv_x}{d\Delta\varphi}\right|_{\alpha \approx 0} \approx \prod_{j=1}^{n-1} \sin(\alpha + j\theta). \tag{10.B.7}$$

If $k = 1$, the second term in the product in (10.B.6) cancels the right term, producing a form similar to (10.B.7) except $\sin[\alpha + (n-1)\theta]$ is replaced by $\sin[\alpha - \theta]$, equivalent to a phase change of $n\theta = \pi$ in that term and resulting in the expression being multiplied by -1. This occurs each time we advance k by 1 so we can write

$$\left.\frac{dv_x}{d\Delta\varphi}\right|_{\alpha \approx 0} \approx (-1)^k \left[\prod_{j=1}^{n-1} \sin(\alpha + j\theta)\right] = \pm C_x. \tag{10.B.8}$$

Thus the same sign occurs at the zero crossing whenever k is advanced by 2, that is, for a change in the lock point of $2\theta = 2\pi/n$. For biphase, as we saw in Section 10.3, this is every 180°. For quadraphase there is a lock point each 90°, and there is a lock point each 45° for eight phase. Thus changes between the allowed phases in a phase-coded signal will not change the tuning voltage, and we can lock to the carrier in spite of the lack of a spectral line there.

Just as a decoded biphase signal is available from one of the PDs in Fig. 10.7 (or Fig. 10.B.1 when $n = 1$), so too can decoded QPSK be obtained from two of the four PDs in Fig. 10.B.1 when $n = 2$. These outputs are proportional to the projections of the four possible vectors in Fig. 10.8 on the horizontal and vertical axes, respectively. In both cases, which PD or PDs provide the decoded signal(s) is arbitrary, so provision must be made for selection or control of the proper output.

Values of C_x are easily computed from (10.B.8) and are shown in Table 10.B.1 for several values of n.

10.C APPENDIX: SYMBOL CLOCK RECOVERY

Once a digital signal has been demodulated and changed to binary levels, a clock is needed that is synchronized with the rate of change of the symbols.

TABLE 10.B.1. Factor in K_p for Several Values of n

n	C_x
2	1
4	0.5
8	0.066
16	0.00052

For a QPSK signal, for example, two binary streams will be generated and the values of the two streams at any given time represent a four-level signal. While each binary stream will change less often than the symbol rate, the basic (highest) frequency in each stream will equal the symbol frequency. A clock at this frequency is required for use in sampling the bit streams. Since the bit patterns of the streams are essentially random, there is no spectral line in these streams at the desired bit rate. As with carrier recovery, one can be created. A circuit for doing this is shown in Fig. 10.C.1.

The incoming bit stream is represented at in Fig. 10.C.1a. In the time domain, it consists of a series of data bits of unknown (random) value, usually 0 or 1 with equal probability. In the frequency domain, the power spectral density S has the shape characteristic of a single pulse of width T_b. There is no discrete spectral line since the stream is random. In order to develop a spectral line, the stream is delayed by T_d, approximately equal to half of T_b, and then enters an Exclusive-OR gate having an inverted output (ExNOR) with the undelayed version. (An analog multiplier can also be used.) During the delay period after the beginning of the pulse, the output from the ExNOR is the product of the two adjacent bits and is unknown (random).

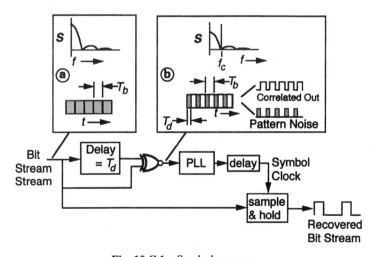

Fig. 10.C.1 Symbol recovery.

But, after T_d has passed, a one is produced since both inputs are from the same pulse. The resulting output at Fig. 10.C.1b can be decomposed into a pulse stream of frequency $1/T_b$ and a stream of random pulses. The random pulses produce "pattern noise" while the regular pulse stream is characterized by a discrete line at the bit rate. A PLL is employed to reproduce the bit frequency while filtering out the noise. A narrow filter could be employed to do this but would not be able to accommodate changes in bit rate. The PLL output, after an appropriate delay, can be used to drive the sample-and-hold circuit to sample the original pulse stream.

10.D   APPENDIX: ADPLL BY PULSE ADDITION AND REMOVAL, ADDITIONAL MATERIAL

10.D.1   Implementation of the Increment/Decrement Circuit

The $\div M$ to $\div 2$ interface is actually mechanized as shown in Fig. 10.D.1. When counting up, if the counter exceeds its modulus (capacity), it outputs a carry signal, which is an increment command to the $\div 2$. Conversely, when counting down, it can output a borrow signal, which is a decrement input.

10.D.2   Stability

Since the analog first-order loop is inherently stable, we look for stability problems due to the sampling effect that has been added by the digital implementation, such as we found with the simple loop of Section 10.5.1.

We can identify three sampling processes: clocking at frequencies F_M and F_N and phase detection. Phase detection with an ExOR gate can be considered a sampling process because phase information is only acquired at the transitions of the output waveform (twice a cycle); no phase information is available between transitions. The sampling process at frequency f_s will not

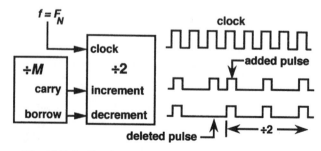

Fig. 10.D.1   Implementation of increment/decrement.

appreciably affect stability if

$$f_s \gg \omega_L = K. \tag{10.D.1}$$

From (10.40), this requires that

$$MN \gg 2, \tag{10.D.2}$$

in order that $F_M \gg \omega_L$, and that

$$MN \gg 2F_M/F_N, \tag{10.D.3}$$

in order that $F_N \gg \omega_L$ cycle/rad.

From (10.40) to (10.42), the lowest input sampling frequency is

$$2f_{in} \geq 2(f_c - F_H) = \frac{F_N}{N} - \frac{F_M}{MN} = \frac{\omega_L}{2}\left(\frac{F_N}{F_M}M - 1\right). \tag{10.D.4}$$

This is much greater than ω_L if

$$M \gg 2F_M/F_N. \tag{10.D.5}$$

Thus Eqs. (10.D.2), (10.D.3), and (10.D.5) are sufficient to guarantee stability and, since M and N are usually large numbers (to reduce ripple), these conditions should be easily met if F_M is not too much greater than F_N.

10.D.3 Ripple Control

Ripple refers to jitter on the loop output due to variations in instantaneous frequency. Although the loop controls average frequency, it does so by removing pulses so, depending on the sequence in which this is done, different cycles may be of different lengths, even though the average frequency is correct. The loop can be designed so it will not produce a jitter of more than $1/N$ cycles at the output.

Assume initially that the output is at f_c so the PD duty factor is 50%, in the center of its range. The time during which the PD output is at a single state is one quarter of an output cycle (Fig. 3.2b), so, using (10.42),

$$T_{PD} = 1/(4f_c) = N/(2F_N). \tag{10.D.6}$$

We can prevent more than one increment or decrement pulse from occurring at I during this period if we make the duration of the M count at least as long as T_{PD}. Thus we require

$$M/F_M \geq T_{PD}. \tag{10.D.7}$$

Combining these last two equations, we obtain

$$M \geq NF_M/(2F_N). \tag{10.D.8}$$

If we meet this restriction, we limit the unnecessary outputs from the M divider to one in each direction during lock at center frequency. When the frequency is offset from center, increments or decrements are required, but (10.D.8) will limit the number of $\div M$ outputs that are in the "wrong" direction to one because the width of the PD output during the time such outputs are generated is narrower than what is given by Eq. (10.D.6) and therefore Eq. (10.D.7) is still met by that period.

Additional circuitry can also be used for ripple control. The circuit in Fig. 10.D.2a inhibits up and down counting when the two inputs are in phase, the zero-error relationship for this circuit. As the phase moves away from this center value, up or down counting occurs in $\div M$, but for short times when the deviation from center is small. When $v(\varphi_{in})$ and $v(\varphi_{out})$ are in phase, the ExOR output is zero and the $\div M$ is not enabled. If $v(\varphi_{out})$ becomes delayed, as in Fig. 10.D.2b, the enable occurs when the second least significant bit of the counter is 1 and its inverse causes the $\div M$ to count up. Conversely, in the case shown in Fig. 10.D.2c, it counts down. However, it only counts during the enable pulse, which would be narrow near band center, as opposed to counting all the time with the original realization. This scheme, however, reduces K_p and the tuning range[5] by 2.

10.D.4   Second-Order (ADPLL) All-Digital Phase-Locked Loops

Second- and higher order loops are also possible using this general design. Figure 10.D.3 is another representation of the system of Fig. 10.25. This time we have shown F_N as a second input. Previously, we had ignored it because it was a constant. Now, however, we wish to use that point to inject a signal. We have redefined K_p and K_v to accommodate this; but K is not affected by the redefinition and so neither is the loop response.

10.D.4.1   *Transfer Function.* We will feed $f_{out}(s)$ through a second loop like this one and back to F_N to create a loop filter. We have added subscripts to Fig. 10.D.3 to differentiate this main loop (1) from the second loop that is used to generate the filter; F_N now changes from a constant clock to a function of s. The transfer function of the second loop is given by Eq. (2.25b) and we will multiply this by a constant, γ, giving

$$H_2(s) = \frac{\gamma}{1 + s/K_2}. \tag{10.D.9}$$

---

[5]The application note [Troha, 1994] indicates that the range is reduced by $(2 + 1/M)$, perhaps due to the details of how the enable and up/down functions work.

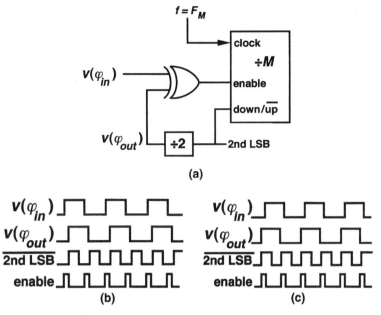

(a)

(b)   (c)

Fig. 10.D.2   Ripple inhibiting circuit.

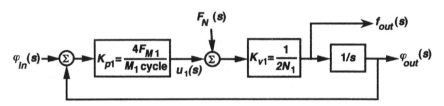

Fig. 10.D.3   Mathematical block diagram rearranged.

This will create a transfer function, between the PD output and f_{out}, of

$$K_{v1}K_{\text{LF}}F(s) = \frac{K_{v1}}{1 - K_{v1}H_2(s)} = K_{v1}\frac{1 + s/K_2}{1 - K_{v1}\gamma + s/K_2} \quad (10.D.10)$$

$$= \frac{K_{v1}}{1 - K_{v1}\gamma}\frac{1 + s/K_2}{1 + s/[(1 - K_{v1}\gamma)K_2]}. \quad (10.D.11)$$

Here we have used Eq. (2.22) with a minus sign on the feedback because the standard diagram has a minus at the summer, but this implementation does not. By the way Eq. (10.D.10) is written, we are effectively abstracting K_v from the response to use for VCO gain constant. From this, and our definition of K_{v1} in Fig. 10.D.3, we can see the following equivalencies to our standard parameters for a loop with a lag-lead filter:

$$K_{p1} = \frac{4F_{M1}}{M_1 \text{ cycle}}, \qquad (10.D.12)$$

$$K_{v1} = \frac{1}{2N_1}, \qquad (10.D.13)$$

$$\omega_z = K_2, \qquad (10.D.14)$$

$$\omega_p = K_2(1 - K_{v1}\gamma) = K_2\left(1 - \frac{\gamma}{2N_1}\right), \qquad (10.D.15)$$

$$K_{LF} = \frac{1}{1 - K_{v1}\gamma} = \frac{1}{1 - \gamma/(2N_1)}, \qquad (10.D.16)$$

$$K_1 = K_{p1}K_{LF}K_{v1} = \frac{2F_{M1}}{M_1 N_1(1 - K_{v1}\gamma) \text{ cycle}}$$

$$= \frac{2F_{M1}}{M_1 N_1(1 - \gamma/(2N_1)) \text{ cycle}}, \qquad (10.D.17)$$

$$\omega_n^2 = K_1\omega_p = \frac{2F_{M1}K_2}{M_1 N_1 \text{ cycle}}. \qquad (10.D.18)$$

Using K_1, ω_p, ω_z, and ω_n from above, ζ and α can be obtained from Eqs. (4.9) and (6.5). (That may be more efficient than expanding those expressions directly in terms of the circuit parameters.)

If we want an integrator-and-lead filter, we set $\gamma = 1/K_{v1} = 2N_1$ and obtain a filter transfer function, from (10.D.10), of

$$K'_{LF} F'(s) = \frac{1 + s/K_2}{s/K_2} = K_2\frac{1 + s/K_2}{s}, \qquad (10.D.19)$$

where primes are used to designate this particular choice of γ. The terms K_{v1} and ω_z are still given by (10.D.13) and (10.D.14). By comparing $K_{p1}K_{v1}K'_{LF}F'(s)/s$ to Eq. (4.23), we find

$$\omega_n^2 = K_2 K_{p1}K_{v1} = \frac{2F_{M1}K_2}{M_1 N_1 \text{ cycle}}, \qquad (10.D.20)$$

which is the same as Eq. (10.D.18). The damping factor is

$$\zeta = \frac{1}{2}\frac{\omega_n}{\omega_z} = \sqrt{\frac{F_{M1}}{2K_2M_1N_1 \text{ cycles}}}.$$ (10.D.21)

Responses for either type of loop (with lag-lead or integrator-and-lead filter) can be obtained from the equations in Table 4.1 or the graphs of Chapters 6 and 7 using the parameters computed above.

Notice, in Eq. (10.D.11), that, if γ should exceed $1/K_v$, the pole would move to the right half plane and the filter would be unstable. However, because we are using numbers rather than analog components, the value of γ can be set precisely so we can obtain an integrator-and-lead filter without drifting into the right half plane. In fact, here is the first time we have met a true type 2 loop, one that really does have a second pole at zero rather than at some frequency small enough to be approximated as zero.

10.D.4.2 *Realization.* The multiplying factor γ can be realized by taking the output from the second loop before the final $\div N$, effectively multiplying the output frequency by N_2, the value of N in loop 2. A $\div L$ circuit is also placed at the input to the second loop, giving $\gamma = N_2/L$. The final structure is shown in Fig. 10.D.4.

10.D.4.3 *Hold-in Range.* The center frequency depends on the center value of $F_{N1}(s)$, which depends on F_{N2}. From Fig. 10.D.4, we can see this to be

$$f_c = \frac{\overline{F_{N1}(s)}}{2N_1} = \frac{F_{N2}}{4N_1}.$$ (10.D.22)

The hold-in range could be set by either loop. From the expression for the output frequency (Fig. 10.D.4), a change in output frequency of Δf_{out} is accompanied by

$$\Delta f_1 = \Delta f_{\text{out}} 2N_1 - \Delta F_{N1} = \Delta f_{\text{out}} 2N_1 - \gamma \Delta f_{\text{out}} = \Delta f_{\text{out}} (2N_1 - \gamma).$$ (10.D.23)

The corresponding phase change in loop 1 is

$$\Delta\varphi_1 = \frac{\Delta f_1}{K_{p1}} = \frac{\Delta f_{\text{out}} M_1 (2N_1 - \gamma)}{4F_{M1} \text{ cycle}^{-1}}$$ (10.D.24)

so we can see immediately that, for a type 2 loop, where $\gamma = 2N_1$, there is no phase change at the main-loop (loop 1) phase detector, as we would expect for such a loop.

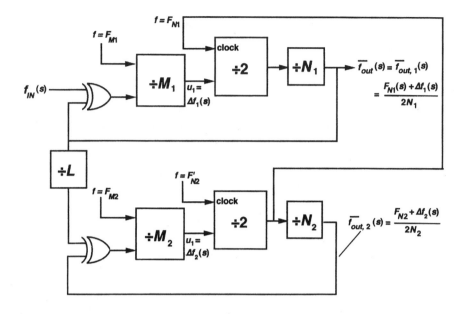

Fig. 10.D.4 Second-order ADPLL using pulse removal.

Referring again to Fig. 10.D.4,

$$\frac{\Delta f_2}{2N_2} = \frac{\Delta f_{\text{out}}}{L}, \tag{10.D.25}$$

$$\Delta \varphi_2 = \frac{\Delta f_2}{K_{p2}} = \frac{\Delta f_{\text{out}} M_2 N_2}{2 F_{M2} L} \text{ cycle}^{-1}. \tag{10.D.26}$$

Neither $\Delta \varphi_1$ nor $\Delta \varphi_2$ can exceed one-quarter cycle, leading to a maximum value of Δf_{out} given by the smaller of the frequency changes with $\Delta \varphi = \frac{1}{4}$ in Eqs. (10.D.24) and (10.D.26),

$$\Delta f_{\text{out}} \le \min \left(\frac{F_{M2} L}{2 M_2 N_2}, \frac{F_{M1}}{(2N_1 - \gamma) M_1} \right). \tag{10.D.27}$$

10.M APPENDIX: MATLAB SIMULATIONS

10.M.1 Higher Order Responses Using MATLAB

While it is difficult to solve for the transient response for third- and higher order loops using classical or transform techniques, we can easily obtain their responses using MATLAB. The program Ord3spcl will produce any of the curves in Fig. 10.A.4, 10.A.6, 10.A.7, and 10.A.8 and more. It uses Eq. (10.A.8) and the error response $1 - H$ derived from it. As before, the ramp error response is obtained by taking the step response of the integral of $1 - H$, and the integration is performed simply by shifting the elements of the numerator vector.

While Ord3spcl is instructive, in that we can see how MATLAB can be used to produce the same results that were obtained in Section 10.A, we would like to be able to get the transient and frequency responses for higher order loops without finding H. During the design process, we will usually work with the open-loop singularities and gain, so we would like to obtain closed-loop responses given the open-loop poles and zeros and gain. The program Open2cls permits us to do this. We begin by specifying column vectors containing the coefficients of power of s (descending order, as before) for the open-loop poles and zeros. Then we use a MATLAB function zp2tf to convert these vectors to the numerator and denominator vectors we have used before—except these are for the open-loop transfer function G. However, we can easily obtain the closed-loop H and $1 - H$, from them as follows:

$$H = \frac{\text{out}}{\text{den}} = \frac{G}{1 + G} = \frac{\text{nopen} / \text{dopen}}{1 + \text{nopen} / \text{dopen}} = \frac{\text{nopen}}{\text{dopen} + \text{nopen}}, \quad (10.\text{M}.1)$$

$$1 - H = 1 - \frac{\text{nopen}}{\text{dopen} + \text{nopen}} + \frac{\text{dopen}}{\text{dopen} + \text{nopen}}, \quad (10.\text{M}.2)$$

where

$$G = \frac{\text{nopen}}{\text{dopen}}. \quad (10.\text{M}.3)$$

As a starting point, Ord3spcl has been written to apply most directly to the special third-order loop described in Section 10.A, and a gain value is employed that places unity gain at the geometric mean of the zero and added pole, as in Section 10.A. We can interactively chose r and then can apply a multiplier to the gain to shift it relative to the nominal value. The resulting closed-loop time and frequency responses can be plotted as can the open-loop response. By varying the gain multiplier we can observe the effects of inaccuracies in setting up the optimum loop described in Section 10.A.

As before, we shift the numerator vector elements of $1 - H$ to the right to obtain the ramp response but, since we do not have the numerator explicitly (having derived it from G), we use matrix multiplication to accomplish the shift.

Open2cls

```
%Closed-Loop Responses from Open Loop Parameters
%Reference is special 3rd Order as in Section 10.A.
%  Comments apply to that case in particular.
% Dynamically choose r and gain multiplier
% Modify program to add poles and zeros

r2deg = 180/pi; % for use in changing radians to degrees
% w -> Frequency points from 10^-1 to 10^1. Modify to shift
                        plots.
  w = logspace(-1,1,201);
% t -> Time points from 0 to 10. Modify to shift plots.
  t = linspace(0,10,200); % Time points

r = 1; kmult = 1; typgr = 'b';
while (r >= 0)
   WL = 1; % default unity-gain frequency
   % r is ratio of pole freq. to WL for nominal gain.
   % More generally it is the square root of the ratio of
                        added pole to zero.
   r0 = input('Enter freq. ratio, r=(zero/pole)^0.5, negative
                        to quit: ');
   if isempty(r0), % if user enters blank, old value retained
      disp(r);
   else
      r = r0;
   end % if
   if (r >= 0),
      z = [-1/r];            % open loop zero at WL/r
      p = [0;0;-r];          % open loop pole at WL*r plus two at
                              zero
      kmult1 = input('Enter gain multiplier: ');
            % ratio to K giving max phase margin
      if ~isempty(kmult1), kmult = kmult1; end;
      k = r*kmult % k = r for unity gain at geometric mean of
                        pole, zero
      [nopen, dopen] = zp2tf(z,p,k);  % open-loop num and den
                              from poles, zeros, k
      den = nopen + dopen;   % closed-loop denominator
      out = nopen;           % numerator for output step response
      er = dopen;            % numerator for error step response
      % integration of er by shifting right:
      rer = er*[0 1 0 0 ; 0 0 1 0 ; 0 0 0 1 ; 0 0 0 0];

      disp('');
      disp('b=Bode plot (open loop)');
      disp('o=Step Output');
      disp('e=Step Error');
```

```
    disp('r=Ramp Error');
    disp('f=Frequency Response');
    typgrl = input('Enter b, o, e, r, or f. ', 's');
    if ~isempty(typgrl), typgr = typgrl; end;
    clg;
    if strcmp(typgr,'b')
        bode(nopen,dopen); % Bode plot of open-loop transfer
                            function
    elseif strcmp(typgr,'e')
        step(er,den,t); % error response to a step
        grid;
    elseif strcmp(typgr,'o')
        step(out,den,t); % output response to a step
        grid;
    elseif strcmp(typgr,'r')
        step(rer,den,t); % error response to a ramp
        grid;
    else % 'f'
        hout = freqs(out, den, w);   % frequency response to
                            output
        magout = 20*log10(abs(hout));    % magnitude of hout
        phout = r2deg*angle(hout);   % phase of hout
        her = freqs(er, den, w);     % frequency response to
                            error
        mager = 20*log10(abs(her));  % magnitude of her
        pher = r2deg*angle(her); % phase of her
        subplot(211), semilogx(w, magout, 'r', w, mager,
                            '--r')
        grid
        title('Frequency Response Magnitude. solid = output
                            dash = error')
        xlabel('Omega / Omega-n')
        ylabel('dB')
        subplot(212), semilogx(w, phout, 'g', w, pher, '--g')
        grid
        title('Frequency Response Phase. solid = output  dash
                            = error')
        xlabel('Omega / Omega-n')
        ylabel('degrees')
    end  % if typgr
  end  % if r
end  % while
```

With appropriate modifications, we can use Open2cls to obtain responses corresponding to arbitrary open-loop configurations. Open2c4 is an example of such a modification to accommodate a fourth-order loop with two poles at zero frequency and two other real poles, pole1 and pole2, plus one zero. The finite poles are input interactively, as is the zero indirectly. The matrix that is used for integration in obtaining the ramp response is increased in size; a 5×5 matrix is required to shift the coefficients of s^0 through s^4. The time and frequency ranges have been set for a particular set of frequencies, to be given below, so the t and w vectors will require modification for different loop bandwidths.

Here is a suggestion for the beginning of an investigation of the properties of the kind of loop represented by Open2c4. Enter $r = 4$ followed by pole1 = 1000. (We are entering the negative of the pole frequencies for simplicity—for reasons of stability we certainly cannot have poles with positive real parts.) This will set the zero (= pole1 / r) to 250. (Frequency units are in radians second.) Enter pole2 = 4000 and a gain constant of 0 dB. Enter b for plot type and thus obtain a Bode plot. Note the frequency where excess phase shift is the least and the gain change necessary to give 0 dB there, thus giving maximum phase margin. Retain the same r, pole1 and pole2 as before by using the Return or Enter key in response to a request for value. Change the gain to the value that will give maximum phase margin and then check the Bode plot to ensure that unity gain occurs at minimum phase lag. At the same time note the gain margin (at $-180°$ phase). Then check the error response and observe the degree of overshoot. Now change the gain to leave about 3 to 5 dB of gain margin and observe how the overshoot changes.

Now change the pole2 frequency to 1000, placing the two poles on top of each other. Adjust gain for maximum phase margin again and compare the overshoot to what previously occurred at maximum phase margin.

Open2c4x is similar to Open2c4, but the zero frequency is entered explicitly. Step through the default values by pressing the return or enter key. You should get the same defaults as in Open2c4, which are just slightly different from what you entered above. Now try entering a pair of complex poles corresponding to a 2-pole (1 complex pair) Butterworth low-pass filter. For the first pole, enter $1414 + 1414*i$ and for the second enter $1414 - 1414*i$. Compare the resulting performance to the case with two poles at 2000 rad/sec. Both filters have the same high-frequency gain, so they will be equally desirable for that parameter where it is important (e.g., many frequency synthesizer applications). Compare Bode plots. Which can give the greatest phase margin?

10.M.2 Simulation of the ADPLL Using MATLAB

The program ADPLL simulates the loop in Fig. 10.18. It will plot N_{out} and show where the output was sampled, as a result of the input transition, to produce a new value of N_2'. It will also show instantaneous frequency versus time and the average frequency for the simulation. Try Example 10.1 using ADPLL. Use Phin = 0.5 and Tmaximum E-4 (10^{-4} seconds).[6] Set the initial values of Nout and N2 to their theoretical steady-state values, which will be displayed when the program is run and which should minimize the initial transient. Note how the sampled values of Nout vary above and below the theoretical steady-state value. Compare the average output frequency to the input frequency. You might want to try a longer simulation[7] to see if the difference is reduced.

Then use an offset of Noff = 100 with an input frequency of 600 kHz. (Use q = 4, nv = 12, Fclock = 8.192E6). Set the initial values of Nout and N2 to the given steady-state values. Now retain those initial values and halve the input frequency. (This simulates stepping the input to 300 kHz after a steady-state lock at 600 kHz.) You should obtain a false lock. The program will also be given a new set of initial values that correspond to the 300-kHz input frequency. Change to those values and note that a correct lock is obtained. You might want to experiment with how far off the initial conditions can be without producing false lock. How do the instantaneous frequency excursions about the average change with input frequency?

Note that the minimum frequency is given as 256 kHz, to keep $C < 2$ in Eq. (10.32) for stability. Set Tmaximum = 2E-4 and try an input frequency of 250 kHz, setting the initial conditions at steady-state values. Is the response stable? Is the average output frequency correct? Increase the input frequency to 260 kHz. (Leave the other conditions alone.) This meets our stability criterion but what do you see?—perhaps the results of quantization effects. Try 270 kHz. That should be better. You might repeat this with Noff = 0. This puts the sampled value of Nout closer to midrange and allows you to watch the instability grow.

---

[6] 6E-5 seconds if you are using MATLAB V.3.5.
[7] 5E-4 seconds is about the limit with MATLAB V.4.

PART 2

PHASE LOCK IN NOISE

CHAPTER 11

PHASE MODULATION BY NOISE

We have previously studied how the phase-locked loop (PLL) responds to sinusoidal phase modulation of the input signal. This chapter will again consider the response of the PLL to phase modulation but, this time, the modulation will be noiselike.

11.1 REPRESENTATION OF NOISE MODULATION

Figure 11.1a shows a noise waveform as a function of time. The random variable, y, could represent voltage v, current i, phase φ, or frequency f. Variable y is described by its probability density, Prob(y), which is a Gaussian function of y with zero mean (otherwise the mean is removed from y and considered to be a second variable) and with a variance, or mean square value, σ_y^2 (Fig. 11.1b). If $y = v$ or if $y = i$, and a 1-Ω system is assumed for convenience, σ_y^2 is power. Mean square phase σ_y^2 and mean square frequency σ_y^2 can also be called power by analogy, as we shall see.

If the noise waveform passes through a narrow filter (Fig. 11.2), the output waveform, as viewed on an oscilloscope, will appear to be a sine wave at the frequency of the filter passband. The narrower the filter, the smaller will be the spread, or fuzziness, of the waveform. Also, the narrower the filter, the more slowly will be the amplitude and phase change. This is because changes in the time domain imply modulation sidebands in the frequency domain. These are separated from the spectral center by the modulation frequency, and a narrow filter allows only a small spread of modulation sidebands and so inhibits high-frequency variations of amplitude or phase (frequency).

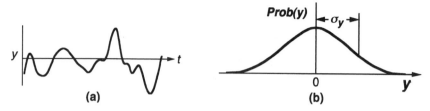

Fig. 11.1 Noise: (a) as a function of time and (b) probability density.

As the bandwidth is narrowed, the amplitude and phase of the signal in that band change ever more slowly. Thus, for any given duration of observation, a filter bandwidth can be chosen that is narrow enough that the amplitude and phase, both of which are unknown except by their statistics, remain essentially constant. Thus, the random waveform can be represented by a continuum of adjacent passbands (e.g., ideal rectangular contiguous passbands), which are sufficiently narrow that the power in each can be considered to represent a single sine wave. Since these sine waves have differing frequencies and unknown phase, we do not know how they combine. We know only that their mean square values, "powers," add; that is, values of σ^2 add, as is usual for random variables. This is an important point. We cannot add volts, amperes, radians, or hertz. We can add volts squared, amperes squared, radians squared, and hertz squared. We refer to these squared variables by the generalized name "powers."

If the center frequency of the filter in Fig. 11.2 is moved or scanned, as in a spectrum analyzer, the average power coming through the filter might be a function of the filter center frequency such as is illustrated in Fig. 11.3. If adjacent filters pass the same mean power, we expect twice as much power from a filter that covers the total band of two individual filters, three times the power from a filter that covered three times the band, and so forth. Thus, over any frequency range where the power is independent of the frequency, we can express the power in a filter bandwidth δf at frequency f' as

$$P_y(f') = \delta f S_y'(f'), \qquad (11.1)$$

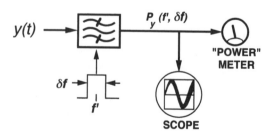

Fig. 11.2 Noise through a narrow filter.

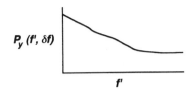

$P_y(f', \delta f)$

f'

Fig. 11.3 Frequency distribution of noise power in bandwidth δf vs. f.

where $S'_y(f')$ is a constant. But, unless $S'_y(f')$ *is* flat in the region of interest, its value will change[1] somewhat as δf changes. However, the relative change will be smaller as δf becomes smaller [barring discontinuities in $S'_y(f')$]. Thus, in the limit where δf becomes differential, $S'_y(f')$ becomes a constant $S_y(f')$ and is called the power spectral density of y at frequency f':

$$S_y(f') \equiv \lim_{\delta f \to 0} \frac{P_y(f')}{\delta f} = \lim_{\delta f \to 0} \frac{E[y^2(f' - \delta f/2 < f < f' + \delta f/2)]}{\delta f}. \quad (11.2)$$

Here $E[y^2(\ )]$ is the expected value of y^2 in the frequency band indicated. Then the power, or mean square value, of y, $E(y^2) = \sigma_y^2$, over any frequency range can be obtained by summing the differential contributions:

$$E(y^2)\big|_{f_1}^{f_2} = \int_{f_1}^{f_2} S_y(f)\, df. \quad (11.3)$$

If $y = \varphi$, then Eq. (11.3) gives the variance of phase represented by the phase power spectra density[2] (PPSD) $S_\varphi(f_m)$ over a range of modulation frequencies from $f_m = f_1$ to $f_m = f_2$,

$$\sigma_\varphi^2\big|_{f_1}^{f_2} = E(\varphi^2)\big|_{f_1}^{f_2} = \int_{f_1}^{f_2} S_\varphi(f_m)\, df_m. \quad (11.4)$$

We can describe frequency modulation by a frequency power spectral density in the same manner as is done for phase modulation. The relationship between frequency deviation at a frequency f_m and phase modulation at the same frequency [see Eq. (7.7)] leads directly to the relationship between the two densities at the same modulation frequency:

$$\frac{S_f(f_m)}{S_\varphi(f_m)} = \frac{S_f(f_m)\,\delta f_m}{S_\varphi(f_m)\,\delta f_m} = \frac{[\Delta f(f_m)]^2}{[\Delta \varphi(f_m)]^2} = f_m^2. \quad (11.5a)$$

[1]When $S_y \sim f$, there is no error due to large δf is S_y is taken in the middle of δf.
[2]The power spectral density $S_y(f)$ is also the Fourier transform of the autocorrelation function of $y(t)$.

We obtain the variance of frequency deviation in a manner similar to Eq. (11.4) and relate it to S_φ by Eq. (11.5a):

$$\sigma_f^2|_{f_1}^{f_2} = \mathbf{E}(f^2)|_{f_1}^{f_2} = \int_{f_1}^{f_2} S_f(f_m)\, df_m = \int_{f_1}^{f_2} f_m^2 S_\varphi(f_m)\, df_m. \quad (11.6a)$$

Similarly, in different units,

$$\frac{S_\omega(f_m)}{S_\varphi(f_m)} = \frac{S_\omega(\omega_m)}{S_\varphi(\omega_m)} = \omega_m^2 \quad (11.5b)$$

and

$$\sigma_\omega^2|_{f_1}^{f_2} = \int_{f_1}^{f_2} \omega_m^2 S_\varphi(f_m)\, df_m = \frac{1}{2\pi} \int_{\omega_1}^{\omega_2} \omega_m^2 S_\varphi(\omega_m)\, d\omega_m. \quad (11.6b)$$

Note that S_φ is in radians squared per hertz (rad^2/Hz) in each of the last four equations, which can be seen from consideration of the units.

11.2 PROCESSING OF NOISE MODULATION BY THE PHASE-LOCKED LOOP

The output and input phase can be related by Eq. (4.3) written for $s = j\omega$:

$$\varphi_{\text{out}}(\omega) = H(\omega)\, \varphi_{\text{in}}(\omega). \quad (11.7)$$

We write the mean squared value of the output phase as

$$\sigma^2[\varphi_{\text{out}}(\omega)] = \sigma^2[H(\omega)\varphi_{\text{in}}(\omega)] = |H(\omega)|^2 \sigma^2[\varphi_{\text{in}}(\omega)]. \quad (11.8)$$

Here $|H(\omega)|^2$ can be brought outside because it is not a function of time, the variable over which the squared value is being averaged.

Since noise modulation can be represented as modulation by an infinite set of sine waves, each at its own frequency f' and having power $S(f')\, df$, the response of the loop to each of these differential oscillations is the same as its response to a discrete sinusoidal modulation at the same frequency. Thus, in response to an input phase power spectral density $S_{\varphi,\text{in}}(f')$, the output phase power spectral density will be

$$S_{\varphi,\text{out}}(f') = S_{\varphi,\text{in}}(f')|H(f')|^2. \quad (11.9)$$

Similarly, the phase error $S_{\phi,e}(f')$ will be described by a phase power

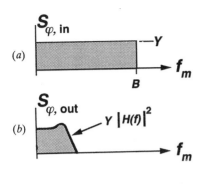

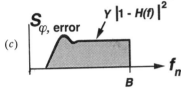

Fig. 11.4 Phase power spectral densities, output at (*b*) and error at (*c*) in response to input at (*a*).

spectral density

$$S_{\varphi,e}(f') = S_{\varphi,\text{in}}(f')|1 - H(f')|^2. \tag{11.10}$$

Figure 11.4 illustrates the reaction of the loop to a limited band of white phase noise at the input according to Eqs. (11.9) and (11.10). It is important that σ_e^2, the integral of $S_{\varphi,e}(f_m)$, be small enough to permit linearity to be assumed at the phase detector (e.g., $\sigma_e^2 \ll 1 \text{ rad}^2$).

Of course, as in the case of discrete signals, the same relationships hold for frequency modulation:

$$S_{f,\text{out}}(f') = S_{f,\text{in}}(f')|H(f')|^2, \tag{11.11}$$

$$S_{f,e}(f') = S_{f,\text{in}}(f')|1 - H(f')|^2. \tag{11.12}$$

11.3   PHASE AND FREQUENCY VARIANCE

The phase power, or variance, of the output, or of the error, in response to an input phase modulation, or to modulation originating at the voltage-controlled oscillator (VCO), is obtained by integrating the corresponding values of S_φ over the frequency range of interest. Often this range covers all frequencies, limited by the response of the circuit or the fall-off of the input modulation. Since both $S(f)$ and $H(f)$ are often represented well by straight lines on a log-log plot, it is helpful to have formulas for integrating under such curves.

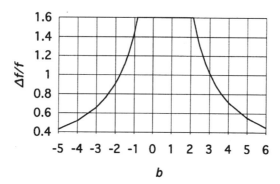

Fig. 11.5 Relative averaging range for a 1-dB error in power vs. slope of the density curve.

The integral from f_1 to f_2 under a curve that has value $S_y(f_1)$ at f_1 and a slope such that $S_y(f) = Kf^b$ is[3]

$$P_y|_{f_1}^{f_2} = \frac{f_1 S_y(f_1)}{b+1}\left[\left(\frac{f_2}{f_1}\right)^{b+1} - 1\right] = \frac{f_2 S_y(f_2)}{b+1}\left[1 - \left(\frac{f_1}{f_2}\right)^{b+1}\right], \quad (11.13)$$

unless $b = -1$. If $b = -1$,

$$P_y|_{f_1}^{f_2} = f_1 S_y(f_1)\ln(f_2/f_1) = f_2 S_y(f_2)\ln(f_2/f_1). \quad (11.14)$$

We can use these equations to determine the error that occurs if we simply multiply a frequency range by the density in the center of the range, rather than using a differential width and integrating. Figure 11.5 implies that the error is greatest where the absolute slope is greatest. The error causes a high estimate except between $b = 0$ and $b = 1$, where the estimate is low. At slopes of 0 and 1, there is no error.

11.4 TYPICAL OSCILLATOR SPECTRUMS

Figure 11.6 shows typical shapes of phase power spectral density in oscillators. Here S_φ is plotted in decibels relative to 1 rad^2 per hertz (dBr/Hz). At high modulation frequencies (region 1 in Fig. 11.6), S_φ is flat. This is white phase noise and results from amplification of the active device's input noise by its gain and noise factor. At lower frequencies, flicker noise predominates (region 2) and S_φ takes on a slope of -3 dB/octave ($= -10$ dB/decade,

---

[3]This equation is represented by a straight line of slope b on a log-log plot,

$$\log[S_y(f)] = \log(K) + b\log(f).$$

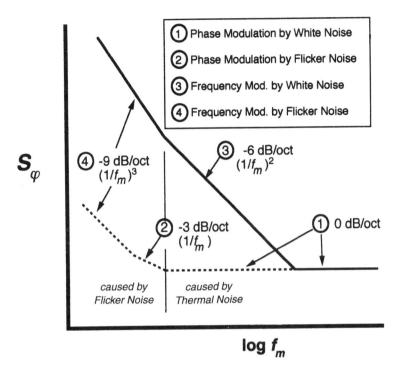

S_φ

① Phase Modulation by White Noise

② Phase Modulation by Flicker Noise

③ Frequency Mod. by White Noise

④ Frequency Mod. by Flicker Noise

④ -9 dB/oct $(1/f_m)^3$

③ -6 dB/oct $(1/f_m)^2$

② -3 dB/oct $(1/f_m)$

① 0 dB/oct

caused by Flicker Noise

caused by Thermal Noise

log f_m

Fig. 11.6   Typical shapes of oscillator phase power spectral densities.

$b = -1$). Below a frequency that is related to the oscillator's frequency f_{out} and the loaded Q, namely $f_m = f_{out}/(2Q)$, the spectrum rises, taking on an additional -6 dB/octave (-20 dB/decade, $b = -2$), for a net of -9 dB/octave (-30 dB/decade, $b = -3$) (region 4). This profile (regions 1, 2, and 4) belongs primarily to high-Q oscillators such as crystal oscillators. For most oscillators, $f_m = f_{out}/(2Q)$ occurs at a frequency where the phase noise is still white. As a result, a -6 dB/octave region (3) is created between regions 1 and 4.

Since $S_\varphi = S_f/f_m^2$, white frequency noise (flat S_f) causes an S_φ that falls at -6 dB/octave due to the $1/f_m^2$ term. That is why the -6 dB/octave region is so named—and similarly for the -9 dB/octave region.

Example 11.1   Integrating Noise Density

a. What is the variance of phase σ_φ^2 between 1 and 10 kHz if the phase power spectral density S_φ is 2×10^{-5} rad²/Hz at 100 Hz and falls at -30 dB/decade throughout the region?

At the low-frequency end of the region, S_φ is 30 dB lower than at 100 Hz, that is 2×10^{-8} rad²/Hz. We use Eq. (11.13), placing the variables into the equation in the order shown there, to get

$$\sigma_\varphi^2 \big|_{10^3\,\text{Hz}}^{10^4\,\text{Hz}} = \frac{(1000\ \text{Hz})(2 \times 10^{-8}\ \text{rad}^2/\text{Hz})}{-3+1}\left[\left(\frac{10^4\ \text{Hz}}{10^3\ \text{Hz}}\right)^{-3+1} - 1\right]$$

$$= 0.99 \times 10^{-5}\ \text{rad}^2.$$

b. What is the variance of frequency?

Using Eq. (11.5a),

$$S_f(1\ \text{kHz}) = S_\varphi(1\ \text{kHz})(10^3\ \text{Hz})^2$$

$$= (2 \times 10^{-8}\ \text{rad}^2/\text{Hz})(10^6\ \text{Hz}^2) = 0.02\ \text{Hz}^2/\text{Hz}.$$

Since $S_f = S_\varphi f_m^2$ and since S_φ has a slope of -30 dB/decade ($b = -3$), S_f has a slope of -10 dB/decade ($b = -1$). Therefore we use Eq. (11.14) to obtain

$$\sigma_f^2 \big|_{10^3\,\text{Hz}}^{10^4\,\text{Hz}} = (1000\ \text{Hz})(0.02\ \text{Hz}^2/\text{Hz})\ln\left(\frac{10^4\ \text{Hz}}{10^3\ \text{Hz}}\right) = 46\ \text{Hz}^2.$$

11.5   LIMITS ON THE NOISE SPECTRUM—INFINITE VARIANCES

We can rewrite $S_y(f_1)$ in terms of the $S_y(f_x)$ at some fixed frequency f_x:

$$S_y(f_1) = S_y(f_x)\left(\frac{f_1}{f_x}\right)^b. \tag{11.15}$$

This allows us to write Eq. (11.13) in terms of a fixed density level $S_y(f_x)$ as the starting frequency f_1 changes.

$$P_y\big|_{f_1}^{f_2} = \frac{f_1 S_y(f_x)(f_1/f_x)^b}{b+1}\left[\left(\frac{f_2}{f_1}\right)^{b+1} - 1\right] = \frac{S_y(f_x)}{f_x^b(b+1)}\left[f_2^{b+1} - f_1^{b+1}\right]. \tag{11.16}$$

From this we see that, if integration begins at $f_1 = 0$, b must be greater than -1 to maintain a finite integral. At the limit where $b = -1$, $f_x S_x$ is constant (since $S_x = K/f_x$). Thus Eq. (11.14) shows that P_y is the same in each octave or each decade or in any other set of ranges for which $f_2 = Kf_1$. Therefore, as one integrates a curve of this slope, beginning at lower and lower starting frequencies, the integral increases uniformly for each decade, for example, of decrease in f_1. For curves with steeper negative slopes the increase under such a process is more rapid. So how is one to interpret these seemingly possible infinities?

If an infinite answer is obtained to a practical question, the problem has probably not been formulated correctly or the answer has not been interpreted properly. For example, Fig. 11.7 shows the open-loop phase power spectral density for a certain (real) oscillator at 296 MHz.

Suppose that a simple PLL is created using this VCO with a unity-gain frequency f_L at the same frequency where the open-loop S_φ bends, about 12 kHz (just to make the problem simpler). Then, with the loop closed, the loop gain will cause the slope to change from -9 dB/octave to -3 dB/octave (effective high-pass) to the left of 12 kHz (approximately). But Eq. (11.14) will give an infinite value for $\sigma_{\varphi e}$ if we integrate to zero frequency. We might think that the slope must flatten at some low frequency, but modulation frequencies with periods as long as a year have been measured [Allan, 1966] so the continuation of the noise slope to such low frequencies had better not cause a problem in a practical application because it does happen—a true lower limit for this slope is not known. Suppose then that we compute the frequency $f_x = c/T_x$ below which this noise must terminate in order that $\sigma_{\varphi e}$ be no greater than 1 rad^2. By Eq. (11.13) the contribution from frequencies above 12 kHz (to infinity) is 6×10^{-7} rad^2 and, by Eq. (11.14), the contribution from frequency f_x to 12 kHz, which we set equal to 1 rad^2 is

$$6 \times 10^{-7} \ln(12 \text{ kHz } T_x/c) = 1, \tag{11.17}$$

so

$$T_x = 8.3 \times 10^{-5} \sec \exp(1/6 \times 10^{-7}) = 10^{7.2 \times 10^5} \sec \approx 10^{7.2 \times 10^5} \text{ years.} \tag{11.18}$$

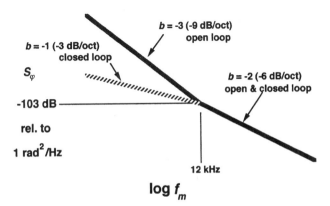

Fig. 11.7 Phase power spectral density for a 240- to 352-MHz oscillator at 296 MHz.

This is $100\ldots00$ years, where the number of zeros is 720,000. Changes in phase that occur with periods this long will have a greatly attenuated effect during the period when we will be concerned about any oscillator.

The real question of interest is not what is the variance of phase error but what is the variance over some period, perhaps a year or ten years or maybe a thousand years? The mathematical formulation of *that* problem will include a factor that attenuates the effect of $S_\varphi(f_m)$ for $f_m \ll c/T_x$. The actual expression for the variance of phase during a period T_x is [Egan, 1981, p. 263]

$$\sigma_\varphi^2 = \int_0^{f_{max}} \left[1 - \left(\frac{\sin \pi f_m T_x/c}{\pi f_m T_x/c} \right)^2 \right] S_\varphi \, df_m. \tag{11.19}$$

For $\omega_m \ll c/T_x$, the integrand here becomes approximately $S_\varphi[\pi f_m T_x/c]^4/3$, which we obtain by taking the first two terms in the series for a sine. Since $[\pi f_m T_x]^4$ has a $+12$ dB/octave-of-f_m slope, the net slope for the expression is $+9$ dB/octave at low frequencies and thus the contribution for $f_m < c/T_x$ is small and the integral is finite.

11.6 POWER SPECTRUM

The spectrum of a cosine that is sinusoidally phase modulated with a peak phase deviation of m at a frequency f_m can be written (see Appendix, Chapter 20)

$$\cos(\omega_c t + m \sin \omega_m t)$$
$$= J_0(m)\cos \omega_c t + J_1(m)[\cos(\omega_c + \omega_m)t - \cos(\omega_c - \omega_m)t]$$
$$+ J_2(m)[\cos(\omega_c + 2\omega_m)t - \cos(\omega_c - 2\omega_m)t] + \cdots \tag{11.20}$$

11.6.1 Spectrum for Small *m*

For a small enough modulation index m (peak phase deviation in radians), we have $J_0(m) \approx 1$, $J_1(m) \approx m/2$, and we can ignore the higher order Bessel functions, $J_i(m)$ with $i > 1$. More specifically, for $m \leq 0.1$: $J_0(m)$ is within 0.25% of 1; $J_1(m)$ is within 0.12% of $m/2$, and $J_2(m) < 0.024 \, J_1(m)$. Therefore, for small phase deviations, the spectrum of the modulated cosine becomes

$$\cos(\omega_c t + m \sin \omega_m t) \approx \cos \omega_c t + \frac{m}{2}[\cos(\omega_c + \omega_m)t - \cos(\omega_c - \omega_m)t],$$
$$\tag{11.21}$$

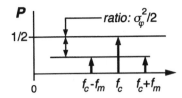

Fig. 11.8 One-sided power spectrum of PM sinusoid, small modulation index.

as illustrated in Fig. 11.8. The carrier power is undiminished, 0.5, corresponding to a sinusoidal amplitude of 1. (We assume a 1-Ω system in converting between voltage and power.) The sidebands' powers are lower by $(m/2)^2$, but this is written in terms of the root mean square (rms) value, $\sigma = m/\sqrt{2}$. We call the spectrum one-sided because all the power is shown at positive frequencies. In a two-sided spectrum, such as is used in Fourier transform theory, half of the power is at negative frequencies. The ratio of sideband to carrier level is the same in either case.

If the phase modulation, rather than being discrete, should represent S_φ in a narrow bandwidth δf, then the relative sideband amplitudes would also be

$$L_\varphi \, \delta f = \sigma_\varphi^2/2 = S_\varphi \, \delta f/2, \tag{11.22}$$

so

$$L_\varphi = S_\varphi/2. \tag{11.23}$$

The φ subscript just indicates that L is due to phase noise.

11.6.2 Single-Sideband Density

Here σ_φ^2 is the mean square phase deviation in radians squared and L is the single-sideband (SSB) power spectral density relative to the carrier, a dimensionless ratio that is often expressed in decibels relative to the carrier per hertz. It is termed single sideband because it gives the relative power of each sideband separately. As can be seen from Eq. (11.23), the relative power in the sum of both sidebands equals S_φ.

In decibels, Eq. (11.23) can be expressed as

$$L|_{\mathrm{dBc/Hz}} = S_\varphi|_{\mathrm{dBr/Hz}} - 3 \text{ dB}. \tag{11.24}$$

One would measure $S_\varphi(f_m)$ by observing the output of a phase detector, as in Fig. 11.9, while L would be observed directly on the spectrum analyzer. The displays that would be observed in each case are illustrated in Fig. 11.10.

As long as the analysis bandwidth of the spectrum analyzer is narrow enough that S_φ and L are nearly constant across that bandwidth, we can obtain an accurate value for S_φ and L by dividing the observed power by δf

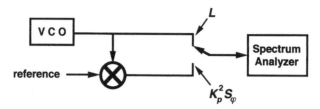

Fig. 11.9   Measuring L and S_φ.

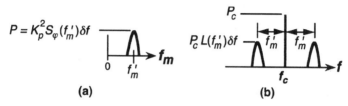

(a)   **(b)**

Fig. 11.10   Spectrum analyzer display of phase power density and power spectral density; δf is the analysis bandwidth in the spectrum analyzer.

and, in the case of S_φ, also by K_p^2. Thus a decrease of the analysis bandwidth by a decade will cause the displayed density to drop by 10 dB, as illustrated in Fig. 11.11. Note, however, that the proper bandwidth to use is the "noise bandwidth." This is the bandwidth that gives the proper ratio between the density and the displayed level. If the analyzer had an ideal rectangular filter, the noise bandwidth would be the same as the -3 dB bandwidth, but, with real filters, there is generally some difference between the two bandwidths. Noise bandwidth will be discussed in some detail in Chapter 14. Other corrections may be required also. Analog spectrum analyzers typically read noise power about 2.5 dB low relative to the power of a sine wave [Hewlett Packard, 1989]. If a spectrum analyzer can display density, that mode should be used both for convenience and because it can include appropriate corrections. Otherwise, consult the instrument's manual.

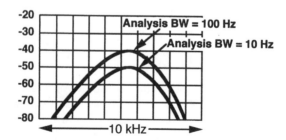

Fig. 11.11   Spectrum analyzer display of noise power at two different analysis bandwidths.

Example 11.2   Spectral Display of Noise Sidebands   If Fig. 11.11 represents an expansion of part of Fig. 11.10b and if the discrete carrier power is displayed at a level of 0 dB (without saturating the analyzer), then the peak value of L displayed in Fig. 11.11 is

$$L = 10^{-4}/100 \text{ Hz} = 10^{-5}/10 \text{ Hz} = 10^{-6}/\text{Hz} \qquad (11.25)$$

or, in terms of dB,

$$L|_{dB} = -40 \text{ dBc/Hz} - 20 \text{ dB} = -50 \text{ dBc/Hz} - 10 \text{ dB} = -60 \text{ dBc/Hz}. \qquad (11.26)$$

Here we have expressed the density in terms of various bandwidths to indicate what happens when the analysis bandwidth changes. Unlike the effect on noise, changing the analyzer's analysis bandwidth has no effect on the display of the discrete carrier level.[4]

Example 11.3   Spectral Display of Phase Noise   If Fig. 11.11 represents an expansion of Fig. 11.10a and if 0 dB is equivalent to 0.01 V rms and the phase detector gain is $K_p = 0.1$ V/rad, then 0 dB represents 0.1 rad rms [i.e., 0.01 V/(0.1 V/rad)]. With a 100-Hz analysis bandwidth, 0 dB represents a noise density of

$$S_\varphi = (0.1 \text{ rad})^2/100 \text{ Hz} = 10^{-4} \text{ rad}^2/\text{Hz} \Rightarrow -40 \text{ dBr/Hz}. \quad (11.27)$$

Thus the peak of -40 dB would represent a density of

$$S_\varphi \Rightarrow -40 \text{ dBr/Hz} - 40 \text{ dB} = -80 \text{ dBr/Hz}. \qquad (11.28)$$

Repeating this process for the 10-Hz analysis bandwidth will produce the same density.

11.6.3   When Is the Modulation Small?

We showed that the approximation of Eq. (11.23) is valid if $m \ll 1$ rad. For the noise power spectrum of Fig. 11.10 the equivalent restriction would be $\sigma_\varphi^2 \ll 1$ rad^2. If that restriction does not hold, then L will not have a shape so similar to that of S_φ because the carrier at f_c will diminish, $S_\varphi(f_m)$ will

---

[4]Actually, even "discrete" signals have finite widths, but we speak of them as being discrete when their power is concentrated in a bandwidth small enough to be considered zero for the problem at hand. The level of a discrete will not change as long as practically all of its power is contained within the analysis bandwidths used.

cause sidebands of L not only at $f_c \pm f_m$ but at $f_c \pm if_m$, for various i, and the magnitudes of the first sidebands will not be linearly proportional to m. We might be tempted to think that, since S_φ is the density in a differential bandwidth, it is small enough; but further reflection should convince us that dividing the spectrum into smaller frequency segments will not make a fundamental difference. The more segments we have, the more multiple sidebands are produced to add to each other. If σ_φ in Fig. 11.10a is larger than a radian, we can expect to see multiple sidebands on each side of the carrier in Fig. 11.10b.

The criterion is more difficult to apply in the case of typical oscillator spectrums, such as shown in Fig. 11.6, because we have shown that σ_φ^2 grows without known limit as the lower frequency of integration decreases. The criterion that we will apply to such spectrums is [Shoaf et al., 1973, p. 43]

$$\int_{f_1}^{\infty} S_\varphi(f_m)\, df_m \ll 1 \text{ rad}^2, \tag{11.29}$$

where f_1 is the lowest modulation frequency at which we employ the small modulation index approximation.

We can rationalize this as follows [Egan, 1981, p. 87]. Consider modulation at frequencies much lower than f_1, where the phase variance is appreciable, as belonging to the carrier. This will broaden the line width of the carrier, which is being modulated by the frequencies that are greater than f_1, and the broadening will result in smearing of the spectrum. However, the smearing will not be noticeable for $f_m > f_1$ because S_φ changes little across frequency increments that are small compared to f_1 at offsets that are greater than f_1. For example, if f_1 is 1 kHz, modulation frequencies much lower than this will broaden the "carrier," perhaps spreading power over 1 Hz or so. Phase modulation at a 1-kHz rate will produce sidebands on the carrier at offsets of ± 1 kHz. If the carrier is 1 Hz wide, then the ± 1-kHz sidebands will actually spread over a few hertz at ± 1 kHz. Usually the magnitude of S_φ does not change much over a few hertz in the vicinity of 1 kHz, so the smearing of this power over a small frequency increment, along with similar smearing from adjacent frequencies, will not make a noticeable difference.

When the criterion of Eq. (11.29) is not met, we can expect differences in the shape of $S_\varphi(f_m')$ and L at offsets from the carrier of f_m'. As S_φ climbs indefinitely at lower frequencies, leading to unbounded σ_φ^2, the power spectrum will remain finite near the spectral center.

11.6.4   Script $\mathscr{L}$

The symbol L is not commonly used. Rather, $\mathscr{L}$ has been in common use for single-sideband power spectral density. Unfortunately, carelessness in usage has caused it to be often taken as equal to $S_\varphi/2$, regardless of whether this was justified by Eq. (11.29). Therefore, $\mathscr{L}$ was redefined by IEEE Standard 1139 [Allan et al., 1988] to be simply equal to $S_\varphi/2$. As a result of its prior usage and its revised usage, $\mathscr{L}$ is somewhat ambiguous, meaning perhaps $S_\varphi/2$ and perhaps L. (Of course, in most cases of interest, they will be the same.) Therefore, we will not use $\mathscr{L}$ in this text, except where Eq. (11.29) has been met so that $\mathscr{L} = L = S_\varphi/2$.

11.7   FREQUENCY MULTIPLICATION AND DIVISION

We briefly considered devices that multiply or divide frequency in Chapter 10 (Fig. 10.9). Multiplying frequency by M automatically multiplies frequency, and phase, deviation by M. Thus it increases S_φ, $S_{\dot{\varphi}}$, σ_φ^2, and so forth by a factor of M^2 ($20 \log_{10} M$). Conversely, frequency division by N decreases these variables by $20 \log_{10} N$.

11.8   OTHER REPRESENTATIONS

Figure 11.12a is identical to Fig. 11.4b. It shows phase power spectral density S_φ in radians squared/hertz or dBr/Hz versus modulation frequency f_m. Figure 11.12b shows the single-sideband spectral density L, usually given in decibels relative to carrier per hertz. This differs from the representation in Fig. 11.10b in three ways (not counting the fact that they represent different spectral shapes). First, only one sideband is shown—there is no loss of information, however, because of symmetry. Second, no powers are given, only the level of the sideband density relative to the carrier power or total power [which are approximately the same under the restriction of

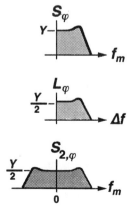

Fig. 11.12   Other representations.

Eq. (11.29)]. Third, the abscissa shows offsets from spectral center rather than actual frequencies. While Fig. 11.10*b* represents what one would observe on a spectrum analyzer, the information is usually plotted as shown here. In both Fig. 11.12*a* and Fig. 11.12*b* the horizontal axis is often logarithmic (of course 0 cannot be shown when it is). At Fig. 11.12*c* is shown the two-sided phase power spectral density. This uses negative frequencies, as in Fourier transforms, so the power is distributed over twice the frequency range, causing the density to be only half as great as at Fig. 11.12*a*. This has the advantage of causing Fig. 11.12*c* to look very much like a spectrum analyzer display (Fig. 11.10*b*) with the carrier level set at 0 dB except that the analyzer's display would be centered on the carrier frequency rather than on zero. The density level then also is the same as at Fig. 11.12*b*, rather than being twice as great, as at Fig. 11.12*a*. However, this uniformity does not seem to be worth the potential confusion. If the two-sided representation were used also for Fig. 11.12*b*, the similarity would disappear again so it depends on using two different representations, one for the power spectrum and another for the modulation density, at the same time. We will generally avoid the use of the two-sided spectrum.

11.S   APPENDIX: SPREADSHEETS FOR INTEGRATING DENSITIES

Many practical noise density plots can be approximated by straight-line segments on semilog paper (decibels vs. log frequency; refer to Sections 11.3 and 11.4). Enter the coordinates at the ends of these segments into the spreadsheet IntPDens and it will give the total integrated power. Companion spreadsheets give specifically the phase deviation or frequency deviation when the segments described represent L_φ [Eq. (11.29) assumed].

IntPDens is an Excel spreadsheet that integrates relative (to carrier) power density to give relative power, using Eqs. (11.13) and (11.14). The rms value of the integrated density is presented for one sideband and for both identical sidebands. We could also enter absolute power density in decibels relative to 1 W/hertz (dBW/Hz) and thus obtain the answers relative to 1 W plus rms values in volts in a 1-Ω system.

IntPhNs also converts the density to the phase deviation it represents. Root mean square phase deviation is given in radians and degrees. The spreadsheet is shown in Fig. 11.S.1. The modulation frequency and SSB density corresponding to the vertices of the segments are entered in columns B and C over the sample data shown there. The bottom line is to be copied as often as required, depending on the number of segments.

Operation of the spreadsheet can be understood from the formulas contained in typical cells. Cell D9 obtains the value of *b*, using = (C9-C8)/ 10/LOG(B9/B8). Cell E9 chooses either Eq. (11.13) or (11.14) to give the

	A	B	C	D	E	F	G	H
1		**INTEGRATED PHASE DENSITY**						
2		Enter SSB density representing phase noise to obtain phase deviation.						
3		ENTER DATA		Do not enter data below. Copy last line OK.				
4		Mod.	SSB		Integrated			
5		Freq.	density		Segment	sum	RMS Phase	
6		Hz	dBc/Hz	slope	rad^2	rad^2	rad	degrees
7		1.00E+2	-89					
8		1.00E+4	-109	-1	1.16E-06	1.16E-06	0.00108	0.0617
9		1.00E+5	-139	-3	1.246E-07	1.284E-06	0.00113	0.06493
10		2.00E+6	-140	-0.08	4.06E-08	1.325E-06	0.00115	0.06595
11								

Fig. 11.S.1 IntPhNs.

integral, depending on how close b is to -1:

```
" = IF(OR(D9<-1.0001,D9<-0.9999),
B8*10 ^(C8 / 10) / (1+D9)*((B9 / B8) ^(1+D9)-1),
B8*10 ^(C8 / 10)*LN(B9 / B8))*2".
```

The factor of 2 at the end comes from Eq. (11.23) and is required to convert from L_φ to S_φ. Cell F9 contains = E9 + F8 to sum the integral under the current segment with those from previous segments. Cells G9 and H9 contain = SQRT(F9) and = G9*180 / PI(), respectively.

IntFNs converts the single-sideband relative power density to frequency density and then integrates to give mean square frequency deviation. Root mean square deviations in radians/second and hertz are also given. The spreadsheet is shown in Fig. 11.S.2. It is similar to IntPhNs but it contains one more column to convert SSB density L_φ to frequency power spectral density S_f before the slope is taken. This follows Eq. (11.5a) and is represented in cell D9 by = C9 + 3 + 10*LOG(B9). Here +3 serves the same purpose as the *2 did previously, to double the density to represent phase

	A	B	C	D	E	F	G	H	I
1		**INTEGRATED FREQUENCY DENSITY**							
2		Enter SSB density representing phase noise to obtain frequency deviation.							
3		ENTER DATA		Do not enter data below. Copy last line OK.					
4		Mod.	SSB	Freq. Dens.		Integrated			
5		Freq.	density	Hz^2/Hz		Segment	sum	RMS Frequency	
6		Hz	dBc/Hz	in dB	slope	Hz^2	Hz^2	Hz	rad/sec
7		1.00E+2	-89	-46					
8		1.00E+4	-109	-26	1	0.0006294	0.0006294	0.03548	0.22292
9		1.00E+5	-139	-36	-1	2.899E-07	0.0006297	0.03549	0.22298
10		2.00E+6	-140	-10.9794	1.923	2.736E-06	0.0006324	0.03556	0.22346
11									

Fig. 11.S.2 IntFNs.

power spectral density rather than SSB density, but we add 3 because we are still working in decibels at this point. Since the doubling takes place prior to integration here, cell F9 does not contain the $*2$ that the equivalent cell E9 did before.

PROBLEMS

11.1 Complete the following table for the loop in Fig. P11.1. Note that the two conditions specified under "At A" for a given ω_m (e.g., lines one and two) do not necessarily occur simultaneously.

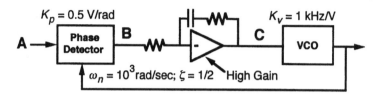

Fig. P11.1

ω_m	At A		At B	At C
10 rad/sec	$\phi_{rms} = 10^{-2}$ rad		___ V rms	___ V rms
10 rad/sec	$S_\phi = 10^{-4}$ rad^2/Hz		___ V^2/Hz	___ V^2/Hz
2500 rad/sec	$\phi_{rms} = 0.01$ rad		___ V rms	___ V rms
2500 rad/sec	$S_\phi = 0.001$ rad^2/Hz		___ V^2/Hz	___ V^2/Hz
700 rad/sec	$S_\phi = 2 \times 10^{-5}$ rad^2/Hz		___ V^2/Hz	___ V^2/Hz

11.2 What is the mean square phase in the region between 1 and 3 kHz if the phase power spectral density at 1 kHz is $S_\varphi = 10^{-4}$ rad^2/Hz and it slopes down at -9 dB/octave?

11.3 (a) Derive

$$\int_{f_1}^{f_2} S_\varphi \, df = \frac{f_1 S_\varphi(f_1)}{b+1}\left[\left(\frac{f_2}{f_1}\right)^{b+1} - 1\right].$$

(b) Show by algebraic manipulation of the right side above that

$$\int_{f_1}^{f_2} S_\varphi \, df = \frac{f_2 S_\varphi(f_2)}{b+1}\left[1 - \left(\frac{f_1}{f_2}\right)^{b+1}\right].$$

CHAPTER 12

RESPONSE TO NOISE MODULATION

The loop processes phase noise just as it processes discrete (single-frequency) phase modulation. We studied this concept in Chapter 7. Here we will apply those same responses to the power spectral density S, which was defined in Chapter 11.

12.1 PROCESSING OF REFERENCE PHASE NOISE

Figure 12.1 indicates where the input phase power spectral density $S_{\varphi,\text{in}}$ and the output and error responses to it are located relative to loop components. This is a special case of Fig. 7.5 wherein the phases are modulation density profiles rather than being at a discrete modulation frequency. The responses in Chapter 7 apply equally well to S_φ as they do to φ. However, since the modulation phase of the noise is unknown, the phase of $H(\Omega)$ (e.g., in Fig. 7.7) is of no value and, based on Eqs. (11.9) and (11.10), squared values of $|H(\omega)|$ and $|1 - H(\omega)|$ are of interest. When the latter are given in decibels, the squaring process is automatic since, for example,

$$|H(\omega)|_{\text{dB}} = 20\log_{10}|H(\omega)| = 10\log_{10}|H(\omega)|^2 = |H(\omega)|^2_{\text{dB}}. \quad (12.1)$$

Thus, for example, Fig. 7.6, 7.8, or 7.9 could be applied to determine $S_{\varphi,e}$ or $S_{\varphi,\text{out}}$. The value of $S_{\varphi,\text{in}}$, in decibels relative to 1 $\text{rad}^2/\text{hertz}$ (dBr/Hz) would be added to the appropriate response in dB to determine the desired output in dBr/Hz.

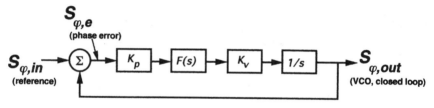

Fig. 12.1   Phase noise on the reference and responses.

The value of $S_{\varphi,\text{out}}$ is important in many applications. Often the purpose of the phase-locked loop (PLL) is to improve $S_{\varphi,\text{out}}$ relative to $S_{\varphi,\text{in}}$. One reason that $S_{\varphi,e}$ is important is that it must be small if the assumption of linearity of the loop is to be valid. For example, with a sinusoidal phase detector, the integrated $S_{\varphi,e}$ should be small compared to $(\pi/4\text{ rad})^2$.

Example 12.1   Loop Response to Reference Phase Noise, Standard Plots
Given: $S_{\varphi,\text{in}} = -90$ dBr/Hz; $\alpha = 0$; $\omega_m/\omega_n = 0.7$; $\zeta = 1$.
Find: $S_{\varphi,e}$ and $S_{\varphi,\text{out}}$.
From Fig. 7.8, $|H(\alpha = 0;\ \omega_m/\omega_n = 0.7;\ \zeta = 1)| \Rightarrow -3$ dB, so

$$S_{\varphi,\text{out}} = -90\text{ dBr/Hz} - 3\text{ dB} = -93\text{ dBr/Hz}.$$

From Fig. 7.9, $|[1 - H](\alpha = 0;\ \omega_m/\omega_n = 0.7;\ \zeta = 1)| \Rightarrow 0$ dB, so

$$S_{\varphi,e} = S_{\varphi,\text{in}} = -90\text{ dBr/Hz}.$$

By the same figure, $S_{\varphi,e}$ would be -99.7 dBr/Hz if α were to equal 1.

As long as the conditions of Eq. (11.29) are met, the same graphs can be used in the same manner to determine $L_{\varphi,e}$ or $L_{\varphi,\text{out}}$ from $L_{\varphi,\text{in}}$, since L differs from S by a constant under those conditions.
The same information can be obtained approximately from the Bode gain plot. This is often more convenient and is especially important where the loop is higher than second order, in which case the plots in Chapter 7 would not apply. Figure 12.2 shows a tangential approximation to typical crystal oscillator noise as $S_{\varphi,\text{in}}$ and a log plot of $|G(\omega_m)|$ for a loop.
Since

$$S_{\varphi,\text{out}}(\omega_m) \approx S_{\varphi,\text{in}}(\omega_m) \quad \text{for } \omega_m \ll \omega_L$$

and

$$S_{\varphi,\text{out}}(\omega_m) \approx S_{\varphi,\text{in}}(\omega_m)|G(\omega_m)|^2 \quad \text{for } \omega_m \gg \omega_L$$

[see Sections 4.1 and 7.4.1], $S_{\varphi,\text{out}}(\omega_m)$ can be roughly determined from this

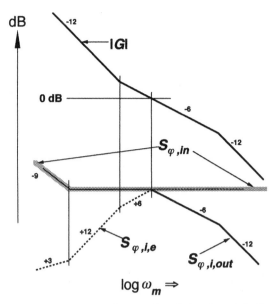

Fig. 12.2 Effect of loop on input phase noise. Subscripts i indicate responses to noise at loop input (to PD). Numbers are slopes in decibels per octave.

information. As shown by $S_{\varphi, i, \text{out}}$ in Fig. 12.2, $S_{\varphi, \text{out}}$ follows $S_{\varphi, \text{in}}$ until the unity-gain point and, beyond that, the decreasing values of $|G|_{\text{dB}}$ are added to $S_{\varphi, \text{in}}|_{\text{dB}}$.

Similarly, since $S_{\varphi, e}(\omega_m) \approx S_{\varphi, \text{in}}(\omega_m)$ for $\omega_m \gg \omega_L$ and $S_{\varphi, e}(\omega_m) \approx S_{\varphi, \text{in}}(\omega_m)/|G(\omega_m)|^2$ for $\omega_m \ll \omega_L$ [see Section 7.4.2], $S_{\varphi, e}(\omega_m)$ follows $S_{\varphi, \text{in}}$ for ω_m greater than ω_L but is reduced by $|G|$ at lower frequencies. This is shown as $S_{\varphi, i, e}$ in Fig. 12.2.

Practically, the response plots are often drawn largely by modifying slopes of $S_{\varphi, \text{in}}$ based on the slopes of $|G|$. Overshoot or attenuation near the unity-gain point might be estimated from the response curves for second-order loops, using Fig. 5.16 to convert computed phase shift to equivalent ζ.

Example 12.2   Loop Response to Reference Phase Noise, Bode Plot   In Fig. 12.2, 100 Hz modulation frequency occurs between the filter zero and unity gain. Unity gain is at 500 Hz. The phase power spectral density (PPSD) of the reference input at 100 Hz is 10^{-6} rad^2/Hz. What are the PPSDs for the voltage-controlled oscillator (VCO) output and for the error at the phase detector (PD) at $f_m = 100$ Hz?

The loop voltage gain at 100 Hz is approximately 500 Hz/100 Hz = 5. The phase error at the phase detector will equal the input phase suppressed by this amount to $(10^{-6}\text{rad}^2/\text{Hz})(\frac{1}{5})^2 = 4 \times 10^{-8}$ rad^2/Hz. The output phase will essentially equal $S_{\varphi, \text{in}}$, 10^{-6} rad^2/Hz.

12.2 PROCESSING OF VCO PHASE NOISE

Figure 12.3 applies to phase noise originating in the loop's VCO. In this case $S_{\varphi,\text{in}}$ is the phase noise that would appear at the output of the VCO if the loop were open. With the loop closed, what is actually seen at the output of the VCO is $S_{\varphi,e}$, the combination of $S_{\varphi,\text{in}}$ and the phase variations that result from feedback arriving on the tuning voltage. And $S_{\varphi,e}$ is also the error seen by the phase detector, so it must again be maintained small for linearity.

Figure 12.4 shows a tangential approximation to typical VCO noise $S_{\varphi,\text{vco}}$ and the same gain plot as in Fig. 12.2. The phase noise at the loop output

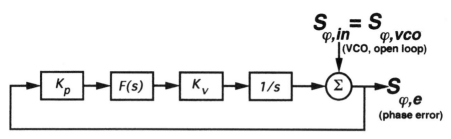

Fig. 12.3   Phase noise on the VCO and responses.

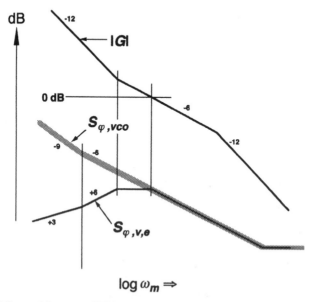

Fig. 12.4   Effect of loop on VCO phase noise. Subscripts v indicate responses to noise at VCO input. Numbers are slopes in decibels per octave.

$S_{\varphi,v,e}$ equals $S_{\varphi,\text{vco}}$ reduced by the loop gain at low frequencies, where the gain is significant.

Example 12.3   Loop Response to VCO Phase Noise   What would be the answers in Example 12.2 if the phase noise originated in the VCO?

The numbers would be the same but the correspondence between the general loop diagram and the system is different. Now the loop (VCO) output would be the error signal, 4×10^{-8} rad^2/Hz, which remains after the loop has canceled most of the original VCO noise. This is also the phase error at the PD, as it was before. In Example 12.1, the phase detector error was the difference between the noisy reference input and the VCO output phase, which closely tracked the input at low ω_m. In this example, the difference signal is generated at the VCO and passed to the PD. Since the other PD input is noiseless, the phase detector error has the same magnitude as the phase noise from the VCO.

12.3   HARMFUL EFFECT OF PHASE NOISE IN RADIO RECEIVERS

As an example of why we may wish to minimize phase noise, consider the application of the local oscillator (LO) in a radio receiver. Its job is to convert the frequency of the received signal to a new frequency. To do this, it enters a mixer along with the received signal and the mixer's output, the intermediate frequency (IF), is at the desired frequency sum or difference. Just as the frequencies combine, so do the corresponding phases. Therefore, any phase variation in the LO will be added to the phase of the signal. Figure 12.5

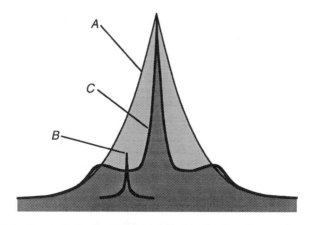

Fig. 12.5   Spectral display illustrating desensitization due to phase noise.

depicts at A the spectrum analyzer display of a strong signal that has been converted by a noisy LO. A weak signal is shown at B and it is masked by the phase noise sidebands on the strong signal. If a PLL can reduce the phase noise on the LO so it is as shown at C, then the weak signal will be discernible above the sidebands on the strong signal. The process by which sidebands on a strong signal increase noise in the vicinity of another signal is one form of receiver "desensitization," sometimes called "reciprocal mixing."

Example 12.4 Desensitization How small a signal can be seen above the noise if the signal is separated from a second, -10 dBm signal by 20 kHz and the relative sideband density L_φ of the LO is -80 dBc/Hz at $f_m = 20$ kHz? The detection Bandwidth is 1 kHz.

In 1 kHz, -80 dBc/Hz produces a noise power of $[(10^{-8}/\text{Hz})(1000 \text{ Hz}) =] \ 10^{-5}$ or, in decibels, $[-80 \text{ dBc} + 10 \text{ dB}\log_{10}(1000) =] -50$ dBc. Thus the noise produced by the -10 dBm signal will be at -10 dBm -50 dBc $= -60$ dBm and that is the weakest signal level that can be seen (if we require 0 dB signal-to-noise ratio).

12.4 SUPERPOSITION

Since noise powers add, the total phase noise can be a combination of phase noise from the reference and phase noise from the VCO (as well as from other sources in the loop). Figure 12.6 shows the closed-loop output noise due to both reference and VCO. These are taken from Figs. 12.2 and 12.4. It also shows their sum, $S_{\varphi,\text{out}}$.

Figure 12.7 shows the total phase noise at the PD output. There is a lack of symmetry between Figs. 12.6 and 12.7. The former shows the sum of an error response and an output response. The latter shows the sum of two

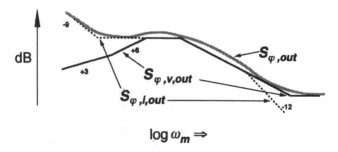

Fig. 12.6 Total phase noise at the output.

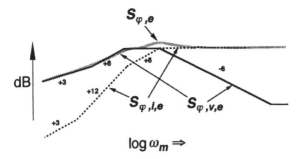

Fig. 12.7   Total phase noise at the phase detector output.

error responses. To understand why this should be so, study the position of the noise inputs and outputs using Fig. 7.5.

12.5   OPTIMUM LOOP WITH BOTH INPUT AND VOLTAGE-CONTROLLED OSCILLATOR NOISE

The value of modulation frequency at which the open-loop phase noise densities of the reference oscillator and the VCO cross is often nearly the optimum loop bandwidth ω_L for minimizing closed-loop output phase noise. At lower frequencies the VCO noise, which is the higher of the two at low frequencies, is attenuated by the loop whereas at higher frequencies the reference noise, which is the higher at high frequencies, is attenuated. The gain in Figs. 12.2 and 12.4 was chosen to produce this "optimum" bandwidth with the results shown in Fig. 12.6. To obtain a more accurate optimum for minimum phase variance, one should write the total integrated noise, using equations of the type given in Section 11.5, and solve for a minimum or plot the sum against ω_L.[1]

Figure 12.8 shows the result with higher gain. The higher gain reduces low-frequency noise due to the VCO but only a small improvement is possible because the $S_{\varphi,\,in}$ is not affected at low frequencies. However, high-frequency $S_{\varphi,\,in}$ that had been attenuated is now passed to the output of the loop. Continued gain increases will further increase high-frequency noise without greatly reducing low-frequency noise. Figure 12.9 shows the opposite effect when gain is too low. Here the amount of low-frequency VCO noise

[1]Even that is not quite optimum because of the straight-line approximations. A truly accurate optimum would require the actual response equations to be used.

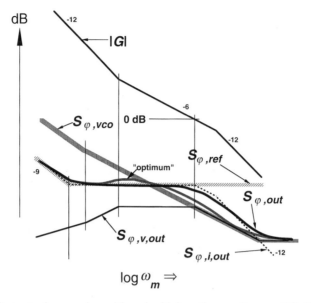

Fig. 12.8 Output phase noise with gain higher than optimum. "Optimum" from Fig. 12.6 is also shown.

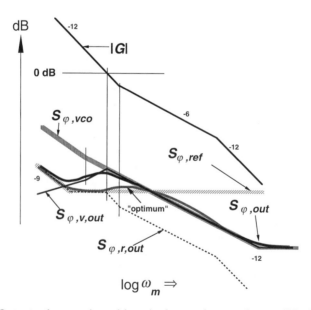

Fig. 12.9 Output phase noise with gain lower than optimum. "Optimum" from Fig. 12.6 is also shown.

that passes through increases, but there is little improvement in high-frequency noise compared to the "optimum."

Example 12.5   Optimum Loop Bandwidth   Estimate the loop bandwidth that will minimize total phase variance for modulation frequencies between 1 and 10^6 Hz if the reference phase noise is flat at -130 dBr/Hz and the VCO PPSD is -40 dBr/Hz at 10 Hz and has a slope of -30 dB/decade.

We estimate the optimum to occur where the PPSD curves cross. The VCO noise must drop 90 dB from its level at 10 Hz to reach the flat reference noise; 90 dB at 30 dB/decade implies 3 decades. Therefore the curves cross at $f_m = 10^4$ Hz.

To check the estimate, we plot the phase variance, using eqs. (11.15) and (11.16). Figure 12.10 shows the contributions from the reference phase noise in curve 1 and from the VCO phase noise in curves 2 and 3. Curve 2 is for a first-order loop and curve 3 is for a type 2 loop with the zero at one-third of the unity-gain bandwidth. Without a zero, the VCO's PPSD continues to climb at lower frequencies because its -30 dB/decade slope cannot be overcome by the loops -20 dB/decade gain slope. If it were not for the restriction that we are integrating noise only above 1 Hz, the integral of the closed-loop VCO noise below f_L would be infinite. When a zero is added, the increased low-frequency gain causes the closed-loop PPSD to fall at frequencies below the zero frequency, and we are no longer dependent on the lower limit of integration to a significant degree. Even if the zero

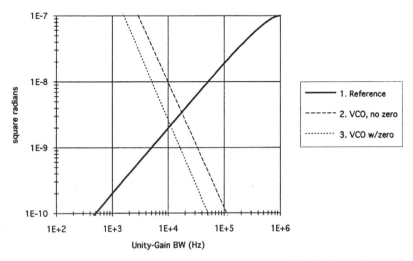

Fig. 12.10   Integrated phase noise, reference, and VCO. This is the variance of the closed-loop VCO phase. The integral is from 1 Hz to 1 MHz.

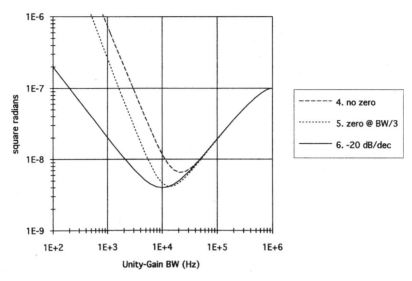

Fig. 12.11   Total phase variance. The sum of reference and VCO noises at the output of the closed loop are shown. One curve is for a first-order loop and the second is for a type 2 loop, both with −90 dBc/Hz open-loop VCO PPSD. Also shown is the variance with −20 dB/decade phase noise.

frequency is moved up to the unity-gain bandwidth, curves 1 and 3 will cross at a frequency that is about 70% of the crossing point show.

The total phase noise is shown in Fig. 12.11. Our estimate was fairly accurate for the type 2 loop. Also shown is the total noise when the VCO has a −20 dB/decade PPSD slope at a level that crosses the reference noise at 10 kHz. We see that the minimum phase variance occurs very close to the estimated point for this case.

12.6   MULTIPLE LOOPS

Sometimes more than one loop is used to minimize phase noise. Figure 12.12 shows an example in which four loops are employed. Oscillator 4 might be a microwave oscillator that is locked to oscillator 3, a surface-acoustic-wave (SAW) oscillator whose frequency has been multiplied to the desired output frequency. The higher Q of the SAW oscillator gives it better phase noise, but the multiplication process increases the noise floor (as well as other phase noise), which is due to thermal noise, so that it exceeds the noise floor of oscillator 4. The loop has an optimum bandwidth near the point at which the noise levels cross. Low-frequency noise modulation is further improved by locking the SAW oscillator to a frequency-multiplied crystal oscillator in a

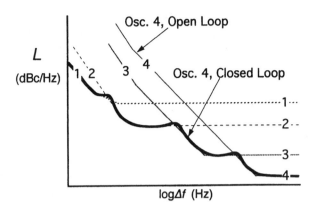

Fig. 12.12   Multiple loops.

similar manner and with similar results. The crystal oscillator may then be locked to an atomic standard that has superior close-in performance. For each of the three loops, the reference has better close-in purity than the VCO because it is of a type that has higher effective Q. The VCO, on the other hand, has better far-out purity (at least after frequency multiplication).

12.7   EFFECTS OF NOISE INJECTED ELSEWHERE

There are many noise sources other than the VCO and the reference. For example, the op-amp in the loop filter may produce enough noise to dominate in certain regions of the output spectrum. In such cases, use the procedures of Section 7.4. While they are described in terms of discrete modulations, the extension to density is straightforward. Remember to use the square of the transfer gains, however, since S is a power density.

Example 12.6   Component Noise   What is the contribution to the noise of the PLL output, at modulation frequencies well above the loop bandwidth, due to the 4-kΩ resistor in Fig. 12.13.

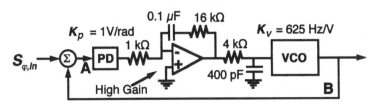

Fig. 12.13

Thermal noise density is -174 dBm/Hz $= -204$ dBW/Hz. This is what the resistor would deliver to a matched load, in which case half of the voltage appears across the load. Therefore, the mean square noise voltage density $\tilde{v}_T^2$ of its equivalent internal noise generator is obtained from the expression for power, $\tilde{v}^2/R$, as

$$(\tilde{v}_T/2)^2 = RN_T = (4000 \ \Omega)10^{-20.4} \ \text{W/Hz},$$

$$\tilde{v}_T^2 = 6.37 \times 10^{-17} \ \text{V}^2/\text{Hz},$$

where $\tilde{v}_T$ is the open-circuit rms thermal noise voltage. Multiplying by $K_v^2 = (625 \ \text{Hz/V})^2$, we obtain $2.5 \times 10^{-11} \ \text{Hz}^2/\text{Hz}$. At high frequencies the voltage is reduced, due to the low-pass filter, by the impedance ratio, $1/(R\omega_m C) = 99{,}471 \ \text{Hz}/f_m$. The phase power spectral density is therefore

$$S_\varphi = S_f/f_m^2 = (2.5 \times 10^{-11} \ \text{Hz})(99{,}471 \ \text{Hz})^2/f_m^4 = 0.25 \ \text{rad}^2 \ \text{Hz}^3/f_m^4.$$

At frequencies well above the loop bandwidth, $1 - H \approx 1$, so this is the value of S_φ at the output. For example, the single-sideband level would be -163 dBc/Hz, corresponding to $S_\varphi = 10^{-16} \ \text{rad}^2/\text{Hz}$, at $f_m = 7071$ Hz, if that is much greater than the loop bandwidth (with which we will not concern ourselves for this example).

Some resistors have troublesome flicker $(1/f)$ noise also, but it is proportional to the direct current (DC) voltage across the resistor. Since the steady-state value of that voltage is zero, $1/f$ noise is probably small, especially at higher frequencies.

12.8   MEASURING PHASE NOISE

There are three basic ways to measure phase noise [Egan, 1981, pp. 263–267]:

- We can measure it directly, using a phase detector.
- We can measure frequency deviation, using a frequency discriminator, and convert S_f to S_φ using Eq. (11.5).
- We can observe the noise sidebands directly using a spectrum analyzer or other receiver, and, if we know the sidebands are due to phase modulation (PM) [rather than amplitude modulation (AM), which we will consider in Chapter 13], use Eq. (11.23) to convert L_φ to S_φ.

In the first two methods the signal is demodulated before the noise is measured, whereas, in the third, the spectrum of the signal is observed directly without demodulation. Direct observation using a spectrum analyzer is often the most convenient method, but it is often less sensitive than the phase detector. That is, with the phase detector, lower noise levels can be measured before the measurement is limited by the noise of the instrumentation.

The discriminator sensitivity has a different dependence on modulation frequency than do the other two devices. As modulation frequency increases, the output from a discriminator increases relative to the output from the others, as is evident from Eq. (11.5) and Fig. 12.14. With the spectrums shown in Fig. 11.6, a phase detector will produce a voltage that increases at lower frequencies where phase noise climbs. Thus, the output from the phase detector will grow relative to the noise of the amplifier or detector that follows it. The discriminator output will be flat in region 3 but will increase with frequency in region 1, so its output will tend to overcome subsequent instrument noise at higher values of f_m. However, noise that precedes the discriminator is partially seen as phase noise by the discriminator. Therefore, it will tend to be the limitation on sensitivity at high values of f_m, climbing along with the noise being measured.

Although the spectrum analyzer usually serves as the final measurement device in all of these methods, we give its name to the method where it is not preceded by a demodulator but rather measures the signal directly. While the similarity of response shapes for the phase detector and spectrum analyzer

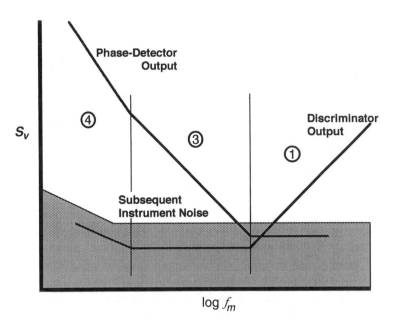

Fig. 12.14 Voltage power spectral density from phase noise measurements in the noise regions of Fig. 11.6. The various curves will shift depending on details of the noise and the measurements.

methods might lead us to expect identical sensitivities, spectrum analyzer performance is limited, in the latter method, by the presence of the strong un-demodulated signal and by requirements placed on it to cover a broad frequency range.

12.8.1 Using a Phase Detector

The test setup is shown in Fig. 12.15. The reference must be maintained at the proper phase relationship to the signal being tested. Ideally the reference is much cleaner than the device under test (DUT). The DUT will often be an oscillator or a PLL. The spectrum analyzer is tuned to the noise modulation frequency. The noise figure of the low-noise amplifier, and, to some degree, the spectrum analyzer, along with K_p of the test phase detector, set a minimum measurable level for S_φ.

The power spectral density S_p, read on the analyzer at f_m, is converted to $S_\varphi(f_m)$ using the PD gain constant,

$$S_\varphi(f_m) = \frac{S_v(f_m)}{K_p^2}. \tag{12.2}$$

Here $S_v = R_0 S_p$, where R_0 is the impedance of the measurement system.[2]

Note that when variables such as S_p, S_v, and K_p are used in an expression, they must all generally be referenced to the same location in the circuit, for example, the output of the phase detector or the input to the analyzer.

If the effective noise density of the amplifier plus analyzer, in the vicinity of the signal frequency, equals N_0, the resulting measured output will be the same as is produced by a PPSD of $S_{\varphi n} = R_0 N_0 / K_p^2$. Levels of S_φ lower than this measurement noise floor cannot be measured accurately. Having considered that, the limitations that we will consider in the rest of this section will be due to additional noise from other sources.

12.8.1.1 Calibration

Measuring K_p Directly. We can ascertain the PD gain constant K_p by observing, on the oscilloscope, the slope of the PD characteristic at the relative phase (e.g., 90° for a balanced mixer) to be maintained during the test. We can do that by changing the phase of one of the inputs to the PD with a phase shifter or line stretcher or by offsetting the reference frequency slightly from the frequency of the DUT and observing the characteristic as the phase changes. We must then maintain the average (DC) voltage during test at a value that will give that measured slope.

---

[2]This is simply an expression of the relationship: power = (rms voltage)^2/(resistance).

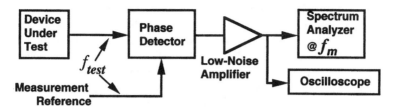

Fig. 12.15   Phase noise measurement with a phase detector.

When the PD Characteristic Is Sinusoidal. *If the PD characteristic is truly sinusoidal* (note Fig. 3.13), K_p can be measured using the spectrum analyzer. When the reference frequency is offset slightly from the frequency of the DUT, the root mean square (rms) voltage from the PD will be, by Eq. (1.10),

$$\tilde{v}_{\text{test}} = \frac{K_p' \text{ rad}}{\sqrt{2}}. \tag{12.3}$$

If we are to make the measurement of S_φ at the highest value of K_p, which is K_p', we can combine the last two equations to obtain

$$S_\varphi(f_m) = \frac{S_v(f_m)}{2\tilde{v}_{\text{test}}^2}. \tag{12.4}$$

The level that we observe during the measurement will be $\sigma_v^2 = S_v B_n$, where B_n is the effective noise bandwidth (to be defined exactly in Chapter 14) of the analyzer. Thus we can write Eq. (12.4) as

$$S_\varphi(f_m) = \left[\frac{\sigma_v^2(f_m)}{\tilde{v}_{\text{test}}^2}\right]\frac{1}{2B_n} \tag{12.5}$$

or, in decibels,

$$S_\varphi(f_m)\big|_{\text{dBr/Hz}} = \left[\sigma_v(f_m)\big|_{\text{dBV}} - 20 \text{ dBV} \log(\tilde{v}_{\text{test}})\right]$$
$$- 10 \text{ dB} \log(B_n/\text{Hz}) - 3 \text{ dB}. \tag{12.6}$$

The ratio of the level observed on the analyzer during measurement to the level observed during calibration, when the signals were offset by f_m, is enclosed in brackets in the last two equations. Thus we set the level read during calibration as a reference and compare measured noise levels to that reference. Then we divide by $2B_n$ to get density.

Example 12.7   Calibration With a Sinusoidal Characteristic   In Fig. 12.15, the reference signal is offset 10 kHz from the output frequency of the DUT. Because the reference is weak compared to the other PD input, the resulting PD output is a good sinusoid at 10 kHz with 0 V average (DC) value. The spectrum analyzer is adjusted so the resulting 10 kHz line is at the top line on the display. Then the reference signal is set to the same frequency as the DUT and adjusted for 0 V DC output. The level of the noise at 10 kHz is 65 dB below the top line. The analyzer's noise bandwidth is 30 Hz. What is S_φ (10 kHz)?

Using Eq. (12.6),

$$S_\varphi(f_m) = -65 \text{ dB} - 10 \text{ dB} \log(30) - 3 \text{ dB} = -83 \text{ dBr/Hz}.$$

By Modulation. We can also calibrate by phase modulating one of the inputs by a known amount. In the absence of a calibrated phase modulator, we can determine the phase deviation by observing the sideband level of the modulated signal using a second spectrum analyzer (refer to Section 11.6).

Example 12.8   Calibration Using Modulation   We phase modulate one of the two signals to be compared and observe the response on the spectrum analyzer. The modulation frequency is 20 kHz and a single line is produced at that frequency on the analyzer. The gain of the analyzer is adjusted to place this line at 0 dB on the display. The spectrum of the modulated signal, as observed directly on a second analyzer, consists of a carrier line and two sidebands, 40 dB smaller and offset 20 kHz on either side. The modulation is removed and the noise level is observed, on the first analyzer, to be at -80 dB. The analysis bandwidth is 100 Hz. What is S_φ(20 kHz)?

Based on Eq. (11.21) and Fig. 11.8, the variance of the modulated signal during calibration is

$$\sigma_\varphi^2 = 2 \times 10^{-40 \text{ dB}/10 \text{ dB}} \text{ rad}^2 = 0.0002 \text{ rad}^2.$$

In a 100-Hz measurement bandwidth this is equivalent to a density of

$$S_\varphi = 0.0002 \text{ rad}^2/100 \text{ Hz} = 2 \times 10^{-6} \text{ rad}^2/\text{Hz},$$

or -57 dBr/Hz. Since the measured noise is 80 dB weaker, its level is

$$S_\varphi(10 \text{ kHz}) = -57 \text{ dBr/Hz} - 80 \text{ dB} = -137 \text{ dBr/Hz}.$$

Calibration can be performed as a function of f_m if necessary.

12.8.1.2 Obtaining a Measurement Reference. Sometimes very stable and accurate sources (e.g., atomic standards) can be compared without locking. Absent that, however, how do we maintain a constant phase relationship between the DUT and the reference? There are several methods.

Comparing Output Phase to Reference Phase. If the DUT is a PLL, we can use the PLL's reference as the measurement reference (Fig. 12.16). Phase noise of the reference will appear at the PLL output multiplied by H. Since the PD subtracts the reference phase from this transformed reference phase, the net result is multiplication of the reference phase by $(H - 1)$, leading to a measured phase of

$$\varphi_m = \varphi_{\text{out},1} - \varphi_{\text{ref}} \tag{12.7}$$

$$= \varphi_{\text{res},1} + \varphi_{\text{ref}}(H - 1), \tag{12.8}$$

where $\varphi_{\text{res }1}$ is the residual phase noise, defined as the phase noise with a perfect reference (i.e., when $\varphi_{\text{ref}} = 0$). The corresponding PPSDs are

$$S_{\varphi,m} = S_{\varphi,\text{out},1} + S_{\varphi,\text{ref}} \tag{12.9}$$

$$= S_{\varphi,\text{res},1} + S_{\varphi,\text{ref}}|1 - H|^2. \tag{12.10}$$

The result can be considered a measurement of $S_{\varphi,\text{res}}$ corrupted by a measurement noise, $S_{\varphi,\text{ref}}|1 - H|^2$, which is small if $f_m \ll f_L$, since H is then approximately 1.

Comparing Output Phase to a Locked Source. Another way to obtain a measurement reference is to lock a clean oscillator to the DUT with a bandwidth narrow compared to the f_m of interest so the locked oscillator does not track the noise of interest. In Fig. 12.17 the measurement point (output) is at the error node of the phase-locked oscillator (PLO) so the

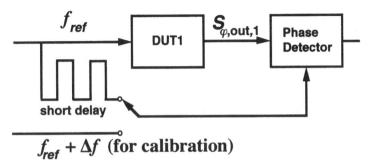

$f_{ref} + \Delta f$ **(for calibration)**

Fig. 12.16 Measurement referred to the DUT's reference. The short delay is used to obtain the correct relative phase at the PD.

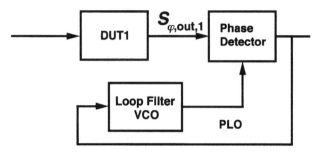

Fig. 12.17 Measurement referred to a PLO.

phase measured there is

$$\varphi_m = \varphi_{res,PLO} + \varphi_{out,1}(1 - H) \tag{12.11}$$

and the PPSD is

$$S_{\varphi,m} = S_{\varphi,res,PLO} + S_{\varphi,out,1}|1 - H|^2. \tag{12.12}$$

If $S_{\varphi,res,PLO}(f_m)$ is small and $|1 - H(f_m)|^2 \approx 1$, then $S_{\varphi,m} \approx S_{\varphi,out,1}$. We can extend the measurements to lower frequencies by measuring $[1 - H(f_m)]$ and using the measured value to cancel its effect mathematically.

Comparing Output Phases of Similar Sources. When a reference that is clean compared to the DUT cannot be obtained, perhaps because the DUT is very clean, two "identical" DUTs are sometimes compared (Fig. 12.18). For example, we might lock two PLLs of identical design to the same reference and compare their output phases. The worst-case noise for each is the measured noise, but often each is assumed to be contributing half of the noise power. If three DUTs having similar noise levels are compared two at a time, the three resulting measurements provide enough information to solve for the noise level of each DUT.

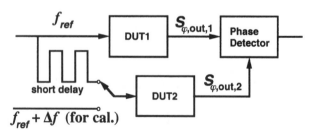

Fig. 12.18 Comparing two DUTs. The short delay is used to obtain the correct relative phase at the PD.

The phase from DUT1 is

$$\varphi_{out,1} = \varphi_{res,1} + \varphi_{ref} H_1 \qquad (12.13)$$

and similarly for $\varphi_{out,2}$. The difference between the two is

$$\varphi_m = \varphi_{out,1} - \varphi_{out,2} \qquad (12.14)$$

$$= \varphi_{res,1} - \varphi_{res,2} + \varphi_{ref}(H_1 - H_2). \qquad (12.15)$$

The corresponding PPSD is

$$S_{\varphi,m} = S_{\varphi,res,1} + S_{\varphi,res,2} + S_{\varphi,ref}|H_1 - H_2|^2. \qquad (12.16)$$

Most of the phase noise of the common reference is either attenuated, at $f_m \gg f_L$, where H is small, or canceled, at $f_m \ll f_L$, where H_1 and H_2 are both very close to one. At $f_m \approx f_L$, however, attenuation is small and good cancellation depends on a good match between the gains of the "identical" PLOs.

Example 12.9   Measurement with a Phase Detector   We are to measure the phase noise of a high-spectral-purity phase-locked oscillator using the setup of Fig. 12.15. The output frequency is a multiple of the reference frequency (see Section 10.1.1), so the PLO and its reference cannot be measured against each other directly. We decide to compare two "identical" PLOs, as in Fig. 12.18. Initially we replace DUT2 by a signal generator set to the same amplitude[3] as DUT2 but offset in frequency by 2 kHz. The output of the balanced-mixer PD, as observed on the oscilloscope, is shown in Fig. 12.19 where the slope at the intended operating point is

$$K_p = 2.5 \text{ V}/0.2 \text{ cycle} = 12.5 \text{ V/cycle}.$$

We then connect PLO2 as DUT2. We connect both DUTs to the same PLL reference and adjust the lengths of the coaxial cables to obtain the required 90° phase shift between the inputs to the test PD. The DC output from the PD is 0 V, corresponding to the operating point where K_p was measured. The spectrum analyzer indicates a power of -55 dBm at 2 kHz when its analysis bandwidth is 10 Hz, which is also approximately its noise bandwidth, so the indicated power density is

$$S_p(2 \text{ kHz}) = \frac{10^{-55/10} \text{ mW}}{10 \text{ Hz}} = 3 \times 10^{-7} \text{ mW/Hz} = 3 \times 10^{-10} \text{ W/Hz}.$$

[3]This is equivalent to calibration by offsetting the reference to one of the PLOs, as in Fig. 12.18, as long as the PLO's amplitude is maintained constant.

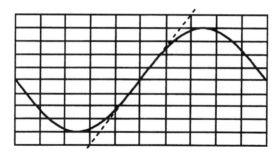

Fig. 12.19   PD calibration output at 0.5 V/division.

In a 50-Ω system, this corresponds to a voltage power spectral density of

$$S_v(2 \text{ kHz}) = 50\Omega \, S_p(2 \text{ kHz}) = 1.5 \times 10^{-8} \text{ V}^2/\text{Hz}.$$

From Eq. (12.2), this implies a PPSD of

$$S_\varphi(2 \text{ kHz}) = \frac{1.5 \times 10^{-8} \text{ V}^2/\text{Hz}}{(12.5 \text{ V/cycle})^2} = 9.6 \times 10^{-11} \frac{\text{cycle}^2}{\text{Hz}} \left(\frac{2\pi \text{ rad}}{\text{cycle}}\right)^2$$

$$= 3.8 \times 10^{-9} \text{ rad}^2/\text{Hz}.$$

This corresponds to -84 dBr/Hz. If we attribute half of the noise to each PLO, each has a PPSD at 2 kHz of 1.9×10^{-9} rad^2/Hz or -87 dBr/Hz.

We attenuate the signal from DUT2 until the noise level on the spectrum analyzer stops decreasing. This occurs at -90 dBm, so we know the noise due to the amplifier, and the analyzer is at least 35 dB below the measured level and, therefore, makes no significant contribution to measurement inaccuracy. Assuming the last term in Eq. (12.16) is small, the noise we have measured is primarily the "residual" noise of the two PLOs.

12.8.2   Using a Frequency Discriminator

One advantage of the use of a frequency discriminator is that it does not require a reference. In a sense it provides its own phase reference. We can see this from a particular implementation of the discriminator (Fig. 12.20) in which the output from the DUT is split and both parts are sent to the phase detector but one path is delayed by T relative to the other. Then the phase detector sees a phase difference of $\varphi = \omega T$. This relationship implies

$$S'_\varphi = S_\omega T^2, \tag{12.17}$$

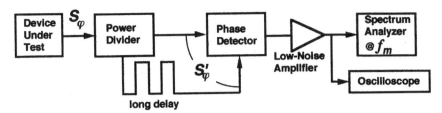

Fig. 12.20   Measurement of S_φ with a frequency discriminator.

where S'_φ is the PPSD measured at the PD and S_ω is the actual frequency PSD. Combining with Eq. (11.5), we obtain the PPSD S_φ of the DUT in terms of S'_φ as

$$S_\varphi = S_\omega / \omega_m^2 = S'_\varphi / (\omega_m T)^2. \qquad (12.18)$$

The maximum measurable deviation is limited by the assumption that K_p remains constant. Therefore, the integrated noise should not be so high that the instantaneous operating point moves from the linear region in Fig. 12.19. If this restriction is met, the sensitivity, by Eq. (12.18), increases with T, and thus with the differential cable length. While $S'_\varphi > S_\varphi$ suggests greater sensitivity with a discriminator than with a phase detector (alone), for S'_φ to exceed S_φ at small values of ω_m, the delay cable must be quite long (roughly 1000 ft at 100 kHz, proportionally longer at lower frequencies). As the delay increases, so to do losses, and additional amplifiers may be necessary along the cable to prevent thermal noise from overwhelming the noise being measured.

12.8.3   Using a Spectrum Analyzer or Receiver

The spectrum analyzer is a swept radio receiver plus a display that indicates the received power versus frequency. Curves A and B in Fig. 12.5 illustrate signals displayed on a spectrum analyzer. It can be used to measure directly the relative sideband amplitudes, from which can be obtained S_φ, by Eq. (11.23) as long as the condition of Eq. (11.29) is met. The spectral purity of the analyzer may be compromised due to other requirements (such as the necessity to be able to sweep over a wide frequency range), so a discrete, single-frequency-at-a-time version may be more sensitive. Of course, a low noise amplifier can be used to improve the sensitivity, but there may then be a problem of overdriving the analyzer so the problem becomes one of dynamic range, sensitivity in the presence of a strong signal. [With the phase detector and discriminator, strong signals are eliminated by cancellation (adjusting for 0 V DC output) and simple filtering, low-passing and possibly high-passing, at the PD output.] We can enhance the dynamic range by using a narrow-band rejection filter (notch filter) at the operating frequency of the DUT to remove most of the power from the center of the spectrum, leaving the sidebands to be measured. Of course, it is necessary then to determine what the total power would have been had it reached the analyzer with the same attenuation as the sidebands, but this is not a fundamentally difficult problem. The spectrum analyzer or receiver, enhanced by filtering when necessary, is the usual choice for measuring the noise floor far from spectral center.

PROBLEMS

12.1 Figure P12.1 shows the phase power spectral density at the input of the loop of Fig. 12.13. Sketch the phase power spectral density at A and at B on a chart such as Fig. P12.1. Use straight-line approximations; indicate slopes and levels.

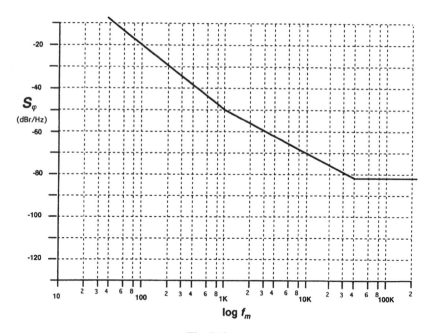

Fig. P12.1

12.2 What is the variance of phase, σ_φ^2, between 100 Hz and 200 kHz indicated by the density S_φ shown in Fig. P12.1? Show separately the calculations for each of the regions defined by different slopes. Give answers in units of radians squared for (a) 100 Hz to 1 kHz, (b) 1 kHz to 40 kHz, (c) 40 kHz to 200 kHz, and (d) the total variance for the three regions combined.

12.3 What equivalent input noise voltage density for the op-amp in Fig. 12.13 will produce $S_f = 10^{-6}$ Hz, at 1 kHz modulation frequency, at the loop output? [The equivalent noise voltage is in series with the positive (noninverting) input of the op-amp. The transfer function from the positive input to the op-amp output is $1 - G_A$, where G_A is the transfer function from the PD output to the op-amp output. Note that, when G_A is real, as when the capacitor impedance goes to zero, it is negative.]

12.4 If the PPSD shown in Fig. P12.1 is due to the VCO and the reference PPSD level is -70 dBr/Hz, what is the approximate unity-gain bandwidth to minimize phase variance at the loop output?

12.5 In Fig. 12.17 the PLO has negligible residual noise. It is a type 2 loop with $f_n = 1$ kHz and $\zeta = 1$. The measured PD sensitivity is $K_p = 0.1$ V/cycle. A spectrum analyzer connected to the PD shows -90 dBm at 800 Hz in a 50-Ω measurement system. The noise bandwidth is 3 Hz. What is S_φ (800 Hz) from the DUT?

CHAPTER 13

REPRESENTATION
OF ADDITIVE NOISE

We have studied, in the two previous chapters, how the loop responds to phase noise. However, many problems require us to determine how the loop acts in the presence of additive noise. Additive noise is noise that accompanies a signal rather than modulation on that signal. So we can apply, to this problem, what we have already learned, we will determine how to represent additive noise as equivalent phase noise. We will see that half of the noise power can be treated as if it were the result of phase modulation.

13.1 GENERAL

As before, we will treat noise as a collection of very narrow-band signals characterized by root mean square (rms) values $\tilde{v}$ and having unknown phase. Figure 13.1 illustrates how we can represent a voltage of any phase as two voltages in mutual phase quadrature, a real component v_r and an imaginary component v_i. The total power (1 Ω assumed) is $\tilde{v}_T^2 = \tilde{v}_r^2 + \tilde{v}_i^2$. Since the phase is random, the power is as likely to be in the real component as in the imaginary component, so each has a mean value of half of the total:

$$P_i = P_r = P_T/2. \tag{13.1}$$

We now wish to represent the real and imaginary noise components as sidebands on some central carrier signal so we can relate these sidebands to modulation of that signal. We will be able to interpret some of the modulation as amplitude modulation (AM) and the remainder as phase modulation

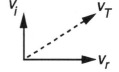

Fig. 13.1 Voltage phasor decomposed into real and imaginary parts.

(PM). The former has only a secondary effect (K_p may increase with amplitude), and we already know how the loop responds to PM. Since phase and amplitude modulation are recognizable by the even or odd sign relationship (see Appendix, Chapter 20) between sidebands (which are offset from the carrier by the modulation frequency), we first consider how to decompose the noise sidebands into even and odd pairs.

Figure 13.2 shows how we can do this. The larger sideband is equivalent to the sum of two new sidebands, which have the same sign. The smaller, symmetrically located, sideband is equivalent to the difference between the same two new sidebands. Thus we can decompose the original pair into an even (same sign) and an odd (different signs) pair. (In fact, we could do this one noise sideband at a time, rather than in pairs, as shown in Appendix 13.A.) As before, each component represents the rms voltage corresponding to the power in a narrow or differential bandwidth.

Again, due to randomness, the likelihood of the even pair is the same as that of the odd pair. Thus the sideband power at any pair of frequencies, located at $f_c \pm \Delta f$, can be equated to eight equally likely, and therefore equally powerful, components. Four are real and four are imaginary. For each of these two types, two form an even pair and two form an odd pair. For each of these pairs, one component is located at $f_c + \Delta f$ and the other is located symmetrically about f_c at $f_c - \Delta f$. We will only treat noise for which $0 < (f_c \pm \Delta f) < 2f_c$ [Blanchard 1976, p. 140]. If the noise power density is symmetrical about the carrier (as it will be with white noise) the total noise power in two differential-width sidebands located at $f_c \pm \Delta f$ is $2N_0 \, df$, where N_0 is the one-sided noise power density. This is divided equally among the eight equally likely components so each will have power $N_0(f) \, df/4$. Thus the total power of the four components at the frequency of each single sideband equals the power of the sideband, $N_0(f) \, df$.

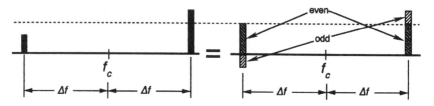

Fig. 13.2 Decomposition of a pair of sidebands into the sum of an even pair and an odd pair.

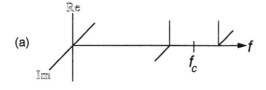

(a)

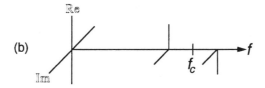

(b)

Fig. 13.3 Decomposition into eight sidebands. Only the positive half of the Fourier spectrum is shown.

The eight components are shown in Fig 13.3.[1] The set at Fig. 13.3a corresponds to AM sidebands on a real (cosine) carrier at f_c. The real pair is produced by a cosine modulation and the imaginary pair by a sine modulation. However, since AM has no first-order effect on a locked loop, the sidebands shown at Fig. 13.3b, which represent half of the total additive noise power, are of primary importance.

13.2   PHASE MODULATION ON THE SIGNAL

Figure 13.3b represents phase modulation (PM) or, equivalently, frequency modulation (FM) sidebands on a real (cosine) carrier as long as $m \ll 1$ rad. The representation is restricted to small m because there would otherwise be multiple sidebands and a reduction in the carrier amplitude. When the noise is true phase modulation, the phase error in the loop must be small if the loop is to be approximately linear. The input phase modulation can be quite strong as long as the VCO tracks it to keep the phase error small. Here, however, even if the phase error is kept small, the input phase deviation must also be kept small for the noise to be accurately modeled as PM.

The real pair in Fig. 13.3b corresponds to sine modulation and the imaginary pair to cosine modulation. If the noise power is symmetrical about the signal at f_c, the total power of the two PM components at each sideband frequency equals $N_0(f_c \pm \Delta f) \, df/2$. Since $L(\Delta f) \, df$ is the ratio of the sideband power at $+\Delta f$ or $-\Delta f$ to the carrier (signal) power P_c, the

---

[1]This can be considered to be the positive half of the Fourier spectrum or to be an analytic signal representation. In the latter case the spectral line at ω represents $\exp(\omega t)$, the real part of which is the signal.

single-sideband power density representing phase noise L_φ is

$$L_\varphi(\Delta f)\, df = L(\Delta f)\, df/2 = N_0(f_c \pm \Delta f)\, df/(2P_c), \qquad (13.2)$$

$$L_\varphi(\Delta f) = L(\Delta f)/2 = N_0(f_c \pm \Delta f)/(2P_c). \qquad (13.3)$$

Based on Eq. (11.23), and under condition that Eq. (11.29) is met, we also have the phase power spectral density in terms of the ratio of additive noise density to signal power,

$$S_\varphi(f_m) = N_0(f_c \pm f_m)/P_c. \qquad (13.4)$$

Thus, whereas for phase noise we have $S_\varphi(f_m) = 2L_\varphi(\Delta f)$, when the noise is additive it is also true that $S_\varphi(f_m) = L(\Delta f)$.

Example 13.1   PPSD from Additive Noise Power Density A signal at a frequency of 10 MHz and power 6 dBm is embedded in noise at a density of -120 dBm/Hz and extending ± 1 MHz from the signal (at which point it is attenuated rapidly by a filter). This is equivalent to a single-sideband density of

$$L(\Delta f < 1\ \text{MHz}) = N_0(10\ \text{MHz} \pm \Delta f)/P_c = (10^{-12}\ \text{mW/Hz})/(10^{0.6}\ \text{mW})$$

$$= 2.5\ 10^{-13}/\text{Hz}$$

$$\Rightarrow\ -120\ \text{dBm/Hz} - 6\ \text{dBm} = -126\ \text{dBc/Hz}.$$

The equivalent phase power spectral density is the same except for units,

$$S_\varphi(f_m < 1\ \text{MHz}) = -126\ \text{dBr/Hz}.$$

The requirement for validity, Eq. (11.29), is, for this case,

$$\int_0^{1\ \text{MHz}} 10^{-12.6}\ \text{rad}^2/\text{Hz}\, df = 10^6\ \text{Hz}\ 10^{-12.6}\ \text{rad}^2/\text{Hz}$$

$$= 10^{-6.6}\ \text{rad}^2 = 2.5 \times 10^{-7}\ \text{rad}^2 \ll 1\ \text{rad}^2.$$

Because of the natural thermal noise floor and because of the equivalence between additive noise and phase noise that was shown above, all oscillator spectrums eventually drop to a flat region, such as those that can be seen in Figs. 11.6 and 12.4.

Even if m does not meet the assumed restriction of smallness, additive noise still produces equivalent phase noise. The zero crossings of the waveform still vary in a noiselike manner. As m becomes larger, however, the theory described above becomes less justifiable as a means of computing the value of the equivalent S_φ from a given L_φ. In such cases we can use a second representation of the spectrum in Fig. 13.3b. We will see results that are the same as if we had simply ignored the limitations on m, but we will have a better understanding of the process.

13.3   AMPLITUDE MODULATION ON A QUADRATURE CARRIER (MULTIPLICATIVE MODULATION)

In this second representation, the components in Fig. 13.3b are represented as AM sidebands on an imaginary (sine) carrier. As in Fig. 13.3a, the sidebands in the same plane as the carrier are even while those in the quadrature plane are odd. And, as before, the former represent cosine modulation and the latter sine modulation. This time, however, suppressed-carrier AM is employed so no carrier signal is required. This is multiplicative modulation. These sidebands exist independently of a desired signal.

Using trigonometric identities, we can determine the spectral components produced by multiplicative modulation. We will represent noise voltages as $\tilde{v}_{mab}$, where a and b are r or i for real or imaginary, and represent the phase (cosine or sine) of the carrier and of the modulating signal, respectively. Modulation of a cosine can be represented as

$$\sqrt{2}\left\{\tilde{v}_{mrr}\cos\omega_m t + \tilde{v}_{mri}\sin\omega_m t\right\}\cos\omega_c t$$

$$= \frac{\tilde{v}_{mrr}}{\sqrt{2}}\left[\cos(\omega_c + \omega_m)t + \cos(\omega_c - \omega_m)t\right]$$

$$+ \frac{\tilde{v}_{mri}}{\sqrt{2}}\left[\sin(\omega_c + \omega_m)t - \sin(\omega_c - \omega_m)t\right]. \quad (13.5)$$

We can identify the right side with the components in Fig. 13.3a. Similarly, the components in Fig. 13.3b can be represented as

$$\sqrt{2}\left\{\tilde{v}_{mir}\cos\omega_m t + \tilde{v}_{mii}\sin\omega_m t\right\}\sin\omega_c t$$

$$= \frac{\tilde{v}_{mir}}{\sqrt{2}}\left[\sin(\omega_c + \omega_m)t + \sin(\omega_c - \omega_m)t\right]$$

$$- \frac{\tilde{v}_{mii}}{\sqrt{2}}\left[\cos(\omega_c + \omega_m)t - \cos(\omega_c - \omega_m)t\right]. \quad (13.6)$$

The power in each component [e.g., $(\tilde{v}_{mii}/\sqrt{2})\cos(\omega_c + \omega_m)t$] on the right side of these equations is $\tilde{v}_{mab}^2/4$ (where $\tilde{v}_{mab} = \tilde{v}_{mii}, \tilde{v}_{mir}$, etc.). But we have seen that each component has power $N_0\,df/4$ so

$$N_0(f_c \pm f_m)\,df = \tilde{v}_{mab}^2. \quad (13.7)$$

This means that the total modulation power at frequency ω_m that is multiplying $\sin \omega_c t$ is

$$\left\langle \left[\sqrt{2} \left\{ \tilde{v}_{mir} \cos \omega_m t + \tilde{v}_{mii} \sin \omega_m t \right\} \right]^2 \right\rangle = \langle \tilde{v}_{mir}^2 \rangle + \langle v_{mii}^2 \rangle = 2 N_0 (f_c \pm f_m) \, df,$$

(13.8)

where $\langle \ \rangle$ indicates mean value, and an average has here been taken over the period of the sinusoids.

By the same process we can show that the modulation power multiplying the $\cos \omega_c t$ has the same expected value. Thus we can write the density of the noise modulation power N_m on either carrier in terms of the noise sideband power density N_0 as

$$N_m(f_m) = 2 N_0 (f_c \pm f_m).$$

(13.9)

13.4   NOISE AT THE PHASE DETECTOR OUTPUT

Now we should answer an important question: What voltage does this noise produce at the output of the phase detector? Assume a sinusoidal phase detector characteristic and that the voltage-controlled oscillator (VCO) is in quadrature with the signal (otherwise there will be a contribution from AM). Then, by the first representation, the phase modulation $S_\varphi(f_m) \, df_m$ will produce noise power

$$\langle u_{1n}^2 \rangle = \left(K_p' \right)^2 S_\varphi(f_m) \, df_m = \left(K_p' \right)^2 N_0 (f_c \pm f_m) \, df / P_c.$$

(13.10)

Let us compare this to the result obtained using the second representation of additive noise. For convenience, we choose the time origin such that the sinusoidal carrier is in phase[2] with the VCO and assume that the mixer acts like a multiplier.

We represent the modulation voltage due to noise in a narrow bandwidth as

$$v_m(t) = (2 N_m \, df)^{0.5} \sin(\omega_m t + \theta).$$

(13.11)

Here θ is an unknown phase and the modulation power is N_m, as before. The PD output due to noise is proportional to the product of the noise voltage, represented by the modulated sine carrier, and the VCO voltage, both at frequency $\omega = \omega_c$. This is

$$u_{1n} = \alpha \left[v_m(t) \sin \omega t \right] \left[V_v \sin \omega t \right],$$

(13.12)

[2] We do this for simplicity. The same results can be obtained by assuming an arbitrary angle but the expressions are longer.

where V_v is the VCO signal amplitude and α indicates the phase detector's efficiency as a multiplier. Combining the last two equations, the noise power at the PD output is

$$\langle u_{1n}^2 \rangle = \alpha^2 \left\langle \left[(2N_m \, df)^{0.5} \sin(\omega_m t + \theta)\langle \sin \omega t V_v \sin \omega t \rangle \right]^2 \right\rangle. \quad (13.13)$$

The first sine is the noise modulation. The other two are the noise carrier and the VCO. The product of those two is first averaged to eliminate the second harmonics, a process that occurs at the PD output. This leaves a DC component of 0.5. Then the mean squared value of the modulation is taken to obtain

$$\langle u_{1n}^2 \rangle = \alpha^2 N_m \, df V_v^2 / 4 \quad (13.14)$$

$$= \alpha^2 N_0 (f_c \pm f_m) \, df V_v^2 / 2. \quad (13.15)$$

To put this in more useful form we must obtain a value for α. We do that by writing the PD output voltage due to the desired signal and relating α to its signal strength. The output produced by the VCO and in-phase signal has an average value (i.e., removing the high-frequency terms as usual)

$$\langle u_{1s} \rangle = \alpha \langle V_v \sin \omega t (2P_c)^{0.5} \sin \omega t \rangle = \alpha V_v (2P_c)^{0.5} / 2, \quad (13.16)$$

where V_v is the VCO amplitude. Note that this is independent of the amount of noise. Since the voltage from a sinusoidal phase detector, when the signals are in phase, is K_p' radians, $\langle u_{1s} \rangle$ is equal to K_p' radians so

$$\left(K_p' \right)^2 = (\alpha/\text{rad})^2 V_v^2 P_c / 2 \quad (13.17)$$

or

$$\alpha^2 = 2 \left(K_p' \text{ radians} \right)^2 / \left(V_v^2 P_c \right). \quad (13.18)$$

Substituting this value into (13.15) gives the mean square noise voltage as

$$\langle u_{1n}^2 \rangle = \left(K_p' \text{ radians} \right)^2 N_0 (f_c \pm f_m) \, df / P_c. \quad (13.19)$$

Notice that we have obtained the same expression for $\langle u_{1n}^2 \rangle$ using either representation [compare to Eq. (13.10)].

The assumption that the phase detector operates, relative to the signal, as a multiplier with some efficiency factor α is fulfilled if the phase detector is a balanced mixer used in the normal fashion for mixers, with a relatively powerful local oscillator (LO) and total input at the signal port that is small enough that amplitudes are preserved in the mixing process (perhaps 10 dB or more below the LO power).

13.5 RESTRICTIONS ON THE NOISE MODELS

Here we discuss some conceptual limitations on the models that we have developed. We have indicated that the PM model cannot be used when the additive noise is too great. The AM (multiplicative) representation, on the other hand, does not have this restriction. Equation (13.19) does not depend upon N_0 being small. Since we have formulated all of our previous theory for PM, we interpret even u_{1n} from (13.19) in those terms. It is the voltage that would be produced by a phase deviation (from quadrature) of

$$\varphi = u_{1n}/K_p' \tag{13.20}$$

if φ were small enough that

$$\varphi \approx \sin \varphi. \tag{13.21}$$

When the reference is modulated by a large phase modulation that the loop does not track (outside the loop bandwidth), u_1 is smaller because (13.21) does not hold. The true sinusoidal phase detector (PD) characteristic has a more limited output than its ideal linear approximation. However, with additive noise, Eq. (13.20) is true regardless of whether Eq. (13.21) holds. Equation (13.19) was developed without reference to any sinusoidal characteristic and without reference to any desired signal. The terms K_p' and P_c were introduced in Eq. (13.19) only as a means of replacing the "new" variable α with variables that we are already using. In other words, (13.20) must apply to the phase error if our linear theory is to apply accurately to phase modulation, but that restriction is not necessary for the linear theory to apply to additive noise.

In using the multiplicative representation with the phase-locked loop (PLL), however, we have assumed a constant VCO phase. If the VCO phase should change, relative to the sine noise carrier in Eq. (13.6), the noise voltage produced by it at the output of the phase detector would be attenuated, multiplied by the cosine of the phase shift. However, the total noise power out of the phase detector would not have decreased because the contribution from the noise modulation on the cosine carrier would replace the reduction of noise from the sine carrier.[3] Or, equivalently, the sine carrier could be reestablished at a new phase angle to match that of the VCO. However, the model assumes that the VCO signal and the noise carriers are sinusoids and any phase modulation of them (as when their phase is allowed to change) violates this assumption. We know, however, that the phase of the VCO will change—that is why we are studying the effects of

---

[3]The same is true for our first model, PM on a signal. If the VCO phase is $\Delta\theta$ from quadrature, $\langle u_1^2 \rangle$ in Eq. (13.10) will be reduced by a factor $\cos^2 \Delta\theta$. However, the equivalent AM component would be $\langle u_1^2 \rangle$ from Eq. (13.10) multiplied by $\sin^2 \Delta\theta$ so the total value of the two components would be independent of $\Delta\theta$.

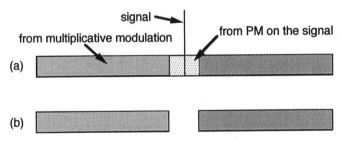

Fig. 13.4   Conceptualization of a combined noise model.

noise—but, if the change is slow enough and if it is uncorrelated with the noise, then we can use the model. For example, if we could move the phase of the noise carriers by only a small amount and infrequently, we could stay close to the assumed conditions while having a small overall effect on the accuracy.

The relative slowness and independence of the noise carrier is true for most of the noise when the bandwidth of the noise is quite wide compared to that of the loop. Most of the noise will not be followed by the phase of the VCO, and that noise will be represented almost perfectly by the multiplicative modulation model. This cannot be true of the noise with effective modulation frequencies that are within the loop bandwidth, however, and that noise can be of great significance because it is what the loop follows with least attenuation. So the noise power that is within the loop bandwidth must be small compared to the signal. If that is so, the signal (assuming it is steady) will stabilize the VCO output.

Since the equivalent phase noise within the loop bandwidth must be small, we could use the first representation for that part of the noise, while possibly using the second for the rest of the noise, as suggested by the division of noise shown in Fig. 13.4a. Suppose we identify a band of noise near the signal that is wider than the (two-sided) loop bandwidth but still has small modulation index. Then we can treat that as phase (and amplitude) modulation on the signal and the rest of the noise, shown in Fig. 13.4b, can be treated as multiplicative noise. Since the modulation index of the close-in noise is small, the phase modulation of the VCO will be small and, since the noise being tracked is close to the signal (small modulation frequency) compared to the remaining noise, the VCO will move slowly compared to the frequencies of the remaining noise. Thus, as long as Eq. (11.29) applies to the equivalent phase noise within the loop bandwidth (i.e., as long as the modulation index of the output of the loop is small), Eq. (13.4) accurately describes the equivalent phase noise, both within and without the loop bandwidth.

Example 13.2 u_1 Due to Additive Noise   A -30-dBm signal is embedded in white noise of density -100 dBm/Hz. A loop with a 1-kHz unity-gain bandwidth is locked to the signal. What is the noise voltage density at the output of the phase detector in the vicinity of 10 kHz if $K_p = 0.1$ V/rad?

From Eq. (13.4), $S_\varphi = -100$ dBm/Hz $+ 30$ dBm $= -70$ dBc/Hz $\Rightarrow -70$ dBr/Hz. In a 1-kHz bandwidth, this is $\sigma_\varphi^2 = (10^{-7}$ rad^2/Hz$)(1000$ Hz$) = 10^{-4}$ rad^2, which is much less than 1 rad^2. Therefore the representations discussed above apply. The loop will not track noise appreciably at 10 times its bandwidth so the error phase power spectral density (PPSD) at 10 kHz is the same as the input PPSD, and the 10-kHz voltage power spectral density at the PD output is

$$S_{u1} = S_\varphi K_p^2 = (10^{-7} \text{ rad}^2/\text{Hz})(0.1 \text{ V/rad})^2 = 10^{-9} \text{ V}^2/\text{Hz}.$$

13.6   DOES THE LOOP LOCK TO THE ADDITIVE NOISE?

When the signal is truly phase modulated, the VCO attempts to follow the modulated signal, and voltage at the phase detector output indicates its lack of success in doing so. The same is not true with additive noise. Even though we may represent the additive noise as an equivalent phase noise for computational purposes, the error signal required to maintain lock is generated by phase detection between the (unmodulated) signal and the VCO output.

We can discover one difference between the responses with true phase modulation and those with equivalent phase modulation by considering what happens when the angle between the VCO and signal moves away from quadrature (in an uncorrelated fashion, i.e., not in a locked loop). With true phase modulation, the detected signal decreases due to the loss of sensitivity of the phase detector at other phase angles. Not so with the equivalent noise. The magnitude of u_{1n} is unaffected by the phase angle.

However, within the bandwidth of a locked loop, u_{1n} will be reduced (see Fig. 11.4c) because the loop will track out the *equivalent* modulation. In this case, phase detection between the signal and the modulated VCO output, which has variance $\sigma_{\varphi n}^2$, produces a voltage that tends to cancel the voltage produced by detection between the VCO output and the additive noise. Note, however, that $\sigma_{\varphi n}^2$ is also the variance of the phase error and, while u_{1n} may be greatly reduced, that does not imply that $\sigma_{\varphi n}^2$ has been reduced. It is the existence of this noise-induced phase modulation at the loop output that causes u_{1n} to be reduced but which also produces a noiselike phase error. Within the loop bandwidth, u_1 is a direct measure of neither the additive noise nor $\sigma_{\varphi n}^2$ but, rather, of a combination of the two.

If the signal and accompanying additive noise pass through a limiter that removes amplitude variations, however, the AM component is stripped and the signal at the limiter output is a true phase-modulated signal. Figure 13.5 illustrates how maximum sensitivity to changes in phase is attained when the

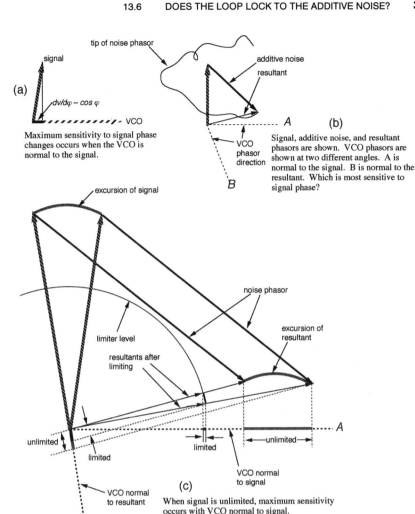

Fig. 13.5   Tracking the signals in additive noise. In this figure the noise phasor is very large for purposes of illustration. Limiting would not be maintained for much of the time with such large noise.

VCO is in quadrature to the signal in the presence of additive noise, but, once the signal is limited, maximum sensitivity occurs when the VCO maintains quadrature to the resulting phase-modulated signal. This supports our contention that the phase error should be considered to be between the VCO and a phase-modulated signal (not just its carrier), but, in the presence of additive noise, it is between the VCO and the actual signal (not the equivalent phase-modulated signal). When the VCO is at an angle that gives best sensitivity to signal changes, it is most likely to respond to those changes and maintain lock. Thus, in the presence of additive noise, the VCO tends to lock to, and maintain quadrature with, the signal and, in the case of phase nose, it tends to lock to, and maintain quadrature with, the modulated signal, not with its unmodulated carrier.

Let us summarize the answer to the question posed in the title of this section.

- With a signal that is phase modulated by noise, the loop locks to the total resultant noise-modulated signal.
- With a signal in additive noise, the loop locks to the signal, not to the equivalent phase-modulated signal, in the sense that it is the error between the signal and the VCO output that must be minimized for proper tracking.
- With a signal in additive noise, the loop acts as though it locks to the equivalent phase-modulated signal in the sense that it tends to reduce the noise at the phase detector output as if that were true.

13.7   OTHER TYPES OF PHASE DETECTORS IN THE PRESENCE OF NOISE

With our first representation of additive noise, in terms of PM, we need give no unusual consideration to the type of phase detector employed. However, when the additive noise is large such that we represent it as multiplicative modulation, there is some variation in its influence as the phase detector type changes [Blanchard, 1976, pp. 145–152]. The development above was for a phase detector that acts like a multiplier. As noted, this applies to balanced mixers with relatively weak signals and strong LOs, even square-wave LOs that switch the diode bridge in a manner similar to strong sinusoids.

13.7.1   Triangular Characteristic

If the input and VCO signals are both square waves, we have shown that a triangular characteristic is produced. If the input signal is a square wave in broadband noise, and we assume an infinite bandwidth in order to pass all the harmonics necessary to maintain a square signal, the equivalent phase noise, relative to that from the sinusoidal PD, will be 4 dB greater for the same signal power. (This is reduced to 1 dB difference, if the square wave has the same peak amplitude as the sine wave in the sinusoidal PD. The 3-dB difference is because a square wave has 3 dB more power than a sine wave with the same amplitude.) The added noise comes from the mixing of LO harmonics with noise that exists at frequencies that are far from the signal and that are assumed to be filtered out when the signal is sinusoidal but that must be passed to maintain the edges of a square wave.

If the signal is squared by passing it through a limiter that follows an input filter, thus allowing the input signal to be filtered, there is no effect as long as the signal-to-noise ratio (S/N) remains high. At the other extreme, where S/N approaches zero, a small degradation is seen in the low-frequency components of u_{1n} (0.3 dB for noise passed through a single-section bandpass input filter; 0.7 dB for an ideal rectangular filter) [Davenport and Root, 1965, Chapter 13, as given in Blanchard, 1976, p. 152]. The low-frequency components tend to be those that the loop follows because they are within its bandwidth.

13.7.2   Sawtooth Characteristic

This type of phase detector triggers on zero crossings (see Section 3.1.1). For purposes of analysis, the signal is assumed to pass through a hard limiter and then a differentiator before triggering the phase detector. As with the triangular characteristic, the performance is similar to that of a sinusoidal detector (mixer) at high S/N while a deterioration is seen at low S/N. By S/N = 0 dB, the peak of the phase detector output characteristic has dropped to about half of its noise-free value and the characteristic has begun to look sinusoidal. The low-frequency S/N deteriorates about 2.9 dB as S/N approaches zero.

13.8   MODIFIED PHASE DETECTOR CHARACTERISTIC WITH NOISE

Pouzet (1972) has shown that the triangular, sawtooth, and bang-bang (two-value) phase detector characteristics all approach sinusoidal shapes and shrink in amplitude with increasing noise as does the sinusoidal amplitude of a limiter followed by a multiplier. Wolaver (1991) has given a technique for computing the modification of phase detector characteristics in the presence of phase noise that applies, in a particularly uncomplicated manner, to the sinusoidal characteristic. We will consider that first. Recall that additive noise did not affect the PD characteristic of an ideal multiplier [Eq. (13.16)] but here we have only phase noise. The average output from the PD, in response to the loop phase error φ_e (between reference signal[4] and VCO) and in the presence of phase noise φ_n, is obtained as the average of the

---

[4]In the case of true phase modulation that is not followed by the VCO, the reference signal is the modulated signal, and φ_n is the difference between its phase and that of the unmodulated VCO. In the case of additive noise, the reference signal is the noiseless signal, and φ_n is the difference between its phase and that of the noisy VCO.

output voltage $v(\varphi_e + \varphi_n)$, weighted by the probability density $p(\varphi_n)$ of the random variable φ_n,

$$\overline{u_1} = \int_{-\infty}^{\infty} p(\varphi_n)v(\varphi_e + \varphi_n)\, d\varphi_n. \qquad (13.22)$$

Assuming probability is an even function of φ_e, we can substitute $p(-\varphi_e)$ for $p(\varphi_e)$, causing (13.22) to become the convolution of the probability density with the phase detector characteristic,

$$\overline{u_1(\varphi_e)} = p(\varphi)^* v(\varphi). \qquad (13.23)$$

The Fourier transform of this equation is

$$\overline{u_1(\omega_e)} = p(\omega) \times v(\omega). \qquad (13.24)$$

Thus the Fourier transform of the average PD characteristic in noise is the product of the transforms of the phase probability density function and the PD characteristic without noise.

In the Fourier frequency domain,[5] $\sin(\varphi)$ is represented by a pair of impulses at $\omega = \pm 1$ so the amplitude of the sinusoid will be multiplied by the transform of the probability density function there. Assuming a Gaussian distribution, the transform will also be Gaussian with variance as shown in Fig. 13.6. Thus the amplitude of the sinusoidal characteristic will be multiplied by the value of a Gaussian at $\omega = 1$.

$$\overline{K_p} = K_p \exp\left(-\frac{\omega^2}{2\sigma_\omega^2}\right) = K_p \exp\left(-\frac{1}{2\sigma_\omega^2}\right) = K_p \exp\left(-\frac{\sigma_\varphi^2}{2}\right), \qquad (13.25)$$

where σ_φ is the rms phase deviation and σ_ω is the corresponding standard deviation in the frequency domain.

We can also see from this development why the triangular and sawtooth PD characteristics approach sinusoidal shape in the presence of noise. The transform of those characteristics, $v(\omega)$, contain many harmonics (at $\omega = \pm n$ rad/sec). When $\sigma(\varphi)$ is small, the Gaussian in the frequency domain $\{\exp[-n^2/(2\sigma_\omega^2)]$ at harmonic $n\}$ will be very broad and will have little effect on the significant harmonics. Therefore, in Eq. (13.24), $\overline{u_1(\omega_e)} \approx v(\omega)$. However, as noise increases and $\sigma(\varphi)$ becomes wider, $\sigma(\omega)$ shrinks, and so, therefore, does the width of the Gaussian. As it does so, it attenuates the harmonics in the PD characteristic but has least effect on the fundamental at

---

[5] We need not relate this Fourier frequency to any frequency defined for the loop. It is merely a variable in the transform domain, which is being employed because it facilitates calculations.

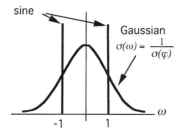

sine

Gaussian

$$\sigma(\omega) = \frac{1}{\sigma(\varphi)}$$

ω

-1   1

Fig. 13.6 Transforms of PD characteristic and phase probability density function.

$\omega = \pm 1$. Thus the frequency-domain representation in Eq. (13.24) approaches that of a sinusoidal characteristic and, therefore, so does the phase-domain characteristic.

Example 13.3 Phase Detector Characteristic Change by Noise A PLL is being used to demodulate a phase-modulated signal. How much will the demodulated output be reduced if additive noise causes the VCO to have an rms phase variation of 0.5 rad? (Assume the strength of the input signal remains fixed.)

The variance is

$$\sigma_\varphi^2 = (0.5 \text{ rad})^2 = 0.25 \text{ rad}^2.$$

By Eq. (13.25), the average PD sensitivity will be multiplied by $\exp(-0.25/2)$. The resulting gain change will be

$$10 \text{ dB} \log_{10}[\exp(-0.125)] = 10 \text{ dB} \ln[\exp(-0.125)]/\ln(10)$$

$$= -(1.25/2.3) \text{ dB} = \underline{-0.54 \text{ dB}}.$$

By applying (13.24) to the fundamental and harmonics of the triangular phase detector characteristic, we can find how the shape of the characteristic changes in the presence of phase noise. The effect on the peak values of both the sinusoidal and triangular characteristics is shown by the curves in Fig. 13.7. Note how the peak value of the triangular characteristic approaches that of the sinusoidal characteristic (even on a relative basis) as both decrease with increasing phase noise. The three data points are from Blanchard (1976), but these represent changes in the triangular characteristics in the presence of additive noise where the abscissa is the equivalent phase noise according to (13.24). In the former case (the curves) one of the two inputs to the phase detector is phase modulated by noise and, in the latter, it is accompanied by additive noise as it passes through a limiter in the

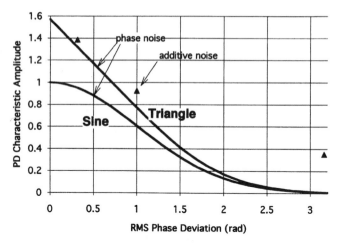

Fig. 13.7 Modification of the amplitude of the characteristics of phase detector in the presence of noise.

process of turning into a square wave. Blanchard (1976) and Stensby (1997) give curves that show changes with additive noise for both triangular and sawtooth characteristics.

By the same process we can determine how the slope of the PD characteristic in the center of the operating range, K'_p, is modified. The results are shown in Fig. 13.8.

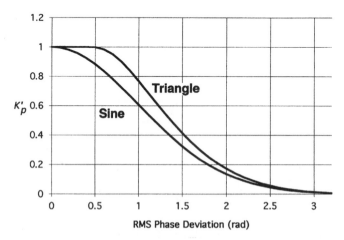

Fig. 13.8 Modification of the slopes of the characteristics of phase detectors in the presence of phase noise.

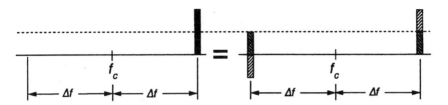

Fig. 13.9 Decomposition of a single sideband into the sum of an even pair and an odd pair.

13.A APPENDIX: DECOMPOSITION OF A SINGLE SIDEBAND

In Section 13.1 we saw how a pair of sidebands can be decomposed into equivalent modulation sidebands on a central carrier. Here we will see how the same basic results can be obtained by treating one sideband at a time.

The equivalence between a single sideband and two sideband pairs is illustrated in Fig. 13.9. Again, due to randomness, the likelihood of the even pair is the same as that of the odd pair. Thus we can equate the power density at any frequency f to eight equally likely, and therefore equally powerful, components. Four are real and four are imaginary. For each of these two types, two are an even pair and two are an odd pair. For each of these pairs, one component is located at f and the other is located symmetrically about f_c at $f_c \pm \Delta f$, where $\Delta f = |f_c - f|$. A component of the total (one-sided) noise power density of value $N_0(f)\, df$ is therefore equivalent to eight components, each with power $N_0(f)\, df/8$. If we just add powers at each frequency, we obtain a total of $N_0(f)\, df/2$ at each frequency. With the contribution from a symmetrically located sideband, the total would be $N_0(f)df$, as desired. However, because of the definite relationship between the even and odd pairs, their powers cannot be added. Rather, due to coherence, the voltages must be added and all of the power appears at the original sideband. If, however, some process, such as limiting, eliminates some of the components, the resulting signal can have power at both sidebands even if there was only power at one of them to begin with.

PROBLEMS

13.1 A one-sided random-noise power spectrum is shown in Fig. P13.1a along with a signal that serves as a "carrier" for the noise. Give the values A through D for the various representations of the noise.

L is the single-sideband relative power spectral density (one sided).

L_φ is that part of L that is due to phase noise.

$S_{1\varphi}$ and $S_{2\varphi}$ are the one-sided and two-sided power spectral densities, respectively.

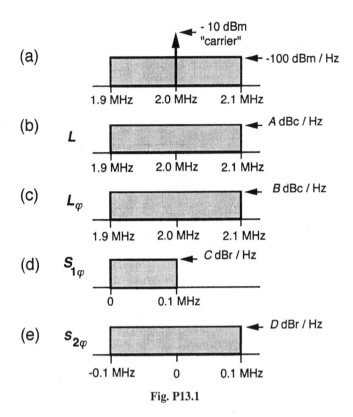

Fig. P13.1

dBm/Hz means dB relative to 1 mW per Hz, dBc/Hz means dB relative to the carrier power per Hz, and dBr/Hz means dB relative to 1 radian² per Hz.

13.2 The noise shown in Fig. P13.2 accompanies the 1-mW carrier into the loop shown. The loop has $\omega_n = 2\pi \times 2 \times 10^5$ rad/sec and $\zeta = 1$.

(a) Sketch the power spectrum at V_1. Show width, shape, and magnitude.

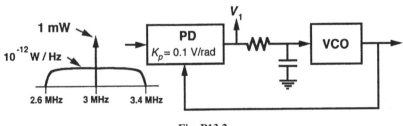

Fig. P13.2

(b) Sketch the phase power spectral density S_φ at the VCO output. Show width, shape, and magnitude.

(c) Sketch the sideband density (as one might see on a spectrum analyzer) at the VCO output. Show width, shape, and magnitude.

HINT: see Example 14.1.

13.3 A triangular PD characteristic can be represented by a series of sinusoids that are the components of its Fourier series. With the linear range between $-\pi/2$ and $+\pi/2$, the spectrum consists of odd harmonics. With $K'_p = 1$ V/rad, the nth harmonic is

$$(-1)^{\frac{n-1}{2}} \left[4 \text{ V}/(\pi n^2) \right] \sin(n\varphi_e).$$

(a) Show that (in the absence of noise) the first five terms give the correct peak amplitude to within 5%.

(b) Find the amplitude in the presence of Gaussian phase noise with an rms value of 1 rad, using the first five terms. [The accuracy will be better than 0.01% in this case. Why is it so much better than in part (a)?]

(c) What is the amplitude of the sinusoidal characteristic for the conditions of parts (a) and (b)?

CHAPTER 14

LOOP RESPONSE
TO ADDITIVE NOISE

In this section we will study the response of the loop to additive noise, using the representations of additive noise developed in the last chapter. Since the analysis will be carried out through an equivalent phase modulation, we could have studied this material before the last chapter. However, we will be considering flat noise spectrums that are characteristic of additive noise, so it is more appropriate to have studied the equivalence first. We will also be able to develop our skills in applying the equivalence. See Example 14.1.

14.1 NOISE BANDWIDTH

The variance, or mean square value, of the phase deviation of the voltage-controlled oscillator (VCO), $\sigma_{\varphi o}^2$, is the integral of $S_{\varphi, \text{out}}$ over all frequency. We saw in Chapter 12 how phase modulation on the reference appeared in a filtered form on the VCO of the locked loop. We saw in Chapter 11 how this $S_\varphi(f_m)$ can be integrated to obtain σ_φ^2. If a flat noise spectrum $S_{\varphi, \text{in}}(f_m) = S_{\text{flat}}$ on the reference is processed by the loop, the phase power spectral density at the output is

$$S_{\varphi, \text{out}}(f_m) = S_{\text{flat}} |H(f_m)|^2. \tag{14.1}$$

The output phase variance is

$$\sigma_{\varphi, \text{out}}^2 = \int_0^\infty S_{\varphi, \text{out}}(f_m) \, df_m \triangleq B_n S_{\text{flat}}, \tag{14.2}$$

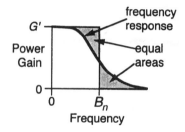

Fig. 14.1 Noise bandwidth.

where B_n is the one-sided[1] noise bandwidth. It is an equivalent bandwidth such that an ideal rectangular filter that passes modulation from $f_m = 0$ to $f_m = B_n$ will produce the actual output phase variance $\sigma_{\varphi,\,\text{out}}^2$.

Comparing the last two equations we see

$$B_n = \int_0^\infty |H(f_m)|^2 \, df_m. \qquad (14.3)$$

The concept of noise bandwidth has application considerably wider than the field of phase-locked loops (PLLs). Figure 14.1 illustrates the concept in general. Some level G' in a frequency response characteristic, usually the highest, is taken as nominal. The noise bandwidth is then the bandwidth B_n of a rectangular filter that has that nominal power gain G' within its passband and zero gain outside the passband and passes the same total power as does the actual filter when they are both excited by a uniformly distributed power density. In general, the right side of Eq. (14.3) equals $G'B_n$ but, for the particular case of a PLL, the maximum gain G' is unity.

If S_{flat} represents additive noise, Eqs. (13.4) and (14.2) can be combined to give

$$\sigma_{\varphi,\,\text{out}}^2 = B_n N_0/P_c. \qquad (14.4)$$

For a first-order loop,

$$B_n = \int_0^\infty \left| \frac{K/j\omega}{1 + K/j\omega} \right|^2 df_m = \int_0^\infty \frac{K}{K + j\omega} \frac{K}{K - j\omega} \, df_m \qquad (14.5)$$

$$= \frac{\text{cycle}}{2\pi} \int_0^\infty \frac{K^2}{K^2 + \omega^2} \, d\omega = \frac{K}{2\pi} \left[\tan^{-1} \frac{\omega}{K} \right]_0^\infty \text{cycles} \qquad (14.6)$$

$$= \frac{K}{4} \text{cycles.} \qquad (14.7)$$

[1] One sided means that only positive frequencies are considered.

Example 14.1   Loop Responses to Additive Noise   The noise shown below accompanies the 10-mW carrier into the loop shown. The loop has $\omega_n = 2\pi \times 10^5$ rad/sec and $\zeta = 1/(2\sqrt{2})$.

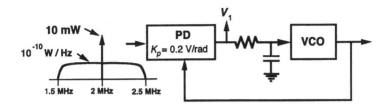

Sketch: (a) The power spectrum at V_1; (b) The phase power spectral density S_φ at the VCO output; and (c) The sideband density (as one might see on a spectrum analyzer) at the VCO output.

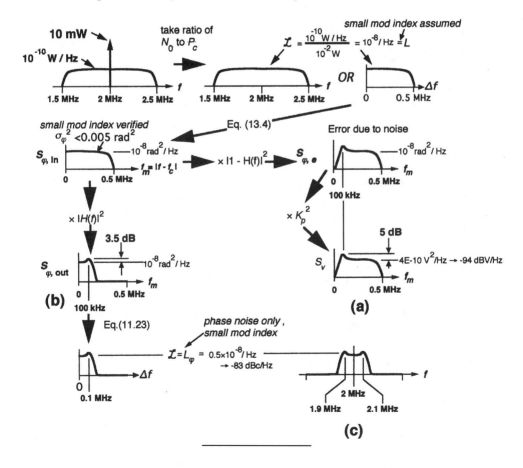

Note that, since K has units of reciprocal seconds (sec^{-1}), the units of B_n are hertz. Since the 3-dB bandwidth of a first-order loop is

$$\omega_L = 2\pi f_L \ \text{rad/cycle} = K, \tag{14.8}$$

B_n can be written in terms of loop bandwidth as

$$B_n = \frac{\pi}{2} f_L. \tag{14.9}$$

Thus the equivalent noise bandwidth extends beyond the 3-dB bandwidth by almost 60% for the single-pole response of the first-order loop.

For the second-order loop we integrate $|H(\omega)|^2$ from Eq. (6.4a) to get

$$B_n = \frac{\omega_n}{4} \left[\frac{1}{2\zeta} + 2\zeta\alpha^2 \right] \frac{\text{cycle}}{\text{rad}}. \tag{14.10}$$

The integration is described in Appendix 14.A. Surprisingly, Eq. (14.10) reduces to (14.7) for $\alpha = 0$. In other words, the addition of a low-pass loop filter has no effect on the noise bandwidth regardless of pole frequency; all of the responses in Fig. 7.8 for $\alpha = 0$, when taken as power gains, have the same area under them. This is illustrated in Fig. 14.2.

We might consider B_n to be the width of a filter on the demodulated signal. For example, we might pass the reference through a phase detector and than place a rectangular filter of width B_n on its output. This would be equivalent to filtering the phase variation with a rectangular filter, if we had a device that could do that, prior to demodulation. Alternately, we could put a rectangular filter on the reference signal before demodulation. It would extend from $f_c - B_n$ to $f_c + B_n$, and thus have a width of $2B_n$, and filter the noise that would produce phase modulation extending from 0 to B_n (assuming small m). These processes are illustrated in Fig. 14.3.

14.2 SIGNAL-TO-NOISE RATIO IN THE LOOP BANDWIDTH

The power of that half of additive input noise to which the loop responds and that passes through the effective filter is

$$P_n = 2B_n(N_0/2) = B_n N_0. \tag{14.11}$$

From this we can write the signal-to-noise (S/N) ratio as

$$\text{S/N} = P_c/P_n = P_c/(B_n N_0). \tag{14.12}$$

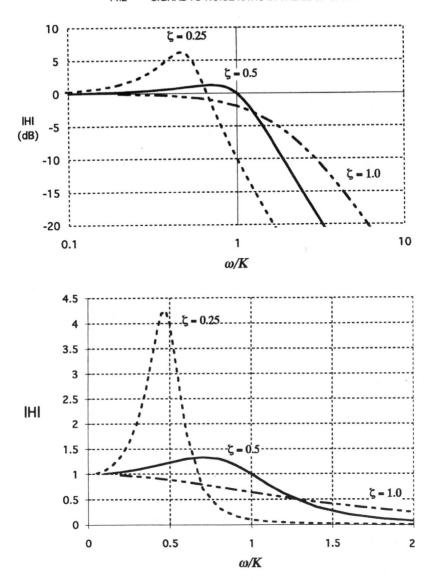

Fig. 14.2 Magnitude of $H(\omega/K)$ for $\alpha = 0$, $\zeta < 1$. These curves, which are also shown in Fig. 7.8, are drawn here to illustrate how the power under them can be the same for all $\omega_p [= K/(2\zeta)^2]$ if K is the same for each.

From Eq. (14.4) this is

$$S/N = 1/\sigma_{\varphi,\text{out}}^2 \qquad (14.13)$$

and is sometimes called signal-to-noise ratio in the loop bandwidth or "in the loop."

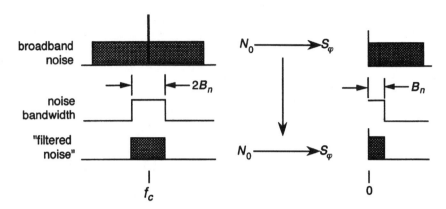

Fig. 14.3   Noise bandwidth, input and modulation equivalents.

14.3   LOOP OPTIMIZATION IN THE PRESENCE OF NOISE

Jaffe and Rechtin (1955) have developed a procedure for optimizing loop parameters in the presence of noise. It is an optimization in the sense that the noise is minimized under a constraint on the loop error when the loop is following certain input transients (step, ramp, parabola). In this section we will give the forms of the optimum transfer functions, which will tell us the general nature of the optimum loops, for each of the three kinds of inputs that they treated. For each kind we will then compute or indicate the parameters that show the tradeoff between response accuracy and output noise.

A theoretical basis, which should aid in extension of the theory to other kinds of inputs, is developed in Appendix 14.B.

14.3.1   The Problem

Optimize the loop transfer function for operation in the presence of noise to give simultaneously (a) *a constrained phase error in the response to the input* and (b) *minimum steady-state output phase noise, subject to that constraint.*

Ideally we would like to minimize both the noise and the error, but the optimum filter for noise would have zero bandwidth and the optimum filter for following the input would have infinite bandwidth. So we accept a certain error in the exactness with which we follow the input and discover the filter that will give the least noise while meeting that constraint.

14.3.2   Measures to Be Used

1. Steady-state noise:

$$\sigma^2_{\varphi,\,\mathrm{out}} = B_n N_0 / P_c. \tag{14.4}$$

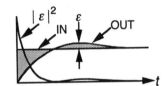

Fig. 14.4 Error in following a step input.

2. Integrated square error of the output $\varphi_{\text{out}}(t)$ relative to the input $\varphi_{\text{in}}(t)$ in the absence of noise: Let

$$\varepsilon(t) = \left| \varphi_{\text{out}}(t) - \varphi_{\text{in}}(t) \right|, \qquad (14.14)$$

as illustrated in Fig. 14.4. Then define E,

$$E^2 \equiv \int_0^\infty |\varepsilon(t)|^2 \, dt = \int_{-\infty}^\infty |\varepsilon(f)|^2 \, df, \qquad (14.15)$$

where $\varepsilon(f)$ is the Fourier transform of $\varepsilon(t)$. The last equation is true by Rayleigh's theorem if $|\varepsilon(t < 0)| \equiv 0$; The value of E^2 is a measure of the inaccuracy with which the output has followed the input.

14.3.3 Optimum Loop for a Phase Step Input

For a phase step, $\Phi(s) = \theta/s$, the optimum loop transfer function is

$$H(s)_{\text{opt}} = \frac{K}{s + K}. \qquad (14.16)$$

This form implies a first-order loop [see Eq. (2.25b)] with open loop $G = K/s$. The integrated error is

$$
\begin{aligned}
E^2 &= \int_{-\infty}^\infty |\Phi(f)|^2 |1 - H(f)|^2 \, df \\
&= \frac{\theta^2}{2\pi} \int_{-\infty}^\infty \frac{1}{\omega^2} \frac{\omega^2}{\omega^2 + K^2} \, d\omega = \frac{\theta^2}{2\pi} \left[\frac{1}{K} \tan^{-1}\left(\frac{\omega}{K} \right) \right]_{-\infty}^\infty = \frac{\theta^2}{2K}. \quad (14.17)
\end{aligned}
$$

Therefore, we can write

$$K = \frac{1}{2} \frac{\theta^2}{E^2}. \qquad (14.18)$$

Thus the loop gain can be chosen to give an acceptable ratio between the phase step θ and the error E in the response to it.

Using Eq. (14.7), the output variance equals

$$\sigma_{\varphi,\text{out}}^2 = \frac{N_0}{P_c}B_n = \frac{KN_0}{4P_c}\text{ cycle} = \frac{N_0}{8P_c}\frac{\theta^2}{E^2}\text{ cycle.} \qquad (14.19)$$

(With θ^2 in rad^2, E^2 in rad^2-sec, and N_0/P_c in rad^2/Hz, $\sigma_{\varphi,\text{out}}^2$ will be in rad^2 and K will be in sec^{-1} without unit conversions.) As we would expect, this shows that a larger tolerated error in response permits a smaller output steady-state phase variance. In practice, one would trade off E^2 against $\sigma_{\varphi,\text{out}}^2$. Then, having chosen these values, one would find K from Eq. (14.19) to get the minimum $\sigma_{\varphi,\text{out}}^2$ and the accepted value of E^2.

14.3.4 Optimum Loop for a Frequency Step Input

For the phase ramp (frequency step) input $\Delta\omega/s^2$, the optimum filter is a second-order loop with an integrator-and-lead filter and $\zeta = 1/\sqrt{2}$,

$$H(s) = \omega_n^2 \frac{1 + \left(\sqrt{2}/\omega_n\right)s}{s^2 + \sqrt{2}\,\omega_n s + \omega_n^2}. \qquad (14.20)$$

Following a procedure similar to that for the phase step, Jaffe and Rechtin (1955) give the expressions that enable us to determine the optimum natural frequency as a function of the ratio of step size to allowed squared error and to determine the variance, due to noise, at the output of the optimum loop,

$$\sigma_{\varphi,\text{out}}^2 = \frac{3}{4\sqrt{2}}\frac{\omega_n N_0}{P_c}\text{ rad-cycle} \qquad (14.21)$$

and

$$E^2 = \frac{(\Delta\omega)^2}{2\sqrt{2}\,\omega_n^3}\text{ rad}^3. \qquad (14.22)$$

We can rearrange this expression to give the natural frequency,

$$\left(\frac{\sqrt{2}\,\omega_n}{\text{rad}}\right)^3 = \frac{\Delta\omega^2}{E^2}, \qquad (14.23)$$

and combine (14.21) and (14.23) to show the tradeoff between error and noise,

$$\sigma_{\varphi,\text{out}}^2 = \frac{3}{8}\frac{N_0}{P_c}\left(\frac{\Delta\omega^2}{E^2}\right)^{1/3}\text{ rad}^2\text{-cycle.} \qquad (14.24)$$

With $\Delta\omega$ and ω_n in radians/second and other units as given above, unit conversions are again unnecessary.

14.3.5 Optimum Loop for a Frequency Ramp Input

For a frequency ramp input, which is a phase parabola having a slope of γt, the optimum loop is of third order. The filter transfer function is defined by

$$KF(s) = 2\omega_3 \frac{[s + (\omega_3 + j\omega_3)/2][s + (\omega_3 - j\omega_3)/2]}{s^2}. \quad (14.25)$$

This has a complex zero pair and two integrators. The resulting closed-loop transfer function is

$$H(s) = \frac{2x^2 + 2x + 1}{x^3 + 2x^2 + 2x + 1}, \quad (14.26)$$

where

$$x = s/\omega_3. \quad (14.27)$$

The closed-loop frequency response of this third-order loop is plotted in Fig. 14.5.[2]

From Jaffe and Rechtin's (1955) results we can also derive

$$\sigma^2_{\varphi,\text{out}} = \frac{5}{12} \frac{N_0}{P_c} \omega_3 \text{ rad-cycle} \quad (14.28)$$

and

$$E^2 = \frac{\gamma^2}{3} \left(\frac{\text{rad}}{\omega_3} \right)^3, \quad (14.29)$$

where γ is the slope of the frequency ramp.

14.4   SPECTRAL SHAPE OF OUTPUT POWER SPECTRUM

The phase power spectral density of Fig. 14.6a results in a power spectrum, shown at Fig. 14.6b, which is given by Eq. (11.23) if the deviation is small enough for Eq. (11.29) to be satisfied. When the deviation increases to the point where that restriction is not well satisfied, we can compute the approximate changes in the spectral shape of P_{sb} by observing the effects of one component of S_φ in a narrow bandwidth.

---

[2] Blanchard (1976, p. 166) suggests approximating the filter with two identical integrator-and-lead filters in series. That results in a response curve that has the correct peak value but crosses 4 dB at 20% higher frequency. He briefly describes an exact filter for which he references Gupta and Solem (1965) for discussion. While we will not discuss it further here, it does not appear that realization should be especially difficult. However, one must consider the stability of the active filter during design.

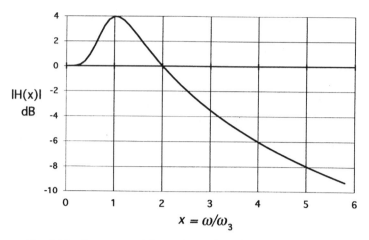

Fig. 14.5 Response of the optimum loop for frequency ramp.

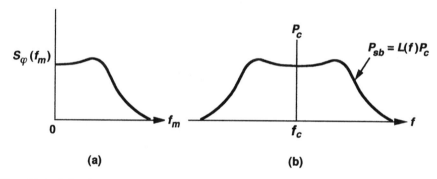

Fig. 14.6 (*a*) Phase power spectral density and (*b*) resulting sideband power spectrum.

While, for small modulation index, the central line has power P_c, which it has also in the absence of modulation, a more careful calculation can be made by considering first what happens if a single modulation component is accounted for. For small m, the power of the central line can be written

$$P_c = PJ_0^2(m_1) \approx P\left[1 - \frac{m_1^2}{4}\right]^2 \approx P\left[1 - \frac{m_1^2}{2}\right], \qquad (14.30)$$

and the power of each of the two first sidebands created by the modulation (Fig. 14.7) is

$$P_1 = PJ_1^2(m_1) \approx P\frac{m_1^2}{4}. \qquad (14.31)$$

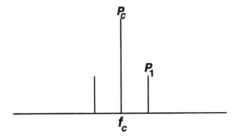

Fig. 14.7   One pair of noise sidebands.

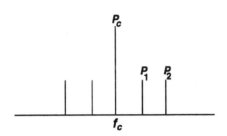

Fig. 14.8   Two pair of noise sidebands.

When a second modulation component is accounted for (Fig. 14.8), the carrier power becomes

$$P_c \approx P\left[1 - \frac{m_1^2}{2}\right]\left[1 - \frac{m_2^2}{2}\right] \tag{14.32}$$

$$\approx P\left[1 - \frac{m_1^2}{2} - \frac{m_2^2}{2}\right], \tag{14.33}$$

and the powers in each of the two first sidebands (on one side of the carrier) are

$$P_1 \approx P\frac{m_1^2}{4}\left[1 - \frac{m_2^2}{2}\right] \tag{14.34}$$

and

$$P_2 \approx P\left[1 - \frac{m_1^2}{2}\right]\frac{m_2^2}{4}, \tag{14.35}$$

respectively.

As we continue this process with more modulation components, we approach a condition similar to modulation by noise. The power in the central line [Eq. (14.33)] becomes[3]

$$P_c = P\left(1 - \frac{1}{2}\sum_i m_i^2\right) \equiv P\left(1 - \sigma_{\varphi,\text{out}}^2\right) \approx P\exp\left(-\sigma_{\varphi,\text{out}}^2\right), \quad (14.36)$$

and the sideband power [Eq. (14.34)] at an offset f_m becomes

$$P_{sb}(f_c \pm f_m) = P\left(1 - \sum_i \frac{m_i^2}{2}\right)\frac{m_m^2}{4}. \quad (14.37)$$

In a differential bandwidth we can write the ratio of this sideband power to total power as

$$L(f_c \pm f_m)\,df_m = \left(1 - \sigma_{\varphi,\text{out}}^2\right)\left[\frac{S_{\varphi,\text{out}}(f_m)}{2}\right]df_m, \quad (14.38)$$

where we have again equated the sum of individual mean square phase deviations to the variance of the phase.

Thus the power of both the central line and the sideband density are reduced by the same factor $(1 - \sigma_{\varphi,\text{out}}^2)$, while the general shape remains that determined by the shape of $S_{\varphi,\text{out}}$. Note that, while this is intended to improve the estimate of sideband levels at higher values of m, it nevertheless depends on the small modulation index approximations used in Eqs. (14.31), (14.32), and so forth, so it is only a valid approximation for $m < 1$ or so.

The total power is not changed by phase modulation. It is still related to the sinusoid's amplitude A by $P = A^2/2$. One check on the accuracy of our approximations is to consider to what degree they have modified this relationship. Before we had taken these reductions into consideration, we could have written the total power as

$$P_{T0} \approx P\left[1 + \int_0^\infty S_{\varphi,\text{out}}(f_m)\,df_m\right] = P\left[1 + \sigma_{\varphi,\text{out}}^2\right]. \quad (14.39)$$

Whether the approximation $L_\varphi = S_\varphi/2$, on which this is based, is accurate or not, the integral of the sideband power produces some additional power and that represents an error. Here that error is $P\sigma_{\varphi,\text{out}}^2$. The question is, does the use of Eqs. (14.37) and (14.38) reduce this error? Repeating the above process, but using the reduced sideband levels from those equations, we obtain

$$P_{T1} \approx P\left[1 - \sigma_{\varphi,\text{out}}^2\right]\left[1 + \int_0^\infty S_{\varphi,\text{out}}(f_m)\,df_m\right] \quad (14.40)$$

$$= P\left[1 - \sigma_{\varphi,\text{out}}^2\right]\left[1 + \sigma_{\varphi,\text{out}}^2\right] = P\left[1 - \sigma_{\varphi,\text{out}}^4\right]. \quad (14.41)$$

[3]In decibels, the change in carrier power is $10\log_{10}\exp(-\sigma_{\varphi,\text{out}}^2) = -8.7$ dB ($\sigma_{\varphi,\text{out}}$/rad).

This an improvement if $\sigma < 1$, and the smaller is σ the greater will be the relative improvement. For example, if $\sigma^2 = 0.1 \text{ rad}^2$, the error in total power in Eq. (14.40) is 10% of the carrier power while in Eq. (14.41) it is only 1%.

14.A APPENDIX: INTEGRATION OF EQ. (6.4a)

We will use indefinite integrals from a table by Bois (1961) and evaluate them from zero to infinite frequency. (A shorter solution can probably be obtained using complex integration in the s plane.) Starting with Eq. (6.4) we obtain

$$|H(\omega)|^2 = \left| \omega_n^2 \frac{1 + j(2\alpha\zeta\omega/\omega_n)}{-\omega^2 + j2\zeta\omega_n\omega + \omega_n^2} \right|^2 = \omega_n^4 \frac{1 + (2\alpha\zeta\omega/\omega_n)^2}{\omega^4 + 2(2\zeta^2 - 1)\omega_n^2\omega^2 + \omega_n^4}.$$

(14.42)

$$B_n = \int_0^\infty |H(\omega)|^2 \, d\omega = \omega_n^4 I_1 + (2\alpha\zeta\omega_n)^2 I_2.$$

(14.43)

For $\zeta < 1$, I_1 and I_2 are each of the form[4]

$$I = \left[k_1 \ln\left(\frac{\omega^2 + k_2\omega + k_3}{\omega^2 - k_2\omega + k_3} \right) + \frac{1}{k_4} \tan^{-1} \frac{k_5\omega}{k_6^2 - \omega^2} \right]_0^\infty$$

(14.44a)

$$= \lim_{\varepsilon \to 0} \frac{1}{k_4} \left\{ \tan^{-1} 0 - \tan^{-1}\left[\frac{k_5(k_6 + \varepsilon)}{-k_6\varepsilon} \right] \right.$$

$$\left. + \tan^{-1}\left[\frac{k_5(k_6 - \varepsilon)}{k_6\varepsilon} \right] + \tan^{-1} 0 \right\}$$

(14.44b)

$$= \frac{1}{k_4} \left\{ 0 + \frac{\pi}{2} + \frac{\pi}{2} + 0 \right\} = \frac{\pi}{k_4},$$

(14.44c)

where k_4 equals k_{41} or k_{42} for I_1 or I_2, respectively. For I_2,

$$k_{42} = 4\omega_n \sin\frac{\cos^{-1} k_7}{2} = 4\omega_n\sqrt{\frac{1 - k_7}{2}} = 4\omega_n\zeta,$$

(14.45)

---

[4]Because of the pole at $\omega = k_6$, evaluate the $\tan^{-1}(\ )$ from $\omega = 0$ to $\omega = k_6 - \delta$ and from $\omega = k_6 + \delta$ to $\omega = \infty$, where δ approaches 0.

where

$$k_7 = 1 - 2\zeta^2. \tag{14.46}$$

For I_1,

$$k_{41} = \omega_n^2 k_{42}. \tag{14.47}$$

Combining Eqs. (14.43) through (14.47), we obtain

$$B_n = \pi \left[\frac{\omega_n}{4\zeta} + \alpha^2 \omega_n \zeta \right] = \frac{\pi \omega_n}{2} \left[\frac{1}{2\zeta} + \alpha^2 2\zeta \right] = \frac{\omega_n}{4} \left[\frac{1}{2\zeta} + \alpha^2 2\zeta \right] \frac{\text{cycle}}{\text{rad}}. \tag{14.48}$$

This is Eq. (14.10).

Bois (1961) does not give Eq. (14.43) for $\zeta \geq 1$. We will not bother with $\zeta = 1$, since it applies to a zero-width interval and we expect continuity, but we will demonstrate that Eq. (14.48) is valid also for $\zeta > 1$. For $\zeta > 1$, Bois gives

$$I = \left[k_1 \tan^{-1}(k_2 \omega) + k_3 \tan^{-1}(k_4 \omega) \right]_0^\infty = \frac{\pi}{2}(k_1 + k_3), \tag{14.49}$$

where, for I_1,

$$k_{11} = \frac{2}{k_5\sqrt{2(k_6 - k_5)}} \qquad k_{31} = \frac{-2}{k_5\sqrt{2(k_6 + k_5)}} \tag{14.50}$$

so

$$k_{11} + k_{31} = \sqrt{2} \, \frac{\sqrt{k_6 + k_5} - \sqrt{k_6 - k_5}}{k_5\sqrt{k_6^2 - k_5^2}} = \frac{1}{\sqrt{2} \, \omega_n^2} \frac{\sqrt{k_6 + k_5} - \sqrt{k_6 - k_5}}{k_5}. \tag{14.51}$$

Here,

$$k_5 = 4\zeta\omega_n^2\sqrt{\zeta^2 - 1} \qquad k_6 = 2(2\zeta^2 - 1)\omega_n^2. \tag{14.52}$$

For I_2,

$$k_{12} = -\frac{k_6 - k_5}{k_5\sqrt{2(k_6 - k_5)}} = -\frac{\sqrt{k_6 - k_5}}{\sqrt{2} \, k_5}$$

$$k_{32} = \frac{k_6 + k_5}{k_5\sqrt{2(k_6 + k_5)}} = \frac{\sqrt{k_6 + k_5}}{\sqrt{2} \, k_5} \tag{14.53}$$

and

$$k_{12} + k_{32} = \frac{\sqrt{k_6 + k_5} - \sqrt{k_6 - k_5}}{\sqrt{2}\,k_5}. \tag{14.54}$$

Substituting Eq. (14.51) and (14.54) each into Eq. (14.49) and then the two resulting versions of (14.49) into (14.43), we obtain

$$B_n = \frac{\pi}{2}\frac{\omega_n^2}{k_5}\left\{\frac{1}{\sqrt{2}}\left[\sqrt{k_6 + k_5} - \sqrt{k_6 - k_5}\right]\right.$$

$$\left. + (\alpha\zeta)^2 2\sqrt{2}\left[\sqrt{k_6 + k_5} - \sqrt{k_6 - k_5}\right]\right\} \tag{14.55}$$

$$= \frac{\omega_n^2}{4k_5}\left\{\frac{1}{\sqrt{2}}\left[\sqrt{k_6 + k_5} - \sqrt{k_6 - k_5}\right]\right.$$

$$\left. + (\alpha\zeta)^2 2\sqrt{2}\left[\sqrt{k_6 + k_5} - \sqrt{k_6 - k_5}\right]\right\} \frac{\text{cycle}}{\text{rad}}. \tag{14.56}$$

Although the equivalence between Eq. (14.48) and (14.56) is not apparent, evaluation for $1 < \zeta \le 3$ showed the same values from each to within a relative error of one part in 10^{15}. Therefore, Eq. (14.55) will be used for all ζ

14.B   APPENDIX: LOOP OPTIMIZATION IN THE PRESENCE OF NOISE

This appendix endeavors to provide a theoretical basis for Section 14.3.

Rather than attempting to repeat Jaffe and Rechtin's (1955) rather lengthy mathematical development, we will base our discussion on two other procedures, minimization using Lagrange multipliers and the Weiner filter, which will be summarized but not derived here. In that way the results will be made available and a theoretical basis provided. However, those who are most familiar with these procedures will undoubtedly achieve a better understanding of the process.

14.B.1   Background

14.B.1.1   *Minimization under Constraint—Use of Lagrange Multipliers.*
To minimize a function $f(x, y, \ldots)$ subject to the constraint that another function has a given value, $g(x, y, \ldots) = g'$, minimize $[f(x, y, \ldots) + \lambda^2 g(x, y, \ldots)]$, where λ^2 is called the Lagrange multiplier. And λ^2 can be selected, perhaps subsequent to the minimization and perhaps implicitly, such that $g(x, y, \ldots) = g'$. Then both the constraint and the minimization will occur simultaneously.

14.B.1.2 Weiner Filter. The Weiner filter is the name sometimes given to a filter that minimizes the mean square error between the actual output and the desired output. The procedure includes safeguards to ensure that the filter is theoretically realizable in that it has no right-half-plane (RHP) poles.

Let $S(f)$ be the phase power spectral density (PSD) of the input (signal plus noise). Since PSD is an absolute value, it will be of the form $S(f) = N_0/P_c + X(f)X^*(f)$, where $X(s)$ is the desired input (equal to the desired output in our development). It is possible to write $S(f)$ in the form $S(f) = [\Psi(s)\Psi(-s)]_{s=j2\pi f}$ where $\Psi(s)$ has only left-hand-plane (LHP) poles and zeros while $\Psi(-s)$ has only RHP poles and zeros.[5] Then the optimum filter for minimum mean square error in reproduction of the desired input has transfer function

$$H(s)|_{\text{opt}} = \frac{1}{\psi(s)}\left[\frac{|X(s)|^2}{\psi(-s)}\right]_+ , \tag{14.57}$$

where $[\ ]_+$ indicates the realizable parts from the Heaviside expansion of $[\ ]$, that is, those parts having LHP poles. In other words, the term in the square brackets is formed by taking the indicated ratio and removing any RHP poles. Here $|X(s)|^2$ means $X(s)X(-s)$ [Truxal, 1955, pp. 469–471].

14.B.2 Explanation of Jaffe and Rechtin's Procedure

We wish to minimize $\sigma_{\varphi,\text{out}}^2$ under the constraint that $E = E_0$, the value that we are willing to accept for the integrated square error. Therefore, according to Section 14.B.1.1, we minimize $[\sigma_{\varphi,\text{out}}^2 + \lambda^2 E^2]$, choosing λ such that $E = E_0$ with the values of the parameters that occur at the minimum. But this is what the Wiener filter would do if the input were $\lambda\Phi(s)$, where $\Phi(s)$ is the Laplace transform of $\varphi_{\text{in}}(t)$. It would minimize the total square error, consisting of the mean square filter response to the input noise plus the integrated square difference between output and input, $\lambda^2 E^2$.[6] The solution is given by Eq. (14.57) with $X(s) \to \lambda\Phi(s)$.

$$H(s)|_{\text{opt}} = \frac{\lambda^2}{\psi(s)}\left[\frac{|\Phi(s)|^2}{\psi(-s)}\right]_+ . \tag{14.58}$$

[5]Symmetry about the real axis is necessary for transforms of real functions of time and symmetry about the imaginary axis corresponds to zero phase shift, which is necessary for an absolute value.

[6]Jaffe and Rechtin (1955) use a more fundamental development, which is related to Wiener filter theory, whereas this presentation attempts to make use of the developed filter theory more directly. Here E is an energy related to the Fourier transform of a single event (e.g., a step). In the usual Wiener filter, E is a power related to the Fourier transform-in-the-limit of a continuous process. In both cases noise power is minimized.

This will give the optimum filter shape, but not all the parameters are determined; λ is still to be chosen. We will probably do that implicitly by writing $\sigma_{\varphi,\,\text{out}}^2$ and E^2 for the loop and choosing a parameter (e.g., ω_n) to give the allowed value of E^2 or $\sigma_{\varphi,\,\text{out}}^2$.

14.B.3   Detailed Calculation for a Phase Step

For a phase step, θ/s

$$X(s) = \lambda\theta/s. \tag{14.59}$$

The two-sided PSD of a signal plus noise is

$$S_2(s) = \frac{N_0}{2P_c} + \frac{\lambda^2\theta^2}{(s)(-s)} = \left(\sqrt{\frac{N_0}{2P_c}} + \frac{\lambda\theta}{s}\right)\left(\sqrt{\frac{N_0}{2P_c}} + \frac{\lambda\theta}{-s}\right). \tag{14.60}$$

The 2 in the denominator of the noise term is because we use two-sided density with Fourier or Laplace transforms. From this we obtain

$$\psi(s) = \sqrt{\frac{N_0}{2P_c}} + \frac{\lambda\theta}{s}. \tag{14.61}$$

We begin substituting into Eq. (14.58):

$$\lambda^2\frac{|\Phi(s)|^2}{\psi(-s)} = -\lambda^2\frac{\theta^2}{s^2}\frac{1}{\sqrt{N_0/2P_c} - \lambda(\theta/s)}$$

$$= \frac{\lambda^2\theta^2}{s\left(\lambda\theta - s\sqrt{N_0/2P_c}\right)} = \lambda\frac{\theta}{s} + \frac{b}{s - \lambda\theta/\sqrt{N_0/2P_c}}. \tag{14.62}$$

The expression on the right, obtained by a Heaviside expansion, has an RHP pole, which is dropped in Eq. (14.58),

$$\lambda^2\left[\frac{|\Phi(s)|^2}{\psi(-s)}\right]_+ = \lambda\frac{\theta}{s}; \tag{14.63}$$

$$H(s)_{\text{opt}} = \frac{1}{\sqrt{N_0/2P_c} + \lambda(\theta/s)}\left(\lambda\frac{\theta}{s}\right) = \frac{\lambda\theta/\sqrt{N_0/2P_c}}{s + \lambda\theta/\sqrt{N_0/2P_c}} = \frac{K}{s + K} \tag{14.64}$$

where

$$K = \lambda\theta/\sqrt{\frac{N_0}{2P_c}}. \tag{14.65}$$

Having obtained this general form, we then write the mean square error E and the phase variance $\sigma_{\varphi,\text{out}}$ in terms of K and choose K to give the best tradeoff between E and $\sigma_{\varphi,\text{out}}$. One could solve for λ but it would only be of value in determining K in terms of E and $\sigma_{\varphi,\text{out}}$, and it is simpler to do that directly.

14.B.4   A Simplified Formula for $H(j\omega)|_{\text{opt}}$

Blanchard (1976, p. 162) gives

$$H(j\omega)|_{\text{opt}} = 1 - \frac{\sqrt{N_0/A^2}}{\psi(j\omega)}. \tag{14.66}$$

Apparently this is equivalent to Eq. (14.58) for the group of waveforms that are considered here, which have Laplace transforms of the form $(k/s)^n$ with $n \le 3$.

PROBLEMS

14.1 Give the one-sided noise bandwidth B_n in Hz for the following:

Loop Filter Type	ω_n (rad/sec)	ζ	K (sec^{-1})
(a) Low pass	1000	0.5	
(b) Integrator and lead	10^4	0.2	
(c) None			10^4
(d) Lag-lead	100	0.5	200

14.2 Find the one-sided noise bandwidth for the loop shown in Fig. P14.2. Give units.

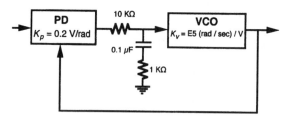

Fig. P14.2

14.3 A loop is to follow a 1-MHz input frequency step while limiting the integrated square difference between input and output phase to a maximum of

$$E^2 = \int_0^\infty [\varphi_{\text{out}} - \varphi_{\text{in}}]^2 \, dt = 0.01 \text{ rad}^2\text{-sec}$$

and minimizing the mean square output phase noise, $\sigma_{\varphi,\text{out}}^2$.

(a) What kind of loop (order, filter type, damping factor) should be used?

(b) The input signal power is 1 mW and the noise density is 1 mW spread evenly (flat) over 1 MHz. What should be ω_n and what will be the value of $\sigma^2_{\varphi,\text{out}}$?

14.4 The VCO output from a PLL is modulated by flat phase noise extending from 0 to 10 kHz.

(a) What is the phase power spectral density S_φ, in rad^2/Hz, that causes a 3-dB reduction in the power of the main signal (carrier)?

(b) How great would be the reduction, in decibels, if S_φ were cut in half?

CHAPTER 15

PHASE-LOCKED LOOP
AS A DEMODULATOR

The effects of noise often determine whether a particular design is usable. In this chapter we will study the loop as a demodulator, much as we did in Chapter 7, but this time we will describe the effects of noise. We will also consider the effects of noise in the carrier recovery circuits of Chapter 10.

15.1 PHASE DEMODULATION

A phase demodulator is pictured in Fig. 15.1. The signal passes through a filter, which limits the noise bandwidth, and enters the phase-locked loop (PLL) as its reference. Because the loop is designed to be too narrow to follow the modulation, the output of the phase detector is the demodulated signal. Also, because the loop does not follow the modulation, m must be small to prevent nonlinear operation.

The power spectrums of the phase-modulated input signal and additive noise are shown in Fig. 15.2. We assume that the signal is in the center of the input filter passband. The corresponding input phase, φ_{in}, consists of signal and noise, which are described by the signal phase power spectral density $S_{\varphi s}(\omega_m)$ and the noise phase power spectral density $S_{\varphi n}(\omega_m)$. We can multiply each by $|1 - H(\omega_m)|^2$ and then by K_p^2 and sum them to give the power spectrum S_{u1} at $\mathbf{x}$, which is illustrated by Fig. 15.3.

The signal spectrums in Fig. 15.2 look like those in Fig. 15.3 because the modulation index m is small and, therefore, Eq. (11.23) holds. Since m is small and the input phase deviation appears, essentially unaltered, as the phase error, u_1 will be proportional to phase. At $\mathbf{x}$ the power spectral density

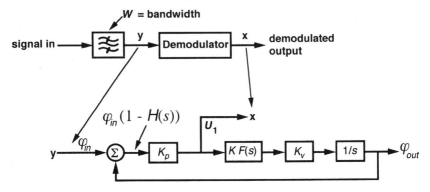

Fig. 15.1 Phase demodulator.

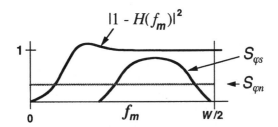

Fig. 15.2 Input power spectrums.

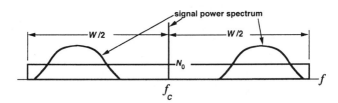

Fig. 15.3 Phase power spectrums and loop response.

of the demodulated signal is

$$S_{s1} = S_{\varphi s}K_p^2|1 - H(s)|^2, \tag{15.1}$$

and the noise power spectral density is

$$S_{n1} = S_{\varphi n1}K_p^2|1 - H(s)|^2 = (N_0/P)K_p^2|1 - H(s)|^2, \tag{15.2}$$

where P is the signal power. We can find the demodulated signal-to-noise ratio (S/N) by integrating the power spectral densities of the signal and noise. These are restricted by the input filter to a noise bandwidth $B_n = W/2$, where W is the equivalent input [radio frequency (RF)] bandwidth. The expression for the demodulated signal power S is

$$S = K_p^2 \int_0^{W/2} |1 - H(f_m)|^2 S_{\varphi s}(f_m) \, df_m = K_p^2 \sigma_{\varphi s}^2, \qquad (15.3)$$

where $\sigma_{\varphi s}^2$ is the mean square phase deviation of the filtered signal. If $|1 - H(f_m)| \approx 1$ for all f_m of significance in the signal, $\sigma_{\varphi s}^2$ is the mean square phase deviation of the original signal. The expression for the demodulated noise power is

$$N = K_p^2 \int_0^{W/2} |1 - H(f_m)|^2 S_{\varphi n} \, df_m = K_p^2 \left(\frac{N_0}{P} \right) \int_0^{W/2} |1 - H(f_m)|^2 \, df_m.$$

$$(15.4)$$

If $|1 - H(f_m)| \approx 1$ for $f_m < W/2$, which implies that the loop bandwidth is small compared to $W/2$ (i.e., $f_L \ll W/2$), this becomes

$$N \approx K_p^2 \left(\frac{N_0}{P} \right) \left(\frac{W}{2} \right), \qquad (15.5)$$

and the signal-to-noise ratio becomes

$$\frac{S}{N} \approx \frac{\sigma_{\varphi s}^2}{(N_0/P)(W/2)} = \frac{2P\sigma_{\varphi s}^2}{WN_0}. \qquad (15.6)$$

If $W/2$ exceeds the signal bandwidth (i.e., $W > 2f_m$ for all f_m), the S/N can be further improved by low passing the output at some video bandwidth, $B_v < W/2$. Then $W/2$ can be replaced by B_v in the expressions above.

The same results could be obtained with a traditional frequency discriminator followed by an integrator, producing a demodulated signal

$$v_d = K_d \int \delta f(\omega_m, t) \, dt = K_d \frac{\delta f(\omega_m, t)}{\omega_m} = K_d \, \delta\varphi(\omega_m, t), \qquad (15.7)$$

where K_d is the detector sensitivity and δf is the frequency deviation. This is not restricted to small modulation index, as is the PLL, but is less sensitive in the presence of noise [Blanchard, 1976].

15.2 FREQUENCY DEMODULATION, BANDWIDTH SET BY A FILTER

A PLL used as a frequency demodulator is pictured in Fig. 15.4. It is in many ways analogous, or complementary, to the phase demodulator. We might expect the signal spectrum to differ from that of Fig. 15.2 somewhat because constant frequency deviation implies greater phase deviation at lower modulation frequencies. For small modulation index, this would cause the signal power to climb near the carrier (or center) frequency. However, this loop is designed to follow the modulation so the restriction to small modulation index is relieved. The phase error can remain small even though the input phase deviates by many radians. At sufficiently high values of m the input spectrum approaches one in which the power is spread evenly between $f_c - \Delta f$ and $f_c + \Delta f$. The demodulated signal spectrum would look like the original information spectrum, perhaps like $S_{\varphi s}$ in Fig. 15.3. Generally the power of the demodulated signal will be given by an expression similar to Eq. (15.3),

$$S = \sigma_{\omega s}^2 / K_v^2, \tag{15.8}$$

where $\sigma_{\omega s}^2$ is the frequency variance of the signal, the mean square frequency deviation.

Other features in Fig. 15.3 will also differ for frequency modulation (FM), however. The loop response will be $H(\omega_m)$, and thus low-pass in nature, and the power created by additive white input noise will not be flat.

Combining Eqs. (11.5b) and (13.4), we obtain the frequency power spectral density due to additive noise as

$$S_{\omega n}(\omega_m) = \omega_m^2 N_0 / P \tag{15.9}$$

so the demodulated spectrums appear as in Fig. 15.5.

For the case where the loop is wide enough that $H(f_m) \approx 1$ for $f_m < W/2$, signal-to-noise ratio can be computed as it was for the phase demodulator. The power in the demodulated signal is

$$S = \frac{1}{K_v^2} \int_0^{W/2} |H(f_m)|^2 S_{\omega s}(f_m) \, df_m = \frac{\sigma_{\omega s}^2}{K_v^2}, \tag{15.10}$$

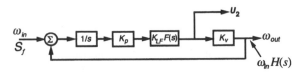

Fig. 15.4 Frequency demodulator.

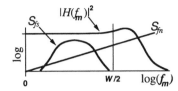

Fig. 15.5 Frequency power spectrums and loop response.

where $S_{\omega s}$ is the frequency power spectral density of the signal and $\sigma_{\omega s}^2$ is its mean square frequency deviation. The detected noise power is

$$N = \frac{1}{K_v^2} \int_0^{W/2} |H(f_m)|^2 S_{fn}(f_m) \, df_m \tag{15.11}$$

$$= \frac{1}{K_v^2} \int_0^{W/2} |H(f_m)|^2 f_m^2 S_{\varphi n} \, df_m \tag{15.12}$$

$$\approx \frac{N_0}{K_v^2 P} \int_0^{W/2} |H(f_m)|^2 f_m^2 \, df_m. \tag{15.13}$$

A Note on Units Equation (15.13) may use mixed units. We have attempted to use radian units for uniformity, but convention sometimes makes this awkward. Noise bandwidths are often in Hz and densities are usually given on a per-hertz basis. We could have used $S_{\omega n}$ in Eq. (15.11), as we did in Eq. (15.10), but Eq. (15.13) would then contain ω_m explicitly, even though the variable of integration is f_m. Some change would have to be made before integrating, even if integration were being done graphically, so we use S_{fn} and f_m. There will be no difficulty in execution as long as we carry units and treat them properly. The integral will have units of Hz^3. This will cancel Hz^{-1} in N_0 and Hz^{-2} in $1/K_v^2$ if K_v is given in Hz/V. If K_v is in (rad/sec)/V, as we have been trying to maintain it, the result must be multiplied by $(2\pi \text{ rad/cycle})^2$, which is, of course, equal to one. We could insert that factor in the equation, but it does not seem worth the increased complexity and might promote the thought that we can depend on answers that are obtained without consideration of units. Note also that, while Eq. (15.8) was written in radian units, it holds equally for Hz units and will be used in that form in some S/N equations to follow.

If $H(f_m) \approx 1$ for $f_m < W/2$, this is

$$N \approx \frac{N_0}{K_v^2 P} \left(\frac{1}{3}\right) \left(\frac{W}{2}\right)^3 = \frac{N_0 W^3}{24 K_v^2 P}, \tag{15.14}$$

and, using also (15.10),

$$\frac{S}{N} \approx \frac{24 P \sigma_{fs}^2}{W^3 N_0}. \tag{15.15}$$

Again we can provide video filtering on the demodulated signal, and the bandwidth of the noise will than be limited to the video bandwidth B_v instead of to $W/2$, giving

$$\frac{S}{N} \approx \frac{3 P \sigma_{fs}^2}{B_v^3 N_0}. \tag{15.16}$$

For sinusoidal modulation with amplitude Δf,[1] $\sigma_{fs}^2 = \Delta f^2/2$, and the above expression can be written

$$\frac{S}{N} \approx \frac{3 P \Delta f^2}{2 B_v^3 N_0}. \tag{15.17}$$

Since we do not want to admit any more noise than necessary, we set B_v equal to the maximum modulation frequency F_m, giving

$$\frac{S}{N} \approx \frac{3 P \Delta f^2}{2 F_m^3 N_0} = \frac{3}{2} \frac{\Delta f^2}{F_m^2} \left[\frac{P}{F_m N_0} \right]. \tag{15.18}$$

The expression in brackets is the S/N for a coherently detected 100% AM signal of modulation bandwidth F_m and carrier power P.[2] The preceding factor, $1.5 (\Delta f/F_m)^2$, is called the FM improvement factor. To take advantage of this factor, $\Delta f/F_m$ must be large and the price that is paid is additional bandwidth. According to Carson's rule, the required input bandwidth for FM is

$$W \geq 2(F_m + \Delta f), \tag{15.19}$$

whereas only $W \geq 2F_m$ is required for narrow-band FM.

The loop that we have been discussing acts like a standard discriminator that multiplies the frequency deviation by a gain factor, $K_d = 1/K_v$ and employs a noise-limiting filter. We shall next consider the use of the PLL to provide not only conversion from frequency to voltage but also filtering.

---

[1] For any sine wave, the root mean square (rms) value is the amplitude divided by $\sqrt{2}$ so the mean square value is this squared.

[2] This assumes a rectangular filter to pass the signal band. Even so, the transmitted power of a 100% amplitude modulation (AM) signal is $1.5P$ because there is power in the sidebands in addition to that in the carrier, as compared to P for the FM signal. On the other hand, a suppressed carrier AM signal can have a transmitted power approaching that of the sidebands alone, $P/2$.

15.3 FREQUENCY DISCRIMINATOR, FIRST-ORDER LOOP

If the first-order loop acts as the noise filter, the noise in Eq. (15.13) becomes

$$N = \frac{N_0}{K_v^2 P} \int_0^{W/2} \frac{f_m^2}{1 + (2\pi f_m / (K \text{ cycle}))^2} \, df_m \tag{15.20}$$

$$= \left(\frac{K \text{ cycle}}{2\pi} \right)^2 \frac{N_0}{K_v^2 P} \left[f_m - \frac{K \text{ cycle}}{2\pi} \tan^{-1} \left(\frac{2\pi}{K \text{ cycle}} f_m \right) \right]_0^{+W/2} \tag{15.21}$$

$$= (K \text{ rad})^2 \frac{N_0}{K_v^2 P} \left[\frac{W}{2} - \frac{K \text{ cycle}}{2\pi} \tan^{-1} \left(\frac{\pi}{K \text{ cycle}} W \right) \right]. \tag{15.22}$$

If the loop bandwidth is wide compared to $W/2$, the first terms of the small-argument expansion, $\tan^{-1} x \approx x - x^3/3$, can be used. The first term in the expanded product equals $W/2$, which cancels the preceding term leaving

$$N = \left(\frac{K \text{ cycle}}{2\pi} \right)^2 \frac{N_0}{K_v^2 P} \left[\frac{K \text{ cycle}}{2\pi} \left(\frac{\pi}{K \text{ cycle}} W \right)^3 \frac{1}{3} \right] = \frac{N_0 W^3}{24 K_v^2 P}. \tag{15.23}$$

This, not surprisingly, is the same as Eq. (15.14).

At the other extreme, let

$$K \text{ cycle} \ll \pi W \tag{15.24}$$

or, equivalently, $f_{-3 \text{ dB}} \ll W/2$, where $f_{-3 \text{ dB}}$ is the cutoff frequency of the first-order loop. Then $\tan^{-1}(\ )$ approaches $\pi/2$ and Eq. (15.22) becomes

$$N \approx K^2 \frac{N_0}{K_v^2 P} \left[\frac{W}{2} - \frac{K \text{ cycle}}{4} \right] \approx K^2 \frac{N_0 W}{2 K_v^2 P} \tag{15.25}$$

$$= \omega_{-3 \text{ dB}}^2 \frac{N_0 W}{2 K_v^2 P}. \tag{15.26}$$

Note that the filtering action is not very good here because the noise power still increases in direct proportion to the bandwidth of the input filter and could become extremely large. This is not surprising when we consider that $H(f_m)$ for a first-order loop never falls faster than f_m^2, which is the rate at which S_{fn} increases [see Eq. (15.9) and Fig. 15.5].

Note that we cannot also make use of the wide input bandwidth to get a higher FM improvement factor because the peak phase error could exceed the linear range, which we will now demonstrate. Set the loop bandwidth f_L equal to the maximum modulation frequency F_m. At this frequency, $\omega_L = K$, the phase error is related to the input phase by Eq. (2.26b), $\varphi_e = \varphi_{in}/(1 - j)$.

Combining this with Eq. (7.7), the peak phase deviation can be written in terms of the input peak frequency deviation as

$$\Delta\varphi_e = \frac{\Delta f}{f_L \sqrt{2}} = \frac{\Delta f}{F_m \sqrt{2}} \tag{15.27}$$

If we restrict the phase deviation by $\Delta\varphi_e < \pi/4$ for linearity, (15.27) requires $\Delta f < 1.1 F_m$, and the FM improvement factor is limited to

$$1.5(\Delta f/F_m)^2 = 1.8. \tag{15.28}$$

To summarize, while a first-order loop can be used as a frequency discriminator, it should not be depended on for noise filtering.

We have seen (Section 7.8.1) that a response similar to that of the first-order loop is available at the phase detector output u_1 when the loop has a low-pass filter. The low-frequency demodulation sensitivity is the same as for the first-order loop but, as illustrated by Fig. 7.17 for $\alpha = 0$, a choice of response shape is available; $\zeta = 0.7$ offers a particularly flat response. This response has the same problems relative to noise filtering, however, since it also falls at only -6 dB/octave at high frequencies. The noise at u_1 is given by Eq. (15.4) and by (15.5) for a loop narrow compared to $W/2$. But this is the same as Eq. (15.25) when $K = K_p K_v$ so S/N is the same as with a first-order loop [Blanchard, 1976, p. 182].

15.4 FREQUENCY DISCRIMINATOR, SECOND-ORDER LOOP

If a second-order loop has a response similar to a first-order loop, we can expect similar results when using it as a frequency demodulator. In order to produce better noise filtering, the loop response must fall faster than -6 dB/octave at high frequencies. This implies that the loop filter must be low pass. (We showed in Section 7.8.1 that this response can also be obtained across the capacitor of a passive lag-lead loop filter.) We will find an expression for the demodulated noise in that case. As usual, from these results we can estimate the performance of higher order loops having similarly shaped responses.

We use Eq. (15.13) again and obtain the expression for $|H(f_m)|^2 f_m^2$ for $\alpha = 0$ (low-pass filter) by manipulating Eq. (6.4). That will enable us to obtain the integral in Eq. (15.12) in terms of B_n, which we have already found. From Eq. (6.4) we can obtain

$$|H(\omega)|^2 = \omega_n^4 \frac{1 + \omega^2(2\alpha\zeta/\omega_n)^2}{\left(\omega_n^2 - \omega^2\right)^2 + (2\zeta\omega_n\omega)^2}. \tag{15.29}$$

From this equation alone we can write the algebraic relationship

$$\left| \omega_m H(f_m)_{\alpha=0} \right|^2 = \left(\frac{\omega_n}{2\zeta} \right)^2 \left[\left| H(f_m)_{\alpha=1} \right|^2 - \left| H(f_m)_{\alpha=0} \right|^2 \right]. \quad (15.30)$$

Substituting this into Eq. (15.12), we obtain

$$N = \left(\frac{N_0}{K_v^2 P} \right) \left(\frac{\omega_n}{2\zeta} \right)^2 \left\{ \int_0^{W/2} \left| H(f_m)_{\alpha=1} \right|^2 df_m - \int_0^{W/2} \left| H(f_m)_{\alpha=0} \right|^2 df_m \right\}.$$

$$(15.31)$$

As W approaches infinity, the integrals become noise bandwidths, B_n. Substituting their values from Eq. (14.10), we obtain

$$N = \left(\frac{N_0}{K_v^2 P} \right) \left(\frac{\omega_n}{2\zeta} \right)^2 \frac{2\zeta\omega_n \text{ cycle}}{4 \text{ rad}} = \frac{1}{8} \left(\frac{N_0}{K_v^2 P} \right) \left(\frac{\omega_n^3}{\zeta} \text{ cycle} \right). \quad (15.32)$$

This response does provide effective (but not necessarily optimum) filtering, justifying our ignoring W when it is much wider than the effective filtering of the loop. We have not obtained an expression for the general lag-lead or integrator-and-lead filter, but Fig. 7.8 illustrates the benefits of the response that we have treated. The noise filtering action is improved by the presence of filter roll off at high f_m and a two-pole Butterworth response is available for $\zeta = 0.707$.

15.5 SUMMARY OF FREQUENCY DISCRIMINATOR S/N

The mean square signal at the voltage-controlled oscillator (VCO) tuning line is σ_{fs}^2/K_v^2 and the noise is given by the various equations developed above. Since those noise expressions contain $N_0/(K_v^2 P)$, the signal ratio can be conveniently written

$$\frac{S}{N} = \gamma \frac{P\sigma_{fs}^2}{N_0}, \quad (15.33)$$

where γ is given in Table 15.1. The equations giving the noise are also indicated in the table. The equations have been written to make it easy to obtain the proper units for γ, which are, as can be seen from Eq. (15.33), Hz^{-3}. The voltage at the point where the demodulated signal is taken is v_D.

The first entry in Table 15.1 applies to the standard discriminator followed by a video filter. The first entry under "Loops" applies to any loop that does not supply filtering but depends on a preceding filter (W wide) or a subsequent video filter (B_v wide).

TABLE 15.1 Frequency Discriminator Characteristics

DISCRIMINATOR TYPE	$\gamma = \left(\dfrac{S}{N}\right)\bigg/\left(\dfrac{P\sigma_{fs}^2}{N_0}\right)$	CONDITIONS
CONVENTIONAL + rectangular filter		
slope $= 1/K_V$ V, f_{in} f_m, $B_V \to V_D$	$\dfrac{3}{B^3}$ where $B = W/2$ or B_V, whichever is much narrower (15.15), (15.16), (15.23)	
LOOPS		
Any wide loop $u_2 = V_D$		$B_n \gg W/2$
First Order $u_1 = u_2 = V_D$ $\left\{\gamma = \dfrac{2}{W\omega_{-3\,dB}^2}\right\}$ see (15.26), (15.25)	$\dfrac{8\pi^2}{(K\,\text{cycle})^2 W}$	
Second Order $u_1 = V_D$ $K = \omega_n/(2\zeta)$ With low pass $(\alpha = 0)$ $u_2 = V_D$ see (15.32)		$B_n \ll W/2$
with lag-lead, $u_1 \to u_2$, $\to V_D$ output across the shunt capacitor (see Section 7.8.1) $(0 < \alpha < 1)$	$\dfrac{4\zeta}{\pi f_n^3}$	

V_D is the discriminator output.

The remaining entries are for the condition wherein the loop provides significant filtering ($B_n \ll W/2$).

The second value of γ is given for the first-order loop and for a particular second-order loop when v_D is taken at the input of the loop filter, rather than at its output. The third value of γ applies to the second-order loop with a low-pass filter when v_D is taken from the loop-filter output and also to the more general second-order loop when v_D is taken across the capacitor in the passive loop filter. The equivalence of these two voltages was established in Section 7.8.1.

Example 15.1 S/N in a Frequency Discriminator A 1-mW signal is accompanied by a broad floor of additive noise of power density 10^{-6} W/Hz. The frequency demodulated signal is obtained from the VCO tuning line of a loop with a low-pass filter. The loop has $K_v = 1$ MHz/V, $\omega_n = 3000$ rad/sec, $\zeta = 0.5$. The signal has an rms frequency deviation of 50 kHz. Find the signal-to-noise ratio of the demodulated signal ($\omega_m \ll \omega_L$).
From Eq. (15.32),

$$N = \frac{1}{8}\left(\frac{10^{-6}\ \text{W/Hz}}{(10^6\ \text{Hz/V})^2 10^{-3}\ \text{W}}\right)\left(\frac{(3 \times 10^3/\text{sec})^3}{0.5}\ \text{cycle}\right)\left(\frac{\text{cycle}}{2\pi}\right)^2$$

$$= 1.7 \times 10^{-7}\ \text{V}^2.$$

The signal power is obtained from (15.8) as

$$S = (5 \times 10^4\ \text{Hz})^2/(10^6\ \text{Hz/V})^2 = 2.5 \times 10^{-3}\ \text{V}^2.$$

The S/N is given by the ratio as 1.5×10^4 or 41.7 dB.
We could also use Eq. (15.33) with γ given by Table 15-1 as $4\zeta/(\pi f_n^3)$,

$$\gamma = \frac{4(0.5)}{\pi[(3000\ \text{rad/sec})(\text{cycle}/2\pi\ \text{rad})]^3} = \frac{5.85 \times 10^{-9}}{\text{Hz}^3}.$$

This would be multiplied by

$$\frac{P\sigma_{fs}^2}{N_0} = \frac{10^{-3}\ \text{W}(5 \times 10^4\ \text{Hz})^2}{10^{-6}\ \text{W/Hz}} = 2.5 \times 10^{12}\ \text{Hz}^3$$

from Eq. (15.30), giving $S/N = 1.5\text{E}4$, as before.

15.6 NOISE IN A CARRIER RECOVERY LOOP

If a carrier recovery loop (for example, Fig. 10.9) receives the same phase noise $S_{\varphi,\text{in}}$ as does the demodulator that it supports, that part of the phase noise that the carrier recovery loop follows will not corrupt the demodulated signal. However, the same cannot be said for additive noise.

In the presence of phase noise φ_n we can rewrite Eq. (10.14) as

$$\varphi_M = M\varphi_{\text{in}} = M(\varphi + \varphi_n) + i2\pi \Rightarrow M(\varphi + \varphi_n). \qquad (15.34)$$

At the output of the carrier recovery PLL, the phase deviation due to noise, at modulation frequency f_m, will be

$$\varphi'_n(f_m) = MH_{\text{CR}}(f_m)\varphi_n(f_m), \qquad (15.35)$$

where H_{CR} is the loop transfer function, and after the output frequency divider it will be

$$\varphi''_n(f_m) = H_{\text{CR}}(f_m)\varphi_n(f_m). \qquad (15.36)$$

Therefore, the signal that is demodulated, by being compared to the recovered carrier, will be accompanied by phase noise,

$$\varphi_{\text{dem}}(f_m) = \varphi_n(f_m) - \varphi''_n(f_m) = [1 - H_{\text{CR}}(f_m)]\varphi_n(f_m). \qquad (15.37)$$

Since $\varphi_n(f_m)$ can be a narrow band of phase noise,

$$\varphi_n^2(f_m) = S_\varphi \, \delta f_m, \qquad (15.38)$$

Eq. (15.37) also implies that the phase power spectral density for the demodulated signal $S_{\varphi,\text{dem}}$, in response to and input density $S_{\varphi,\text{in}}$, is

$$S_{\varphi,\text{dem}} = S_{\varphi,\text{in}}|1 - H_{\text{CR}}|^2. \qquad (15.39)$$

Thus this noise is filtered by the error response of the carrier recovery loop, as discussed in Section 15.1. One would minimize the deleterious effects of this noise by maximizing the bandwidth of the carrier recovery loop. Equivalent performance can be expected from the Costas loop.

We would like to write the additive noise in terms of equivalent phase noise using Eq. (13.4). Unfortunately the relationship between a particular additive noise component and the signal changes as the data changes, and this de-correlates the phase noise at the carrier recovery output from the noise that accompanies the signal. For example, a particular sideband may be in quadrature with the signal, but, when the signal changes phase by 90°, it is suddenly in phase with the signal. Thus the effect of additive noise is not attenuated by the loop because the noise into the demodulator is not coherent with the noise at the output of the carrier recovery loop. In fact, that noise adds to the other noise in the demodulator.

The phase variance, due to additive noise, of the recovered clock in a Costas loop or multiply-and-divide recovery circuit (Fig. 10.9) is

$$\sigma_\varphi^2 = S_L^{-1}/\rho. \qquad (15.40)$$

Here S_L is the "squaring loss," ρ is the signal-to-noise ratio in the recovery loop bandwidth, and the PLL is preceded by a filter that is wide compared to the PLL [Lindsey and Simon, 1973, pp. 57–75]. Here S_L^{-1} is greater than unity and represents the additional noise due to mixing that is inherent in the multiplier. With (small) phase noise, the multiplier acts linearly on the phase deviation, but this is not the case with additive noise.

For a biphase Costas loop, (Fig. 10.7) having a one-pole low-pass filter with a noise bandwidth of B_1,

$$S_L^{-1} = 1 + N_0 B_1/(2P). \tag{15.41}$$

If the filter is an ideal rectangular filter,

$$S_L^{-1} = 1 + N_0 B_1/P. \tag{15.42}$$

For the squaring circuit (Fig. 10.9 with $M = 2$), B_1 is the equivalent low-pass bandwidth (half the actual bandwidth) of a filter that is inserted between the multiplier and the PLL and Eqs. (15.41) and (15.42) hold as for a PLL with the same low pass.

Thus the optimum bandwidth is a tradeoff between wide bandwidth to reduce the effects of phase noise and a narrow bandwidth to minimize the effects of additive noise.

PROBLEMS

15.1 The loop shown in Fig. P15.1 is driven by a frequency-modulated signal, which is embedded in noise as indicated by the power spectrum shown. (Hint: Section 7.8.1 should be useful here.)

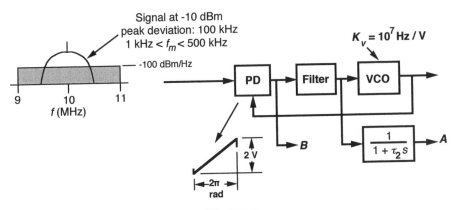

Fig. P15.1

(a) What is the peak signal output at A if the response is flat with modulation frequency?

(b) What is ω_n to give -3 dB response to FM at the maximum modulation frequency? Use a high-gain $[F(s) \approx (1 + \tau_2 s)/s]$ loop with $\zeta = 1/\sqrt{2}$ to frequency demodulate the signal, taking the output at A. Use this design in the remaining parts.

(c) Why use $\zeta = 1/\sqrt{2}$?

(d) How much noise power appears at A (assume 1Ω)?

(e) What is the maximum peak phase error deviation at any single frequency (due to signal modulation)?

(f) At what modulation frequency does that occur?

(g) Approximately how much noise appears at B (assume $1\,\Omega$)?

(h) What is the lowest modulation frequency for which accurate phase demodulation could be obtained at B (the -3 dB frequency)? (This is not related to the FM input spectrum shown.)

CHAPTER 16

PARAMETER VARIATION
DUE TO NOISE

The received signal strength can affect the loop parameters through its influence on the phase detector gain constant K_p. To maintain control of the loop parameters we can employ various techniques to maintain them constant. However, input noise modifies the effectiveness of these techniques. Here we will study the gain control techniques and how they are affected by noise.

16.1 PREVIEW

Before delving into the details let us consider the gain control techniques at a higher level.

16.1.1 Automatic Gain Control

We have seen [Eq. (13.17)] that K_p tends to be proportional to $\sqrt{P_c}$, where P_c is the signal power. If the phase detector were an ideal multiplier, this would be strictly true; but, even with practical balanced mixer phase detectors, it is true as long as the signal is weak relative to the voltage-controlled oscillator (VCO). Since the loop parameters depend on K, and therefore on K_p, they will change as the signal strength changes. One way to maintain control of the design parameters is to vary the gain preceding the phase-locked loop (PLL) to compensate for changes in signal strength. Such an automatic gain control (AGC) subsystem is shown in Fig. 16.1.

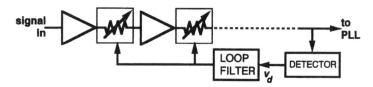

Fig. 16.1   AGC circuit.

Here the amplitude, or the power, of the signal entering the PLL is detected and used to control attenuators that precede the detector. There is a feedback control loop that attempts to maintain the strength of the signal seen by the PLL at a constant level as the input signal level varies. Of course, the degree to which this is accomplished depends on the AGC loop gain, in a manner analogous to the PLL. Gain may be distributed, as shown, to maintain an optimum tradeoff between noise and saturation level in the chain.

16.1.2 Limiter

Another device to control the input amplitude to the PLL is the limiter. The limiter output is a constant-amplitude signal with zero crossings (the time when the sine wave changes sign) synchronized to those of the signal at its input. Figure 16.2 illustrates such a circuit.

In this simple form of a limiter the input sinusoid v_{in} produces an approximately square wave v_d across the back-to-back diodes. As v_{in} rises through zero, v_d is essentially equal to v_{in} because the diodes do not draw appreciable current at low voltages (assume a high-impedance input to the filter). But, as the forward diode drop (≈ 0.7 V for silicon) is approached, the diodes begin to draw current and part of v_{in} is dropped across the resistor. As v_{in} far exceeds the diode forward drop, the diode impedance drops greatly, and v_d stays approximately constant until v_{in} drops toward zero again. The same occurs on the negative half cycle and, as a result, an approximately square wave is produced across the diodes. If a sine wave is required, the square wave is passed through a filter that rejects harmonics of the square wave (ideally there are no even harmonics so the filter must primarily cut off the third harmonic). The fundamental component of the

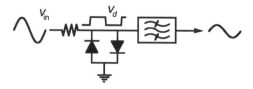

Fig. 16.2   Diode limiter.

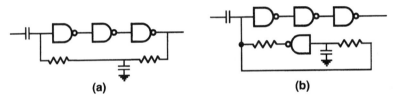

Fig. 16.3 Gate limiter circuit: (a) biased by feedback and (b) separate bias.

square wave is a sine wave with amplitude equal to $4/\pi$ times the amplitude of the square wave. The synchronism between input and output zero crossings is obvious and the amplitude varies little as long as the input signal is sufficiently large. Thus the limiter passes phase (frequency) information but, ideally, strips off amplitude variations.

Digital logic gates, with their inputs properly biased, can perform the limiting function also, particularly if several are placed in series. One advantage of the use of gates is that they operate naturally in a "saturated" mode, where the output levels are fixed. Therefore, as the signal level grows, the gates at the high end of the chain begin to operate in normal fashion.

Sometimes the average output of the chain of gates is fed back to provide bias for the input, as shown in Fig. 16.3a. Again, this forms a closed loop. A net inversion in the forward path (such as obtained from an odd number of NAND gates) is required for negative feedback. It can be useful in placing the input bias at a level that will ensure operation with small signals, since the steady-state direct current (DC) solution places the input bias in the linear region between the one and zero levels.[1] However, this circuit can have very high gain, and it has the potential to oscillate if not properly designed. In addition, it may not work well with certain signals, such as large pulse trains that do not have 50% duty cycle or even large square waves, since they might have to be biased far from their midvoltage to establish the duty factor required by the design. Another configuration, one that establishes the bias in an open-loop fashion, is shown at Fig. 16.3b.

Flip-flops can also serve as limiters if the signal is input to a properly biased clock line. At microwave frequencies, YIG (yttrium iron garnet) devices have been used for limiting.

16.1.3 Driving the Phase Detector Hard from the Signal

Since the output amplitude to a balanced mixer typically is little affected by small changes in the high-level local oscillator (LO) input, why not make the varying signal, rather than the VCO, the high level input? This may be beneficial under some circumstances, but it will only work over a limited

---

[1]Be cautious of possible negative consequences when some types of logic gates are held in this state.

range, between the power level at which the response to the LO flattens and the level at which the mixer's diodes are destroyed. (We might also say that for the simple form of the limiter in Fig. 16.2, but the range there is not necessarily as limited.) In addition, if there are multiple signals at the input, their tendency to produce additional spurious signals in the mixer is enhanced as their levels are increased.

If we do use this method, our first representation of additive noise as phase modulation (PM) on the signal will have no problem (assuming the restrictions on this representation are met) and, of course, there will be no problem with true phase noise. However, our second representation of additive noise as multiplicative modulation will not fare well in this analysis because it depends on representation of the phase detector as a multiplier.

16.1.4 Effects of Variations

These control devices have inherent limitations in the presence of noise. The limitations are due to the difficulty in differentiating between the signal and noise. To maintain K_p' constant, the signal must be maintained constant. If the control device holds the sum of signal and noise constant, then K_p' will decrease as the noise increases relative to the signal. We will characterize the control devices by a parameter η, called the suppression factor, that indicates by how much K_p' is reduced due to the presence of noise. Later we will study how these imperfections affect performance.

16.2 AUTOMATIC GAIN CONTROL

Certain aspects of automatic gain control, in particular the coherent detector and the effect of AGC in aiding acquisition, were previously discussed in Chapter 9.

16.2.1 Types of Detectors

We can choose the detector in Fig. 16.1 from several types. A DC or low-frequency output is generated when current is passed through a nonlinear device such as a diode. All the even-order terms (i.e., those where v is raised to an even power) in Eq. (3.1) generate DC. We can see this in Eq. (3.5), which contains the terms

$$v_d = c\left[\frac{A^2 + B^2}{2}\right]. \tag{16.1}$$

Since we are only considering one signal in an amplitude detector (not two as in the phase detector), let $B = 0$. At low levels (small A), v_d is proportional

Fig. 16.4   Diode amplitude detector.

to A^2, for which reason the detector is called a square-law detector. Since the DC voltage from the nth-order term will be raised to the nth power, all of these terms will change more rapidly with A than will the second-order term. Therefore we can expect this square-law performance to dominate at low signal levels. As the level increases, however, the other terms come into play, and the detector loses its square-law characteristic, commonly taking on a linear characteristic. We can see that, while the detector in Fig. 16.4 might generate a square-law output at low levels, it would become a rectifier at higher levels. When the diode begins operating like a switch, by being in either a relatively high- or relatively low-impedance state during most of the input wave, it allows the capacitor to charge only when v_{in} is positive. Thus the capacitor tends to hold something proportional to the amplitude of v_{in}.

The coherent detector, shown in Fig. 16.5, can be a linear amplitude detector even at low levels. The detected output comes from a balanced mixer whose LO is in phase with the detected signal. By Eq. (3.5), this phase will produce an output proportional to the signal amplitude. The LO for the coherent detector's mixer is produced by the VCO in a PLL, which is locked to the signal being detected. If the PLL gain is high, its input to the phase detector (PD) will be in quadrature with the signal, but the 90° hybrid circuit ensures that quadrature at the PD mixer implies phase alignment at the

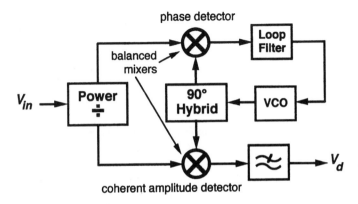

Fig. 16.5   Coherent amplitude detector.

other, amplitude detecting, mixer. Unlike the diode detector, noise or other signals to which the PLL is not locked will produce a non-DC output that can be attenuated by a low-pass filter.

When the coherent detector is used to control an AGC, the simultaneously operating PLL and AGC loops can present a complex analysis problem, but, once proper lock has been achieved, the coherent detector can provide accurate detection. Conversely, it does not provide such detection before the PLL is locked since the output from the balanced mixer, due to the signal, will then be roughly a sinusoid.

16.2.2   Square-Law Detection

With square-law detection the output of the detector is proportional to total power so the AGC circuit will attempt to keep it constant. The power gain of the AGC circuit can be written

$$G_p = \frac{P_o}{N_i + S_i} = \frac{P_o/N_i}{1 + S_i/N_i},$$ (16.2)

where P_o is the total output power, S_i is the input signal power, and N_i is the input noise power. The output signal power is

$$S = G_p S_i = \frac{P_o}{N_i + S_i} S_i = \frac{P_o S_i/N_i}{1 + S_i/N_i}$$ (16.3)

while the noise power output is

$$N = G_p N_i = \frac{P_o}{N_i + S_i} N_i = \frac{P_o}{1 + S_i/N_i}.$$ (16.4)

These relationships are plotted in Fig. 16.6. When the signal power dominates (right side of the figure), the total output power P_o is largely signal power, so, when the AGC controls the gain to achieve constant P_o, it is also controlling for constant signal level, which is what we want. Since the signal is held constant at the output, any change in input signal-to-noise (S/N) is reflected as a change in output noise. Conversely, when the noise power dominates, it is held constant at P_o and the output signal level changes with S/N.

Consider what happens if the noise level is constant. If the maximum gain is not high enough to amplify the noise to the design output level P_o, the gain will go to its maximum value when a signal is absent or very small. As the input signal power increases, it will be amplified linearly until the output power reaches P_o, at which point the gain will begin to decrease to hold the total output power at P_o. The gain-limited condition is represented by the

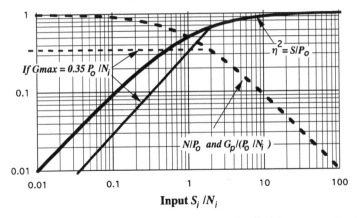

Fig. 16.6 AGC characteristics normalized to P_o, the limiting power level at the output.

straight lines in Fig. 16.6. It occurs at signal levels where the gain G_p is less than the required value, P_o/N_i, implying $S_i < (P_o/G_p) - N_i$.

From here we will consider operation in the region where sufficient gain is available to maintain constant power output (i.e., not within the straight-line regions). We can find the maximum gain required from the AGC circuit from Eq. (16.2) with S_i and N_i set at minimums.

We define a factor η, called the "suppression" factor, that describes the reduction in signal amplitude relative to the level set by the AGC in the absence of noise,

$$\eta^2 \triangleq \frac{S}{P_o}. \tag{16.5}$$

The output noise can be written in terms of η as

$$N = P_o - S = P_o(1 - \eta^2). \tag{16.6}$$

From these two equations the S/N at the output, which must equal the input S_i/N_i, since an AGC provides linear amplification, can be written in terms of η as

$$\rho \triangleq \frac{S}{N} = \frac{\eta^2}{1 - \eta^2} = \rho_i \triangleq \frac{S_i}{N_i}, \tag{16.7}$$

and, in turn, the suppression factor can be written in terms of ρ as

$$\eta^2 = \frac{\rho}{1 + \rho} = \frac{\rho_i}{1 + \rho_i}. \tag{16.8}$$

This is plotted in Fig. 16.10.

16.3 LIMITER

A second way to obtain constant input amplitude is through use of a limiter, which does not allow the voltage to rise above a level A' nor to drop below a level $-A'$. This is illustrated in Fig. 16.7. A sine wave is limited to excursions that are $\pm A'$ relative to its average value. If the wave is large compared to the limits (which we will assume), the output is approximately a square wave of amplitude A'. This is sent through a filter that passes only the fundamental component, the amplitude of which is $(4/\pi)A'$.

16.3.1 Limiting in the Presence of Small Noise

If additive noise is small enough so we can approximate it as phase and amplitude noise on the signal, we can represent the phase noise by Eqs. (13.3) and (13.4). Hard limiting (where a square wave is created as illustrated below) will remove amplitude variations, leaving only the phase modulation. Thus the relative phase power spectral density [e.g., in decibels relative to carrier per hertz (dBc/Hz)] will be reduced by passage through the limiter, but the noise will be all phase noise. See Fig. 16.8. Since this is the only kind that the PLL follows, the improvement in S/N is only apparent so far as the loop is concerned.

The value of L_φ is unchanged in passing through the limiter and thus so is S_φ. Therefore the equations in Section 13.4 still apply. The main change is in the value of P_c. It becomes set at a level determined by the limiter,

$$P_c = [(4/\pi)A']^2/2 = 0.81(A')^2. \tag{16.9}$$

This permits K'_p to be independent of the signal.

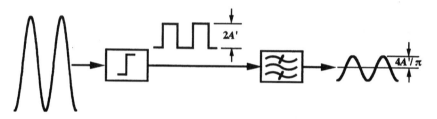

Fig. 16.7 Limiter.

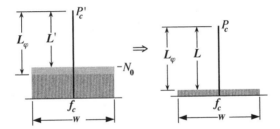

Fig. 16.8 Spectral changes in passing small noise through a limiter. The ordinates are logarithmic (dB).

16.3.2   Limiting in the Presence of Large Noise

If S_φ cannot satisfy the small-modulation-index restriction of Eq. (11.29), we must use the alternate representation of Section 13.3, multiplication of quadrature carriers. The signal is represented by v_c in Fig. 16.9 and the in-phase and quadrature noise voltages are represented by v_I and v_Q, respectively. The mean square values of these components are

$$\langle v_c^2 \rangle = P_c' \tag{16.10}$$

and, integrating (13.9) over $f_m \leq W/2$,

$$\langle v_I^2 \rangle = \langle v_Q^2 \rangle = N_0 W. \tag{16.11}$$

The resultant, v, varies in both amplitude and phase due to the noise. The output of the limiter v_{lim} is the same as v except that its amplitude is constant; thus it consists only of signal and phase noise.

The ratio of signal to noise at the output of an ideal limiter is [Blachman, 1966, pp. 86–87]

$$\frac{S}{N} = \frac{a}{b - a}, \tag{16.12}$$

where

$$a = \pi \rho_i \left[I_0 \left(\frac{\rho_i}{2} \right) + I_1 \left(\frac{\rho_i}{2} \right) \right]^2 \qquad b = 4 e^{\rho_i}, \tag{16.13}$$

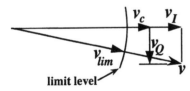

limit level

Fig. 16.9   Carrier and noise.

where ρ_i is the input S/N and I_0 and I_1 are modified Bessel functions of the first kind. The ratio of signal to total power P_o is

$$\eta^2 = \frac{S}{S+N} = \frac{a}{b} = \left\{ \frac{\sqrt{\pi \rho_i}}{2} e^{-\rho_i/2} \left[I_0\left(\frac{\rho_i}{2}\right) + I_1\left(\frac{\rho_i}{2}\right) \right] \right\}^2. \quad (16.14)$$

An approximate formula, accurate to 0.1 dB, is [Tausworthe, 1966]

$$\eta^2 \approx \frac{0.7854\rho_i + 0.4768\,\rho_i^2}{1 + 1.024\rho_i + 0.4768\,\rho_i^2}. \quad (16.15)$$

For high ρ_i, we can use the first term in the large argument expansion [Abromowitz and Stegun, 1964, p. 377] for I_v (see Appendix 16.A),

$$\lim_{z \to \infty} I_v(z) = \frac{e^z}{\sqrt{2\pi z}} \left\{ 1 - \frac{4v^2 - 1}{8z} + \frac{(4v^2 - 1)(4v^2 - 9)}{128 z^2} \right\} - \cdots, \quad (16.16)$$

to give $I_{0,1}(\rho_i/2) \approx \exp(\rho_i/2)/\sqrt{\pi \rho_i} \approx I_0(\rho_i/2)$ so Eq. (16.14) approaches $\eta^2 = 1$, as it must. Using the first two terms in Eq. (16.16), Eq. (16.12) also approaches $2\rho_i$, under this extreme condition, corresponding to the elimination of amplitude noise. For low ρ_i, $I_v(z) \approx (z/2)^v/\Gamma(v+1)$, which approaches [Abromowitz and Stegun, 1964, p. 375] 1 for I_0 and approaches 0 for I_1, so Eq. (16.14) approaches

$$\eta^2 \approx (\pi/4)\,\rho_i. \quad (16.17)$$

That is, the ratio of the output signal to P_o is 1.05 dB lower than ρ_i.

Suppression factors for both the limiter and the AGC are plotted in Fig. 16.10. Since the limiter is not linear, the output S/N can, and does, differ from the input S/N. Because the output is a square wave at the level set by the limiter, the power near the signal frequency, signal plus noise, is constant at P_o. The signal portion is $\eta^2 P_o$ so the noise power is $(1 - \eta^2)P_o$, and the output signal-to-noise ratio is

$$\frac{S}{N} = \frac{\eta^2}{1 - \eta^2}. \quad (16.18)$$

This is plotted in Fig. 16.11. At high ρ_i, the 3-dB increase in S/N is ignorable, insofar as its effect on a PLL, because the noise that remains is entirely phase noise, and thus the improvement is in the type of noise that does not affect the loop anyway. At low ρ_i, the deterioration of S/N as seen by a PLL is not as bad as it would seem from the plot because the additional noise is distributed more heavily toward higher modulation frequencies that

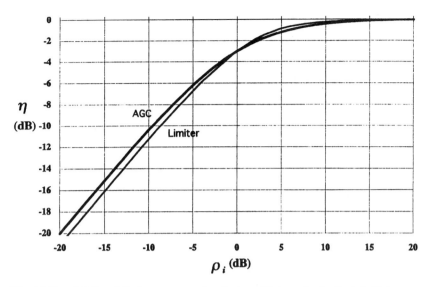

Fig. 16.10   AGC and limiter suppression factors. Note: η in decibels is $20\log_{10}\eta$.

the loop is less likely to follow. The effect seen at the output of a PLL, that is narrow compared to the noise, corresponds to a deterioration of ρ_i, compared to what would occur with the same loop parameters in the absence of the limiter, by only 0.25 dB, if the filter preceding the limiter is a single-pole filter, 0.49 dB if it is Gaussian, and approximately 0.64 dB if it is a rectangular filter [Springett and Simon, 1971]. Because this effect is small (and not simply quantified), and because the AGC has no effect, we will neglect any effect on S/N in a PLL due to either the AGC or the limiter.

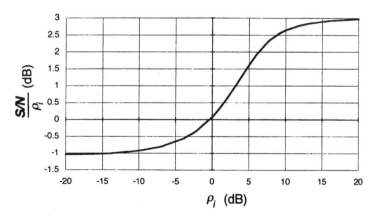

Fig. 16.11   Ratio of output to input S/N vs. input $S/N = \rho_i$ with a limiter (equals 0 dB with AGC). The effect at the output of a PLL is much less than this seems to indicate. It depends also on the type of noise and the distribution of the density.

16.4   EFFECTS OF GAIN VARIATION ON THE SECOND-ORDER LOOP

If $K_p = K_{p0}$ with the signal power at the design level P_0, the value when the signal has been reduced to $\eta^2 P_0$ will be $K_{pn} = \eta K_{p0}$. Therefore, the loop gain will decrease from K_0 to $K_n = \eta K_0$ in the presence of noise. This affects the second-order loop parameters as follows:

$$\omega_{nn}^2 = K_n \omega_p = \eta K_0 \omega_p = \eta \omega_{n0}^2; \tag{16.19}$$

$$\omega_{nn} = \sqrt{\eta}\, \omega_{n0}; \tag{16.20}$$

$$\zeta_n = \frac{1}{2}\left(\frac{\omega_p}{\omega_{nn}} + \frac{\omega_{nn}}{\omega_z} \right) = \frac{1}{2}\left(\frac{\omega_p}{\sqrt{\eta}\, \omega_{n0}} + \frac{\sqrt{\eta}\, \omega_{no}}{\omega_z} \right). \tag{16.21}$$

These effects can be seen in Figs. 4.3 and 4.4 if the various value of K shown there are taken to represent reductions in the effective value of K due to noise. Apparently the noise bandwidth and other bandwidths generally drop as η drops. While the effect of η on the loop parameters can be computed for any particular case, it is instructive to consider one case. It is not uncommon to have ω_n much higher than the geometric mean of ω_p and ω_z, in the manner illustrated by the highest curve in Figs 4.4 and 16.12. (Typically $\omega_p \ll \omega_n$ to keep the bandwidth narrow compared to the acquisition range and $\omega_z \ll \omega_n$ to give phase margin). This restriction,

$$\omega_n \gg \sqrt{\omega_p \omega_z}, \tag{16.22}$$

can be rewritten

$$\frac{\omega_n}{\omega_z} \gg \frac{\omega_p}{\omega_n}, \tag{16.23}$$

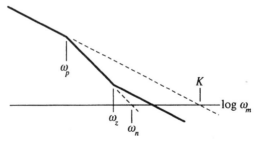

Fig. 16.12   Bode plot of a "high-gain" loop.

and therefore

$$\zeta_n \approx \frac{1}{2}\left(\frac{\sqrt{\eta}\,\omega_{no}}{\omega_z}\right) = \sqrt{\eta}\,\zeta_0. \qquad (16.24)$$

From Eq. (14.10), with (16.22),

$$B_n = \frac{\omega_n}{4}\left[\frac{1}{2\zeta} + \frac{2\zeta}{(1+\omega_z/K)^2}\right]\frac{\text{cycle}}{\text{rad}} \approx \frac{\omega_n}{4}\left(\frac{1}{2\zeta} + 2\zeta\right)\frac{\text{cycle}}{\text{rad}}. \qquad (16.25)$$

Substituting for ζ from Eq. (16.24), we obtain the noise bandwidth when the loop parameters are modified by the noise,

$$B_{nn}\big|_{\omega_n \gg \sqrt{\omega_p \omega_z}} = \frac{\sqrt{\eta}\,\omega_{n0}}{4}\left(\frac{1}{2\sqrt{\eta}\,\zeta_0} + 2\sqrt{\eta}\,\zeta_0\right)\frac{\text{cycle}}{\text{rad}} = \frac{\omega_{n0}}{8\zeta_0}(1 + 4\eta\zeta_0^2)\frac{\text{cycle}}{\text{rad}}.$$
$$(16.26)$$

Thus the noise bandwidth drops with decreasing suppression factor. Since the latter decreases with increasing noise, the noise at the output of the PLL tends to be more constant than it would in the absence of suppression.

Problem 16.1 illustrates how noise can reduce bandwidth and phase margin in a loop fed by an AGC. It is apparent from Fig. 16.12 why the phase margin would decrease when the gain decreases due to the noise. Under what conditions would noise tend to increase phase margin?

16.5   EFFECT OF AUTOMATIC GAIN CONTROL OR LIMITER ON AN OPTIMIZED LOOP

In Section 14.3 we studied the procedure for minimizing $\sigma_{\varphi,\text{out}}^2$ under a constraint on the allowed transient error $|\varepsilon|^2$. We did this by minimizing $\Sigma^2 = \sigma_{\varphi,\text{out}}^2 + \lambda^2 E^2$. Jaffe and Rechtin (1955) also considered how Σ varied with ρ, after the loop had been optimized at $\rho = 0$ dB, in the case where an AGC was employed and where a limiter was employed. Their results, illustrated in Fig. 16.13, show a fairly broad optimum that tolerates considerable variation in ρ.

Example 16.1   AGC Suppression Factor   A signal has power equal to the noise in a 1-MHz rectangular filter passband. The phase variance at the output of a high-gain loop due to the noise is 0.01 rad^2 and the loop

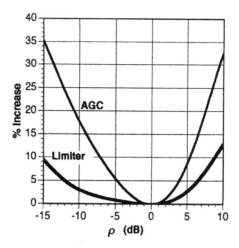

Fig. 16.13 Percent increase in Σ relative to continuously optimized loop as input S/N varies. The loop is optimized for minimum Σ with a frequency step at $\rho = 0$ dB. The percentage by which Σ exceeds the value that would be obtained by reoptimizing for each value of ρ is plotted for loops whose gain is controlled by AGC and by a limiter. [Based on Jaffe and Rechtin (1955, Fig. 10) © 1955 IEEE).]

has $\zeta = 1$ under these conditions. Compare the change in variance when the signal drops 10 dB with and without an AGC circuit.

From Fig. 16.10, a drop from $\rho_i = 0$ dB to $\rho_i = -10$ dB is accompanied by a 7-dB reduction in η. According to Eq. (16.26), B_{nn} is proportional to $(1 + 4\eta\zeta_0^2)$ and, since $\sqrt{\eta}\,\zeta_0$ is the effective value of ζ, this factor equals $(1 + 4) = 5$ when $\rho_i = 0$ dB (since the effective ζ is given as 1 under those conditions). This factor will drop to $(1 + 4 \times 10^{-0.7}) = 1.8$ when η drops 7 dB. Thus the noise bandwidth ratio is $B_{nn}(\rho_i = -10 \text{ dB})/B_n$ $(\rho_i = 0 \text{ dB}) = 1.8/5 = 0.36$. Without the AGC, the 10-dB decrease in ρ_i will cause a 10-dB increase in the phase variance, to 0.01 rad^2 $(10) = 0.1 \text{ rad}^2$. With the AGC, the noise bandwidth simultaneously shrinks to 0.36 of its former value so the variance is 0.036 rad^2.

16.A   APPENDIX: MODIFIED BESSEL FUNCTIONS

Figure 16.14 shows I_0 and I_1 plus points that correspond to the first two of the three terms in Eq. (16.16).

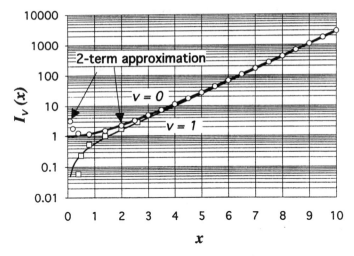

Fig. 16.14 Modified Bessel functions.

PROBLEMS

Assume one-sided noise and bandwidths and additive white noise, unless otherwise indicated, in the following problems.

16.1 Compare phase margins exactly in this problem. Do not use straight-line approximations for gain or phase. The objective is to discover the effects of noise on parameters, including phase margin, and we should not allow it to be masked by approximations. The block diagram in Fig. P16.1 shows parameters representative of the loop when it is tested with a strong input signal ($\rho \gg 1$).

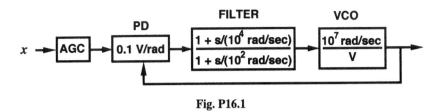

Fig. P16.1

(a) What is the phase margin in degrees?

(b) What is the noise bandwidth in hertz?

When the loop is used with a signal that is accompanied by noise of twice the power level of the signal, an AGC circuit maintains the same total power level at its output as during test. Under these circumstances:

(c) What is the phase margin in degrees?

(d) What is the noise bandwidth in hertz?

(e) If the signal and noise powers at **x** are equal, *approximately* by how much will the phase margin and noise bandwidth change if the AGC is replaced by a limiter (K_{p0} fixed)?

(f) With twice as much noise power as signal power into the AGC, what will be the signal-to-noise ratio at its output in decibels?

(g) Under these same conditions at the input to a limiter, what will be the signal-to-noise ratio at its output in decibels?

16.2 A limiter operates with an input noise level 3 dB below the input signal.

(a) If the noise is additive Gaussian noise, what will be the output signal-to-noise ratio of the limiter in decibels?

(b) If the noise is phase noise, what will be the output signal-to-noise ratio in decibels?

(c) If it is amplitude noise, what will be the output signal-to-noise ratio in decibels?

16.3 A circuit similar to Fig. P16.1 is driven by a signal and additive noise. The S/N at the AGC output measures 17 dB while the phase variance at the loop output is $(20 \text{ mrad})^2$. Another identical circuit is similarly driven through a limiter, and the S/N also measures 17 dB at the loop input (limiter output). What is the variance of phase from that loop?

CHAPTER 17

NONLINEAR OPERATION DUE TO NOISE IN A LOCKED LOOP

We have seen how high levels of noise can affect loop parameters by changing K_p when gain control devices are employed. Even if the input signal should be maintained at constant amplitude, however, noise modulation of the voltage-controlled oscillator (VCO), caused by input noise, can have a similar effect. The phase modulation on the VCO can drive the operating point out of the linear range of the phase detector. This reduces K_p, which affects the loop noise bandwidth, and it, in turn, affects the level of phase modulation on the VCO. In this chapter we consider how to analyze the loop in the presence of such high noise levels. We employ a method that modifies K'_p, based on the noise level, using a factor η as was done with gain control devices. We also study similar effects due to modulation. While this chapter concentrates on noise that is large enough to cause nonlinear operation, we will still assume that the loop is locked. Nonlinear effects large enough to cause loss of lock will be discussed in the next chapter.

17.1 NOTATION

To differentiate the various phase modulations that we consider, we define the input and output signals as illustrated below.

Loop in: $\qquad A\cos[\omega_c t + \phi_i(t)] + \underbrace{v_i(t)\sin \omega_c t + v_r(t)\cos \omega_c t}$ (17.1)

$\qquad\qquad\qquad\qquad\quad\uparrow \qquad\qquad\qquad\qquad\quad \uparrow$

$\quad$ Input phase modulation related $\qquad$ Input additive noise related

$\qquad\qquad\qquad\qquad\quad\downarrow \qquad\qquad \downarrow$

Loop out (VCO): $\quad -B\sin[\omega_c t + \phi_o(t) + \varphi_n(t)]$ $\qquad\qquad\qquad$ (17.2)

407

where $\phi_o(t)$ is the output phase in response to an input phase modulation $\phi_i(t)$; $\varphi_n(t)$ is the output phase response to the additive input noise, represented by $v_i(t) \sin \omega_c t$, that is in phase with the VCO. In cases where the loop is considered linear,

$$\varphi_n(t) \Rightarrow \varphi_o(t). \tag{17.3}$$

Even in cases where the noise level is high enough to cause the loop to be nonlinear, we designate φ_o as the linear response. We may compute it, based on linear assumptions, as a step in computing a more accurate value φ_n that takes the nonlinearity into account. Initially we will use $\phi_i(t)$ as a test modulation so we can see how the response $\phi_o(t)$ is affected by the presence of $\varphi_n(t)$.

17.2   PHASE DETECTOR OUTPUT u_1

The terms of interest that are generated in a balanced-mixer phase detector are obtained by retaining the difference-frequency parts of the product of Eqs. (17.1) and (17.2), as we have done previously.

$$u_1(t) = \alpha B \{ A \sin[\phi_i(t) - \phi_o(t) - \varphi_n(t)]$$
$$- v_i(t)\cos[\phi_o(t) + \varphi_n(t)] - v_r(t)\sin[\phi_o(t) + \varphi_n(t)] \} \tag{17.4}$$

$$= K_p' \Big\{ \sin[\phi_i(t) - \phi_o(t) - \varphi_n(t)]$$
$$- \frac{v_i(t)}{A} \cos[\phi_o(t) + \varphi_n(t)] - \frac{v_r(t)}{A} \sin[\phi_o(t) + \varphi_n(t)] \Big\}. \tag{17.5}$$

The noise terms represent a voltage $u_{1n}(t)$, as described by Eq. (13.19), with the restrictions of Section 13.5 applicable here also, so $u_1(t)$ can be represented by

$$u_1(t) = K_p' \sin[\phi_i(t) - \phi_o(t) - \varphi_n(t)] + u_{1n}(t). \tag{17.6}$$

If the phase error is small, this becomes

$$u_1(t) \approx K_p \{ \phi_i(t) - \phi_o(t) - \varphi_n(t) + n'(t) \}, \tag{17.7}$$

where

$$n'(t) = u_{1n}(t)/K_p \tag{17.8}$$

is the input phase that would produce $u_{1n}(t)$.

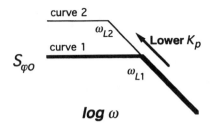

Fig. 17.1 Change in output phase noise with K_p.

Note that the linear relationship between $u_1(t)$ and $u_{1n}(t)$ does not depend on small phase deviation, since it exists in Eq. (17.6). This means that, while $n'(t)$ acts like an equivalent phase modulation, as we can see by Eq. (17.7), the voltage produced at the phase detector output is not diminished by the nonlinear nature of the sine function as is so for true phase modulation.

17.3   CHANGES IN THE OUTPUT SPECTRUM

Consider what happens to the output phase noise $\varphi_o(\omega_m)$ in response to input white additive noise as K_p decreases due, for example, to a reduction in signal strength or to a static phase error. Assume a first-order loop for simplicity. The straight-line approximation of the output phase power spectral density is shown in Fig. 17.1, curve 1. As K decreases, it causes ω_L to decrease. Since the corner between the flat and sloped phase noise regions moves lower, one or the other of these regions must change level. The question is, which one? Since only K_p, and not the rest of $|G(\omega_m)|$, has decreased, and since $u_{1n}(t)$ is not really affected by K_p, the output noise density at high frequencies will remain unchanged, as illustrated in Fig. 17.1, curve 2. Therefore, the low-frequency output phase noise density increases inversely as K_p. This is also because, within ω_L, the output phase moves to cancel u_1, but more phase deviation is required to produce a voltage to counter u_{1n} when K_p is lower. Another way to look at it is that the equivalent phase noise, $n'(t)$ in Eq. (17.8), increases as K_p decreases.

17.4   QUASI-LINEARIZATION METHOD[1]

We will now approximate the nonlinear effects produced by noise as a decrease in K_p.

---

[1]Blanchard (1976, p. 314) credits Develet (1963) with proposing this method for PLL study. Develet, in turn, credits the basis for his method to Booton (1952).

Example 17.1 Effective Gain with Noise   Determine the effective gain of a phase detector (PD) in the presence of uniformly distributed phase noise.

Figure E17.1-1 represents the operating point on a sinusoidal PD characteristic where the average phase is Θ and noise causes equiprobably phase deviations from $-\alpha$ to $+\alpha$. If the noise is sufficiently high in modulation frequency so the loop does not respond to it, the average phase will be the important loop variable.

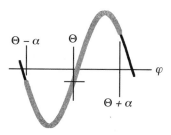

Fig. E17.1-1

For $\alpha = \pi$, $K \Rightarrow 0$ because the average phase is

$$\frac{1}{2\pi}\int_{\Theta-\pi}^{\Theta+\pi} \sin\varphi\,d\varphi = \left.\frac{-\cos\varphi}{2\pi}\right|_{\Theta-\pi}^{\Theta+\pi} = 0.$$

In general,

$$\frac{1}{2\alpha}\int_{\Theta-\alpha}^{\Theta+\alpha}\sin\varphi\,d\varphi$$

$$= \left.\frac{-\cos\varphi}{2\alpha}\right|_{\Theta-\alpha}^{\Theta+\alpha} = \frac{1}{2\alpha}[\cos(\Theta-\alpha) - \cos(\Theta+\alpha)] = \frac{\sin\alpha}{\alpha}\sin\Theta.$$

Therefore the average PD output would be given by the PD characteristic in the absence of noise multiplied by $[\sin(\alpha)]/\alpha$ (illustrated in Fig. E17.1-2), which equals one when $\alpha = 0$ (no noise).

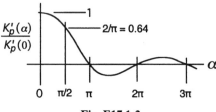

Fig. E17.1-2

17.4.1 Basics

By means of trigonometric identities, Eq. (17.6) can be expanded to

$$u_1(t) = K_p'\{\sin(\phi_i(t) - \varphi_o(t))\cos\varphi_n(t) - \cos(\phi_i(t) - \phi_o(t))\sin\varphi_n(t)\}$$

$$+ u_{1n}(t). \tag{17.9}$$

If we approximate the noise terms by their expected values, we obtain

$$u_1(t) \approx K_p'\{\sin(\phi_i(t) - \phi_o(t))\overline{\cos\varphi_n(t)} - \cos(\phi_i(t) - \phi_o(t))\overline{\sin\varphi_n(t)}\}$$

$$+ \overline{u_{1n}(t)}. \tag{17.10}$$

As long as φ_n and n' have zero mean,[2] the last two terms are zero. If we assume that the output phase noise is a Gaussian random variable (or so approximate it), the mean value of the cosine is

$$\overline{\cos\varphi_n(t)} = \exp\left(-\tfrac{1}{2}\sigma_{\varphi n}^2\right). \tag{17.11}$$

Therefore Eq. (17.10) becomes

$$u_1(t) \approx K_p'\exp\left(-\tfrac{1}{2}\sigma_{\varphi n}^2\right)\sin(\phi_i(t) - \phi_o(t)). \tag{17.12}$$

Thus the effect of the noise is to multiply K_p' by a suppression factor,

$$\eta_n \triangleq \frac{K_p'}{K_{p0}'} = \exp\left(-\frac{1}{2}\sigma_{\varphi n}^2\right). \tag{17.13}$$

The subscript 0 indicates the value of the parameter without noise. We would have obtained the same results had we used the method that gave Eq. (13.25).

We can use the suppression factor as we did previously. However, $n'(t)$ must be multiplied by $1/\eta_n$ in order that $u_{1n}(t)$ not be affected in Eq. (17.8). Thus, under nonlinear conditions due to noise, the effective input phase power spectral density given by Eq. (13.4) is modified to

$$S_\varphi(f_m) = \frac{N_0(f_c \pm f_m)}{P_c\eta_n^2} = \frac{N_0(f_c \pm f_m)}{P_c}\exp\left(\sigma_{\varphi n}^2\right). \tag{17.14}$$

[2] If φ_n has a nonzero mean, it should be included in the mean output phase, rather than the noise. If n' should have a mean (DC) value that would generally be an implementation problem.

17.4.2 Phase Variance in Loop with Integrator-and-Lead Filter

Let us apply this suppression factor to the high-gain loop discussed in Section 16.4. The resulting output phase noise is, from Eqs. (14.4) and (16.26),

$$\sigma_{\varphi n}^2 = \frac{N_0 B_{nn}}{\eta_n^2 P_c} = \frac{1}{\eta_n^2} \frac{N_0}{P_c} \frac{\omega_{n0}}{8\zeta_0}\left(1 + 4\zeta_0^2\eta_n\right)\frac{\text{cycle}}{\text{rad}} \tag{17.15}$$

$$= \exp\left(\sigma_{\varphi n}^2\right)\frac{N_0}{P_c} \frac{\omega_{n0}}{8\zeta_0}\left(1 + 4\zeta_0^2\exp\left(-\frac{1}{2}\sigma_{\varphi n}^2\right)\right)\frac{\text{cycle}}{\text{rad}}. \tag{17.16}$$

We can write the ratio of this phase variance to the variance under the linear assumption as

$$\frac{\sigma_{\varphi n}^2}{\sigma_{\varphi 0}^2} = \exp\left(\sigma_{\varphi n}^2\right)\frac{\left(1 + 4\zeta_0^2\exp\left(-\frac{1}{2}\sigma_{\varphi n}^2\right)\right)}{\left(1 + 4\zeta_0^2\right)}. \tag{17.17}$$

We can easily compute $\sigma_{\varphi 0}^2$ as a function of $\sigma_{\varphi n}^2$ and thus plot Eq. (17.17), as in Fig. 17.2.

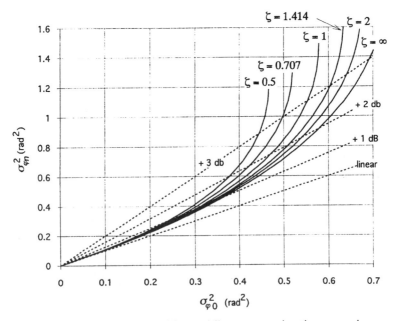

Fig. 17.2 Output phase variance with quasi-linear approximation vs. variance under linear assumption, high-gain second-order loop. Curves become double valued beyond extremes shown, suggesting inaccuracy at larger values. Here $\zeta = \zeta_0$.

17.4.3 Comparison to Other Results

These results have been compared to results using two other methods[3] of analysis for a loop with $\zeta = 0.707$. All three show an increase of about 1 dB over the linear approximation at $\sigma_{\varphi 0}^2 = 0.35$ ($\rho = 4.5$ dB). The quasi-linear approximation is more pessimistic than the others as it rises toward its asymptote at low S/N.

17.4.4 Effect on Phase Error

The reduced effective value of K_p also affects the mean phase error that occurs from various causes that we have previously considered. We can write Eq. (17.12) for a steady phase error, $\Phi_e = \Phi_i - \Phi_o$, as[4]

$$u_1 \approx K'_{p0} \exp\left(-\tfrac{1}{2}\sigma_{\varphi n}^2\right)\sin \Phi_e. \qquad (17.18)$$

We can write the phase error required to tune the oscillator from center frequency simply based on a sinusoidal PD characteristic.[5] Considering also the reduction of the effective gain in the presence of noise [Eq. (17.13)], we can write it as

$$\Phi_e = \sin^{-1}\left[\frac{\omega_{\text{out}} - \omega_c}{K'_0}\exp\left(0.5\sigma_{\varphi n}^2\right)\right]. \qquad (17.19)$$

Similarly, applying also Eq. (16.20), Eq. (9.5) for the error during a frequency sweep would become

$$\varphi_e = \sin^{-1}\left\{\exp\left(0.5\sigma_{\varphi n}^2\right)\left[\dot{\omega}_{\text{in}} - \dot{\omega}_c\right]\left[\frac{1}{\omega_{n0}^2} - \frac{t}{K_0}\right]\right\}. \qquad (17.20)$$

17.5 PHASE MODULATION

17.5.1 Tracking the Carrier

Phase-locked loops (PLLs) are often used to lock to the carrier frequency of a signal that is phase modulated. In such cases, the loop bandwidth is made narrow compared to the modulation frequency so it will not follow the modulation. Nevertheless, the modulation can affect the loop parameters. We will apply the quasi-linearization method to these effects, assuming that the modulation frequency is too high to be followed by the loop.

---

[3]From a JPL report by Lindsay and Trausworthe, reported in Blanchard (1976).
[4]Note that Φ_e is a deviation from the normal 90° phase difference.
[5]The characteristic without noise is expressed by Eq. (1.11). A difference (cos $\leftrightarrow$ sin) between (1.11) and (17.19) is due to a difference in the definition of phases in the two cases.

We write the phase detector output as

$$u_1(t) = K_p \sin(\Phi_e + \phi_i(t)), \tag{17.21}$$

where Φ_e is the difference between the average input phase and the constant output phase, ignoring the nominal 90°, and $\phi_i(t)$ is the phase modulation on the input signal. By trigonometric identity, we have

$$u_1(t) = K_p[\sin \Phi_e \cos \phi_i(t) + \cos \Phi_e \sin \phi_i(t)]. \tag{17.22}$$

So far as tracking of Φ_e is concerned, the effective value of u_1 is the value averaged over the variations in $\phi_i(t)$, since the loop does not follow them. Since the mean value of the sine is zero, this is

$$u_1(t)|_{\text{effective}} = K_p \overline{\cos \phi_i(t)} \sin \Phi_e, \tag{17.23}$$

so we can define a modulation suppression factor

$$\eta_m = \overline{\cos \phi_i(t)}. \tag{17.24}$$

For biphase modulation, where $\phi_i(t) = \pm \Theta$, Eq. (17.24) becomes

$$\eta_{m, \text{biphase}} = \cos \Theta. \tag{17.25}$$

This effect is illustrated in Fig. 17.3. The slope of the sinusoidal phase detector characteristic at the operating points is more shallow than at the center, so changes in the mean phase produce less voltage change.

For Gaussian modulation with phase variance σ_{mod}^2, Eq. (17.24) becomes

$$\eta_{m, \text{Gaussian}} = \exp\left(-\tfrac{1}{2}\sigma_{\text{mod}}^2\right). \tag{17.26}$$

For sinusoidal modulation, Eq. (17.24) becomes

$$\eta_{m, \text{sine}} = \overline{\cos[m \sin(\omega_m t + \theta_m)]} = \overline{J_0(m) + 2J_2(m)\sin(2\omega_m t + 2\theta_m) + \cdots} \tag{17.27}$$

$$= J_0(m). \tag{17.28}$$

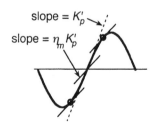

Fig. 17.3   Gain suppression by biphase modulation.

Since $J_0(2.4) = 0$, $\eta_m = 0$ at $m = 2.4$. At that modulation index the carrier disappears and there is nothing to lock to. Similarly, in the case of MPSK digital phase modulation, in which there are M equiprobable states at phases $(m \times 360°/M)$, $m = 0, 1 \ldots M - 1$), the mean value of $\cos \Theta$ is zero and

$$\eta_{m,\mathrm{MPSK}} = 0. \tag{17.29}$$

17.5.2 Phase Modulation with VCO Noise

In Eq. (17.6), we let $\phi_i(t) \Rightarrow \Phi_i + \phi_i(t)$, where Φ_i is the mean value and $\phi_i(t)$ represents input phase modulation, and $\phi_o(t) \Rightarrow \Phi_o$, representing that the VCO is not following the modulation. (This could represent carrier recovery, where the loop does not follow modulation.) We then take the mean value of $u_1(t)$ to determine the effect of simultaneous input phase modulation and VCO phase noise. The effective value of u_1 is thus

$$u_1(t)\big|_{\mathrm{effective}} = K_p' \overline{\sin(\Phi_e + \phi_i(t) - \varphi_n(t))}, \tag{17.30a}$$

where

$$\Phi_e = \Phi_i - \Phi_o \tag{17.30b}$$

is the mean phase error.

By trigonometric identity, this is

$$u_1(t)\big|_{\mathrm{effective}} = K_p'\left\{\sin \Phi_e \overline{\cos[\phi_i(t) - \varphi_n(t)]} + \cos \Phi_e \overline{\sin[\phi_i(t) - \varphi_n(t)]}\right\}. \tag{17.31}$$

The last term has zero mean and the first term can be expanded, again using a trigonometric identity, plus the fact that the mean of a sum equals the sum of the means:

$$u_1(t)\big|_{\mathrm{effective}} = K_p' \sin \Phi_e\left\{\overline{\cos[\phi_i(t)]\cos[\varphi_n(t)]} + \overline{\sin[\phi_i(t)]\sin[\varphi_n(t)]}\right\}. \tag{17.32}$$

Since ϕ_i and φ_n are independent, the means of the products are equal to the products of the means. The last term disappears due to zero means, leaving

$$u_1(t)\big|_{\mathrm{effective}} = K_p' \sin \Phi_e \overline{\cos[\phi_i(t)]}\; \overline{\cos[\varphi_n(t)]} = \eta_m \eta_n K_p' \sin \Phi_e. \tag{17.33}$$

Thus the gain suppression factor in the presence of both noise and modulation is the product of the two individual suppression factors. Similarly, in Eq. (17.18) through (17.20),

$$\exp\left(-0.5\sigma_{\varphi n}^2\right) \Rightarrow \eta_m \exp\left(-0.5\sigma_{\varphi n}^2\right). \tag{17.34}$$

Similarly the phase variance for a high-gain loop with an integrator-and-lead filter, Eq. (17.16), becomes

$$\frac{\sigma_{\varphi nm}^2}{\sigma_{\varphi 0}^2} = \frac{\exp(\sigma_{\varphi n}^2)}{\eta_m^2} \frac{\left(1 + 4\zeta_0^2 \eta_m \exp(-\frac{1}{2}\sigma_{\varphi n}^2)\right)}{\left(1 + 4\zeta_0^2\right)}. \qquad (17.35)$$

A set of curves similar to those of Fig. 17.2 could be plotted for any given value of η_m by the same method that was used to obtain Fig. 17.2. That is, we can solve for $\sigma_{\varphi 0}^2$ in Eq. (17.35) as a function of $\sigma_{\varphi nm}^2$ for any value of η_m. Alternately, we can solve iteratively for a particular value of $\sigma_{\varphi n}^2$, rather than drawing curves.

If φ_n is produced by additive input noise, the effective phase noise density must also be divided by this product to maintain u_{1n} constant.

$$S_{\varphi,\,\text{in}}(f_m) \Rightarrow \frac{N_0(f_c \pm f_m)}{P_c \eta_m^2 \eta_n^2} = \frac{N_0(f_c \pm f_m)}{P_c \eta_m^2} \exp(\sigma_{\varphi n}^2). \qquad (17.36)$$

17.5.3   Distortion of the Demodulated Signal

The clearest way to see how distortion can occur in the demodulated signal is to consider the phase detector as an ideal multiplier, multiplying the modulated input signal by the steady VCO signal. The spectral description of a signal modulated by a sinusoid is given by

$$v_{\text{in}} = A \left(\frac{\overbrace{J_0(m)\sin\varphi_c(t)}^{\text{control signal}} + \overbrace{J_1(m)\{\cos[\varphi_c(t) + \varphi_m(t)] + \cos[\varphi_c(t) - \varphi_m(t)]\}}^{\text{demodulated phase}}}{\underbrace{\begin{array}{c} -J_2(m)\{\sin[\varphi_c(t) + 2\varphi_m(t)] + \sin[\varphi_c(t) - 2\varphi_m(t)]\} \\ -J_3(m)\{\cos[\varphi_c(t) + 3\varphi_m(t)] + \cos[\varphi_c(t) - 3\varphi_m(t)]\} + \cdots \end{array}}_{\text{distortion in demodulated phase}}} \right).$$

$$(17.37)$$

Here $\varphi_c(t)$ and $\varphi_m(t)$ include ωt and a fixed phase θ. The parts of the input signal are marked in Eq. (17.37), to indicate their use after multiplication by the VCO signal, $B \cos \omega_c t$. This is shown in detail in Fig. 17.4. The first term produces a constant; this is the control signal that keeps the loop locked. The second term produces the desired sinusoidal signal. The remaining terms produce distortion. This occurs because of the fundamental nature of the demodulation process with a balanced-mixer phase detector, that is, multiplication.

$y_{in} = \sin[\varphi_c(t) + m\cos\varphi_m(t)]$ where $\varphi_c(t) = \omega_c t + \theta_c$ and $\varphi_m(t) = \omega_m t + \theta_m$

$$= \left\langle \begin{array}{l} J_0(m)\sin\varphi_c(t) + J_1(m)\{\cos[\varphi_c(t) + \varphi_m(t)] + \cos[\varphi_c(t) - \varphi_m(t)]\} \\ - J_2(m)\{\sin[\varphi_c(t) + 2\varphi_m(t)] + \sin[\varphi_c(t) - 2\varphi_m(t)]\} \\ - J_3(m)\{\cos[\varphi_c(t) + 3\varphi_m(t)] + \cos[\varphi_c(t) - 3\varphi_m(t)]\} + \cdots \end{array} \right\rangle$$

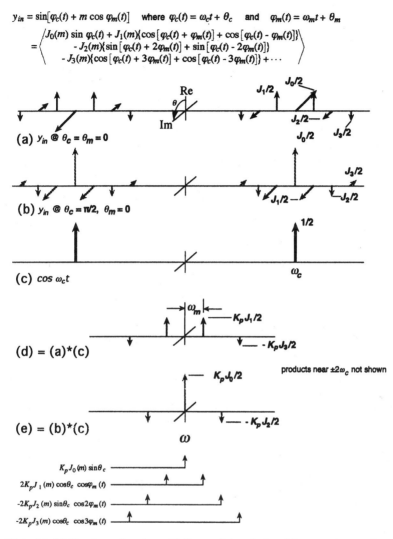

Fig. 17.4 Multiplication of a sinusoidally modulated signal by a cosine at the same carrier frequency. (* represents corresponding convolution in the frequency domain.)

The even harmonics of ω_m (with coefficients J_2, J_4, etc.) appear in the product after multiplication by $B\cos\omega_c t$ only to the degree that the phase relationship between the signal and VCO are not ideal, that is $\theta_c \neq 0$. This is analogous to the zero low-frequency output from a phase detector when one input is a sine and the other is a cosine.

To eliminate the distortion, the components that produce it must be filtered out of the input. Thus the input filter would have to pass signals offset from the carrier by the highest modulation frequency, $\omega_{m,\max}$, but reject signals offset by $3\omega_{m,\,\min}$ or, in case the signal and VCO are out of quadrature, $2\omega_{m,\,\min}$. Since modulation frequencies often cover many octaves, this is not necessarily possible. Moreover, such a filter might be very difficult to realize due to the narrow relative bandwidth or due to drift of the carrier frequency. The distortion can be reduced to an arbitrary level by keeping m small since the higher order Bessel functions drop more quickly as m drops.

Extra distortion terms also appear when the signal is modulated by multiple frequencies. These occur at frequencies that are offset from the carrier by the sums and differences of the modulation frequencies and the harmonics thereof. Such components do not represent distortion in the phase modulation but cause distortion as a natural result of the use of multiplicative demodulation, which treats the spectral components separately.

17.5.4 Demodulation with a Linear Phase Detector Characteristic

A phase detector with a sawtooth or triangular characteristic can be analyzed most efficiently in terms of its response to the modulation, since that is linear. If we try to analyze the sinusoidal phase detector in terms of its nonlinear phase response or if we try to analyze the triangular or sawtooth phase detector in terms of multiplication by an equivalent series of sinusoidal harmonics, we will get the same answers. We will just be less efficient and more prone to confusion.

Because u_1 is linearly proportional to ϕ_i, no distortion is produced (within the range of the phase detector, of course). These types of phase detectors do not need a carrier to lock to either. Figure 17.5 shows possible operating points for the four phases in a QPSK signal (Fig. 10.8).

Since these states are generally equiprobable, the signal has no spectral line at the carrier frequency. Nevertheless, it is obvious from Fig. 17.5a that the average voltage will change as Φ_i, the average value of ϕ_{in}, changes relative to Φ_o. When Φ_e changes, the average voltage will change by the same amount as does the voltage at each state. However, the linear range is reduced by $M = 4$, as illustrated in Fig. 17.5b.

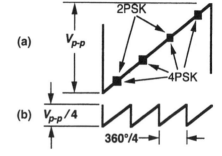

Fig. 17.5 Operating points for (a) QPSK signal (voltage versus phase) and (b) equivalent phase detector characteristic.

17.6   FREQUENCY MODULATION

17.6.1   Phase Error Expected in a PLL

We will now show that, in the absence of noise, a frequency demodulation loop that is made wider than the input half bandwidth $W/2$ should remain locked. By implication, a loop that is used also for filtering, that is, one that is made as wide as the maximum modulation frequency F_m but narrower than $W/2$, may not. Carson's rule states that the bandwidth necessary for a frequency-modulated (FM) signal is

$$W \geq 2[\Delta f + F_m], \tag{17.38}$$

where Δf is the peak frequency deviation and F_m is the highest modulation frequency. If the bandwidth of the PLL is made at least $W/2$ wide, the unity-gain frequency will be approximately

$$f_L \geq [\Delta f + F_m], \tag{17.39}$$

and the peak phase error will be

$$\varphi_e = \frac{\Delta f}{f_m}(1 - H(f_m)) \approx \frac{\Delta f}{f_m G(f_m)}. \tag{17.40}$$

Here the approximation is possible because the modulation frequencies are within the loop bandwidth where $(1 - H)$ can be approximated as $1/G$. Since $G(f_m)$ will be falling at least as fast as $1/f_m$, the value of φ_e will be no larger than what occurs at the unity-gain frequency f_L, so Eqs. (17.39) and (17.40) imply

$$\varphi_e \leq \Delta f/f_L < 1 \text{ rad}, \tag{17.41}$$

where the last inequality comes from Eq. (17.39). Therefore, even with no margin relative to $W/2$, operation is approximately linear.

Noise can cause the effective loop bandwidth to shrink and violate Eq. (17.39). In addition, we will now consider another noise effect, at low S/N, that calls for minimization of W, but consistent with acceptable width for signal linearity. A narrow PLL, but one that is wide enough to prevent excessive phase error, can give an advantage under these conditions. Gardner (1979, pp. 175–196) suggests $1 \leq \zeta \leq 1.5$, no limiter preceding the loop, and experimental optimization.

17.6.2   Standard Discriminator and Click Noise

The expression for S/N in a standard discriminator was given by Eq. (15.18). This can be written in decibels as

$$\left[\frac{S}{N}\right]_{dB} = \left[\frac{S_i}{N_i}\right]_{dB} + 20 \log_{10} m + 1.8 \text{ dB}. \tag{17.42}$$

However, this does not hold at low S/N, as illustrated in Fig. 17.6. This is because of pulse noise, or "clicks."

Figure 17.7 shows a phasor representation of a signal plus noise. At high S/N the noise phasor rotates about the tip of the signal phasor producing the path shown, and the deviation of the phase of the resultant from the signal's phase produces phase noise and, simultaneously, frequency noise. At low enough S/N, however, the trajectory of the noise vector sometimes circles the origin (O). When this happens the phase changes by 2π rad,

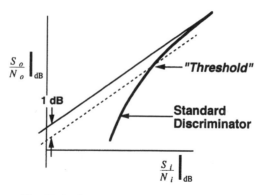

Fig. 17.6 S/N for standard discriminator.

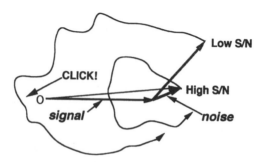

Fig. 17.7   Phasor representation of click noise.

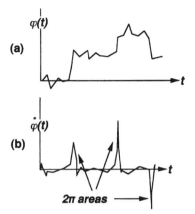

(a)

(b)

2π areas

Fig. 17.8 (*a*) Phase and (*b*) frequency during clicks.

causing a severe perturbation. Figure 17.8*a* illustrates the phase noise with steps corresponding to 2π rad changes when the origin is circled. Figure 17.8*b* shows the derivative of the phase noise, which is the frequency noise. The assumption of a narrow video filter permits these transitions to be approximated as phase steps and corresponding frequency impulses. The implied requirement that B_v be narrow compared to $W/2$, and a similar requirement on B_n of the PLL, is somewhat relieved by the rapidity of these nonlinear transitions.

The effective frequency power spectral density due to the clicks is approximately[6] (see Appendix 17.A)

$$S_{\dot{\varphi}} = 2N\frac{\text{sec-Hz}^2}{\text{Hz}}, \qquad (17.43)$$

where N is the rate of clicks. If the signal is centered in the input filter and unmodulated,

$$N = \frac{\gamma}{2\pi}\text{erfc}\sqrt{\rho}, \qquad (17.44)$$

where

$$\gamma^2 = \frac{\int_{-\infty}^{\infty} (\Omega)^2 |F(\Omega)|^2 \, d(\Omega)}{\int_{-\infty}^{\infty} |F(\Omega)|^2 \, d(\Omega)}. \qquad (17.45)$$

[6]See Schwartz et al. (1966, pp. 144–154) with references to Rice (1963).

Here Ω is the offset from the center of the filter $F(\Omega)$ and

$$\mathrm{erfc}(x) \triangleq \frac{2}{\sqrt{\pi}} \int_x^\infty e^{-u^2} \, du \qquad (17.46)$$

is the complementary error function.

For a rectangular filter,

$$\gamma^2 = \frac{\int_{-W/2}^{W/2} (\Omega)^2 \, d(\Omega)}{\int_{-W/2}^{W/2} d(\Omega)} = \frac{\frac{2}{3}(W/2)^3}{W} = \frac{W^2}{12}$$

$$\gamma = \frac{W}{2\sqrt{3}} \left(\frac{2\pi \ \mathrm{rad}}{\mathrm{cycle}} \right) = \frac{\pi}{\sqrt{3}} \frac{W}{\mathrm{cycle}}. \qquad (17.47)$$

Combining this with Eqs. (17.43) and (17.44), we obtain the frequency power spectral density as

$$S_\varphi = \left(\frac{W}{\sqrt{3} \ \mathrm{Hz}} \mathrm{erfc}\sqrt{\rho} \right) \frac{\mathrm{Hz}^2}{\mathrm{Hz}} = \frac{W}{\sqrt{3}} \mathrm{erfc}\sqrt{\rho}. \qquad (17.48)$$

Some factors not included here are an increase in the number of clicks due to modulation (we might have guessed this from Fig. 17.7) and that clicks decrease the signal's apparent frequency deviation, causing a further deterioration at low S/N.

Adding this noise to what we included in Eq. (15.18) (based on the assumption of rectangular filters), we obtain

$$\frac{S}{N} = \frac{\frac{1}{2}\Delta f^2}{\frac{1}{3}(N_0/P)F_m^3 + S_\varphi F_m} \qquad (17.49)$$

$$= \frac{\frac{1}{2}\Delta f^2}{\frac{1}{3}(N_0/P)F_m^3 + (WF_m/\sqrt{3})\mathrm{erfc}\sqrt{\rho}}. \qquad (17.50)$$

The numerator is the signal power; the first term in the denominator is the filtered input noise; and the second term is noise due to clicks. For comparison to S/N without clicks, we can rearrange this such that the numerator is Eq. (15.17):

$$\frac{S}{N} = \frac{(3P\Delta f^2)/(2N_0 F_m^3)}{1 + \sqrt{3}(W^2/F_m^2)\rho \ \mathrm{erfc}\sqrt{\rho}}. \qquad (17.51)$$

17.6.3 Clicks with a PLL

In a PLL, if the transition is fast enough, the loop will not follow the phase change during the clicks. The phase error will skip a cycle, producing a relatively small phase perturbation at the VCO and resulting in a bipolar frequency transient that contains little energy compared to a frequency transient that has a one-cycle area. This is illustrated in Fig. 17.9. A limiter can be harmful here because it can prevent a desirable drop in input level during a click.

The effects of noise that is severe enough to cause cycle skipping properly belong in the next chapter. However, we will borrow the results from there so we can treat, in one place, the effects of noise in a loop being used for frequency demodulation.

Based on an estimation that the times of the clicks are governed by a Poisson probability law (Blanchard, 1976, p. 340), the frequency power spectral density will be

$$S_{\dot{\varphi}} = 2\overline{F}, \tag{17.52}$$

where $\overline{F}$ is the mean frequency of cycle skipping in hertz. Substituting this into Eq. (17.49), we obtain (still assuming a sharp postdetection filter)

$$\frac{S}{N} = \frac{\frac{1}{2}\Delta f^2}{\frac{1}{3}(N_0/P)F_m^3 + 2\overline{F}F_m}. \tag{17.53}$$

We can use the results of Section 18.2.3 to give us F (with W very wide, which is not optimum). Then the threshold for a PLL demodulator, based on this theory, is better than for a standard discriminator, as shown in Fig. 17.10 [Gupta et al., 1968]. While modulation makes $\overline{F}$ worse, as shown in the figure, it does not much affect the threshold. However, Rice (1963) has indicated a degradation of about 2 dB in one case. Although conditions and results vary widely the potential for improvement with a PLL is evident.

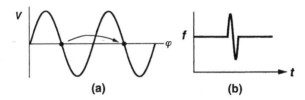

Fig. 17.9 Fast phase step of 2π may cause the phase error in the PLL to skip a cycle (a), resulting in a low-energy frequency transient (b) at the PLL output.

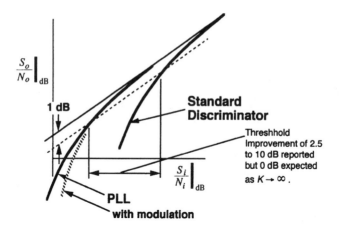

Fig. 17.10 S/N for a discriminator PLL.

17.A   APPENDIX: SPECTRUM OF CLICKS

Here we develop the frequency power spectral density due to clicks [Eq. (17.43)]. The frequency change due to a single click occurring at time T_i can be represented by the function $\Delta f(t - T_i)$. If this frequency deviation is processed linearly by the discriminator and then passed through the video filter F_v, the output voltage due to the click will equal the discriminator gain constant times the convolution of the pulse with the impulse response of the filter,

$$\Delta v_i(t) = K_d \, \Delta f(t - T_i) * F_v(t). \tag{17.54}$$

If the half bandwidth of the filter that precedes the discriminator is wide compared to the video filter bandwidth, $W/2 \gg B_v$, the typical click will have much of its energy at frequencies that are outside of the video bandwidth—that is, the click will be fast compared to $1/B_v$. Then the response of F_v to Δf will be similar to its response to an equivalent impulse. Substituting an impulse with area equal to one cycle for the pulse,

$$\Delta f(t - T_i) \approx 1\text{-cycle } \delta(t - T_i), \tag{17.55}$$

(17.54) becomes

$$\Delta v_i = K_d \text{ cycle } \delta(t - T_i) * F_v(t) \equiv K_d \text{ cycle } F_v(t - T_i). \tag{17.56}$$

The Fourier transform of this is

$$\Delta v_i(f) = K_d F(f) \exp(-j2\pi f T_i) \text{ (Hz-sec)}^{1/2}. \tag{17.57}$$

Because of the randomness of T_i, the spectrum of $\Delta v_i(f)$ has random phase and the transform of the sum of many pulses cannot be written at any frequency except zero, where it represents the sum of frequency changes. (There the sum will be zero if the number of positive and negative clicks are equal). However, the energy spectral density can be written at other frequencies. The variance of the sum of $\Delta v_i(f)$ in a narrow bandwidth δf for n pulses is[7]

$$\sigma_{vi}^2 = nK_d^2|F(f)|^2 \, \delta f \text{ Hz-sec.} \tag{17.58}$$

This is an energy (for a 1-Ω system). The power is obtained by dividing by the elapsed time, which converts the total number of pulses n to the pulse rate N. Dividing also by δf to get spectral density, we obtain

$$S_{v2} = NK_d^2|F_v(f)|^2 \text{ Hz-sec.} \tag{17.59}$$

Note that this is a two-sided density, since we have been dealing with Fourier transforms. This would be the output of the discriminator if it were driven by a one-sided frequency power spectral density of

$$S_{\dot{\phi}} = 2N \sec \frac{\text{Hz}^2}{\text{Hz}}. \tag{17.60}$$

PROBLEMS

17.1 A second-order loop is locked to a signal in the presence of additive noise. The true variance of the output phase is 0.38 rad^2.

    (a) What would be the variance if the loop were linear and had an integrator-and-lead filter with $\zeta = 0.5$? Obtain the value more accurately than can be done from Fig. 17.2.

    (b) What would be the variance if it were linear and had a low-pass filter? There is no formula in the text that will give you the answer directly.

    (c) Again with the low-pass filter, suppose that, at a certain frequency, the phase deviates from quadrature by 20° due to finite gain. What would be the deviation if the noise went away?

17.2 A loop has an integrator-and-lead filter. The filter zero is at 150 Hz and the open-loop gain is unity at 50 Hz.

    (a) What is the phase margin in degrees?

    (b) A binary phase modulation of $\pm 60°$ at a 20-kHz rate is added to the input. What is the phase margin then?

[7]In this section we are using $\delta(t)$ as the traditional symbol for an impulse and δf for a narrow frequency bandwidth.

17.3 A high-gain ($K \gg \omega_n$) second-order loop has a lag-lead filter, a natural frequency ω_n of 500 rad/sec, and a damping factor ζ of 0.5. The input signal power is -10 dBm and the noise power in a band from 0 to 500 kHz is -13 dBm. The input carrier frequency equals the VCO center frequency, 250 kHz. Give answers in units of radians squared.

(a) What is the phase variance $\sigma_{\varphi n}^2$ at the loop output?

(b) What is $\sigma_{\varphi nm}^2$ if the input signal is phase modulated $\pm 45°$ with equal probability of either state? Here, and below, assume the frequencies of the modulation are well beyond the loop bandwidth.

(c) What is $\sigma_{\varphi nm}^2$ if the signal is phase modulated with information that has a noiselike Gaussian distribution with $\sigma = 0.7$ rad?

(d) What is $\sigma_{\varphi nm}^2$ if the input is phase modulated with a sinusoid of peak deviation 45° [$J_0(\pi/4) = 0.85$].

17.4 Develop an equation equivalent to Eq. (17.35), but for a loop with a low-pass filter.

17.5 What values of the following parameters will reduce the effective loop gain by 1.9 dB ($\eta = 0.8$)?

(a) S_i/N_i in a limiter [$B_n \ll W$] in decibels

(b) S/N, $\rho_{L0} = 1/\sigma_{\varphi 0}^2$, in the loop [$B_n \ll W$; integrator-and-lead filter with $\zeta \approx 1$]

(c) $J_0(m)$ for sinusoidal phase modulation

(d) θ with equiprobable binary phase modulation at $\pm \theta$

(e) $\sigma_{\varphi, in}^2$ in radians squared for Gaussian phase modulation

CHAPTER 18

CYCLE SKIPPING DUE TO NOISE

In this section we consider performance of the phase-locked loop (PLL) when the signal-to-noise (S/N) ratio is so poor that it causes cycle skipping. Figure 18.1 shows the phase error of a loop in the presence of significant noise. The noise here has half the power, in the (linear) loop bandwidth, of the signal. For a while the result is phase noise on the voltage-controlled oscillator (VCO), which is reflected in the phase error (at Fig. 18.1a). Then the deviation becomes large enough that the phase skips a cycle, locking again one cycle from the original phase. Over a longer observation period multiple cycle skips will be observed, including events during which two or more cycles are skipped without an obvious pause as a potential lock point is passed (see Fig. 18.1b). Analysis under these conditions is understandably difficult but some useful results have been obtained.

18.1 FOKKER–PLANK METHOD

Viterbi (1966, Chapter 4) employed the Fokker–Plank differential equation to describe the change in the probability distribution $p(\varphi_n, t)$ of the phase error over time. He attempted to solve that equation for time approaching infinity to obtain the steady-state probability distribution of phase error. He did obtain the solution for simple loops and suggested an approximation for more complex loops in which we determine the S/N under linear assumptions, ρ_{L0}, and then determine how a simple loop, with that same ρ_{L0}, would react when analyzed exactly.

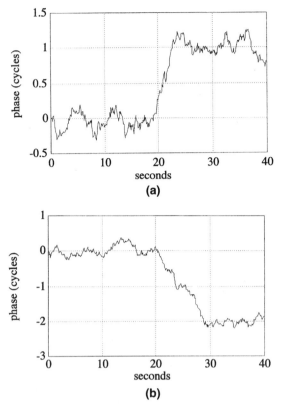

Fig. 18.1   Examples of phase error in the presence of noise: $\rho_{L0} = 2$, $\alpha = 1$, $\zeta = 0.7$, $\omega_n = 1$ rad/sec. (These plots were produced by the software in Appendix 8.M.)

We assume no input modulation and zero phase error so that Eq. (17.6) becomes

$$u_1(t) = -K_p' \sin \varphi_n(t) + u_{1n}(t). \tag{18.1}$$

The differential equation describing the loop performance in the presence of the input noise is (see Fig. 18.2)

$$\omega_{\text{out}}(t) = \frac{d\varphi_n}{dt} = K'\{-\sin \varphi_n(t) + n'(t)\} * f(t), \tag{18.2}$$

where $* f(t)$ represents convolution with the filter impulse response.

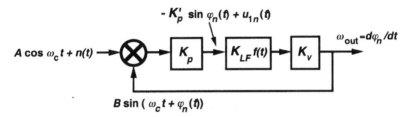

Fig. 18.2   Loop in the presence of noise.

18.1.1   Assumption Regarding the Nature of the Noise

Based on the premise that the VCO phase $\varphi_n(t)$ changes slowly compared to the input additive noise, $u_{1n}(t)$ is assumed to be Gaussian and white. In other words, the frequencies f_m of the modulation on the noise carrier, due to noise at $f_c \pm f_m$, are predominantly high compared to the loop noise bandwidth B_n. This in turn is based on the premise that the input bandwidth W is much wider than twice the loop bandwidth

$$W \gg 2B_n. \tag{18.3}$$

The assumption, regarding the VCO phase changing slowly compared to the input noise, is at least questionable since the most important components of the input noise, those that most affect $\varphi_n(t)$, are those that do not satisfy this restriction. We discussed this problem in Section 13.5 and considered how not meeting the assumed restriction might only slightly modify the picture. Our discussion there, however, included the assumption that the phase error would be small enough to prevent cycle skipping and that does not hold here. One might claim that there is so much noise outside the loop bandwidth that its effect is dominant even though the loop reacts more readily to the noise within its bandwidth. Perhaps the best argument is that the analysis, based on the assumed noise characteristics, produces useful results.

18.1.2   First-Order Loop

For the first-order loop, $f(t)$ in Eq. (18.2) equals 1, so we have

$$\frac{d\varphi_n}{dt} = K'\{-\sin \varphi_n(t) + n'(t)\}. \tag{18.4}$$

This is a Markov process in that the derivative of φ_n at time t depends only on its value at time t. As a result its probability distribution is described by the Fokker–Plank equation,

$$\frac{\partial p(\varphi_n, t)}{\partial t} = -\frac{\partial}{\partial \varphi_n}\left[-p(\varphi_n, t)K' \sin \varphi_n\right] + \frac{K^2}{4}\frac{N_0}{P}\frac{\partial^2 p(\varphi_n, t)}{\partial \varphi_n^2}. \quad (18.5)$$

If we solve this equation, starting with the initial condition that the phase error is zero [i.e., $p(\varphi_n, 0)$ is an impulse at $\varphi = 0$], the solution for a sequence of times, $0, t_1, t_2, t_3$, is shown in Fig. 18.3. It is easy to see that, as time increases, $p(\varphi_n, t)$ diffuses so the ultimate value approaches zero everywhere. This simply reflects the fact that, if there is any chance that the phase error will skip a cycle, as time approaches infinity the chances that it will be in any particular 2π range approach zero.

We have obtained the correct answer to the wrong question. We do not want the probability distribution of the phase error but rather the distribution on the error mod 2π. We want the probability at angle φ to represent probability that the phase is at $\varphi \pm m2\pi$ for all m. To obtain this function, Viterbi (1966) used the related function

$$P(\varphi_n, t) = \sum_{-\infty}^{\infty} p(\varphi_n + m2\pi, t). \quad (18.6)$$

This is like the original function except that each 2π-wide region contains an identical distribution; P begins with impulses at $m2\pi$ for all m. As P diffuses with time, the probability that the phase will jump out of the primary interval $-\pi < \varphi < \pi$ is the same as the probability that it will jump into that interval. As the edges of the distribution at $\pm \pi$ build up, the probability of

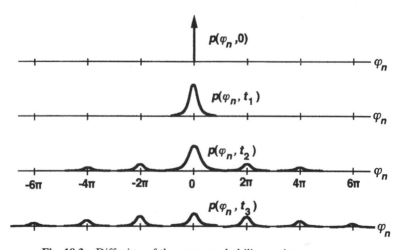

Fig. 18.3 Diffusion of the error probability as time progresses.

the phase moving from $\pi - \varepsilon$ to $\pi + \varepsilon$ is the same as the probability of the phase moving from $-\pi - \varepsilon$ to $-\pi + \varepsilon$. As a result, in the $\pm \pi$ region (18.6) can be written

$$P(-\pi < \varphi_n < \pi, t) = p(\varphi_n, t) \bmod 2\pi, \tag{18.7}$$

which is the desired function. Equation (18.6) is substituted for $p(\varphi_n, t)$ in Eq. (18.5) and the initial condition is set as an impulse of probability density at $\varphi_n = 0$. Boundary conditions are set as

$$P(\pi, t) = P(-\pi, t). \tag{18.8}$$

The probability density is normalized such that

$$\int_{-\pi}^{\pi} P(\varphi_n, t) \, d\varphi_n = 1. \tag{18.9}$$

Under the steady-state condition,

$$\lim_{t \to \infty} \frac{\partial P(\varphi_n, t)}{\partial t} = 0, \tag{18.10}$$

the solution is found to be

$$P(\varphi_n) = \frac{\exp(\rho_{L0} \cos \varphi_n)}{2\pi I_0(\rho_{L0}) \text{rad}} \qquad -\pi \text{ rad} < \varphi_n < \pi \text{ rad}, \tag{18.11}$$

where I_0 is the modified Bessel function of the first kind, order 0, and ρ_{L0} is the S/N in the loop bandwidth under linear assumptions, Eq. (14.12). This distribution is not Gaussian but does have a zero mean. The variance from Eq. (18.11) is

$$\sigma_{\varphi n}^2 = \left[\frac{\pi^2}{3} + \sum_{k=1}^{\infty} (-1)^k \frac{I_k(\rho_{L0})}{I_0(\rho_{L0})} \right] \text{rad}^2. \tag{18.12}$$

In Fig. 18.4 $P(\varphi_n)$ is shown as a function of ρ_{L0}.

For high S/N, Eq. (18.11) becomes

$$P(\varphi_n) \approx \sqrt{\frac{\rho_{L0}}{2\pi}} \, e^{-\varphi_n^2 \rho_{L0}/2}/\text{rad}, \tag{18.13}$$

which is Gaussian with a variance $\sigma_{\varphi n}^2 = 1/\rho_L$, as is the case with a linear analysis. For very low S/N, Eq. (18.11) approaches a uniform distribution

$$p(\varphi_n) \approx 1/(2\pi \text{ rad}), \tag{18.14}$$

indicating that the phase is completely unknown. We can observe these trends in Fig. 18.4.

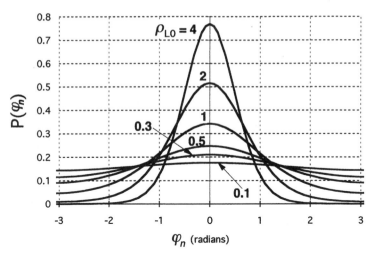

Fig. 18.4 Probability density (per radian) of phase error vs. phase error, first-order loop, or second-order loop with low-pass filter. The parameter $\rho_{L0} = 1/\sigma_{\varphi 0}^2$ is the signal-to-noise ratio in the loop bandwidth in the absence of noise.

18.1.3 Second-Order Loop

The solution for a second-order loop [Viterbi, 1966, Chapter 4; Blanchard, 1976, pp. 302–314] is more difficult because $d\varphi_n(t)/dt$ is not dependent only upon $\varphi_n(t)$, since the second-order loop has memory,[1] and thus $\varphi_n(t)$ is not a Markov process. The solution involves the formation of a two-dimensional vector related to $\varphi_n(t)$. The vector describes the state of the loop, but it is Markov because its derivative depends only on its own value. The second-order Fokker–Plank equation is then used to describe the time dependence of the joint probability function of this two-dimensional variable. The solution has been obtained [Viterbi, 1966, Chapter 4] for $\omega_n \gg \omega_z$, which is similar to a first-order loop, and for a loop with a low-pass filter [Blanchard, 1976, pp. 302–314]. These solutions are again given by Eq. (18.11). In the case with the low-pass filter, the distribution for the frequency error is also obtained. It is

$$p(\dot{\varphi}_n) = \frac{1}{\sqrt{2\pi}\,\sigma_{\dot{\varphi}}} \exp\left(-\frac{\dot{\varphi}_n^2}{2\sigma_{\dot{\varphi}}^2}\right). \tag{18.15}$$

This is Gaussian with a null mean and with variance

$$\sigma_{\dot{\varphi}n}^2 = \frac{K^2\omega_p}{4}\frac{N_0}{P} = \omega_n^2\sigma_{\varphi 0}^2. \tag{18.16}$$

---

[1]Refer to text after Eq. (8.3).

Example 18.1   Phase Error with Large Noise, Exact Solution   A second-order loop with a low-pass filter has a natural frequency of 1 kHz and a damping factor of 0.25 (in the absence of noise). With an input signal strength of 0 dBm and one-sided noise density of -38 dBm/Hz:

a. What is the probability that the phase error will be within $\pm 35°$ of the nominal (90°) value?

From Eq. (14.10), the (linear) noise bandwidth is

$$B_n = \frac{2\pi \times 10^3 \text{ rad/sec}}{4} \left[\frac{1}{2(0.25)} + 0 \right] \frac{\text{cycle}}{\text{rad}} = \pi \times 10^3 \text{ Hz}.$$

From Eq. (14.11), the noise in the loop's (linear) noise bandwidth is

$$P_n = \pi \times 10^3 \text{ Hz} \times 10^{-3.8} \text{ mW/Hz} = 0.5 \text{ mW}.$$

Since the signal strength is 1 mW, the signal-to-noise ratio in the (linear) loop bandwidth is $\rho_{L0} = 2$. To get the exact answer we would integrate Eq. (18.11) from -0.61 to $+0.61$ rad (35° is 0.61 rad), but we can obtain an approximate answer using Fig. 18.4. The average probability density over ± 0.61 rad is approximately 0.47/rad. (This can be obtained by inspection or by properly averaging a number of values over the range.) Multiplying this density by the full width of the range, 1.22 rad, we obtain a probability of *0.57* that the phase is within $\pm 35°$.

[By "exact" we mean that we are applying Eq. (18.11), which is the mathematical solution that applies to the type of loop we are analyzing. Nevertheless, the analysis involves an approximation concerning how the loop's phase is correlated to the noise, which we have discussed, and we have done an approximate integration.]

b. What is the root mean square (rms) frequency deviation of the loop's VCO?

Using Eq. (18.16) with $\rho_{L0} = 1/\sigma_{\varphi 0}^2$ from part (a), we obtain

$$\sigma_{\dot{\varphi} n}^2 = (2\pi \times 10^3 \text{ rad/sec})^2 (\tfrac{1}{2}),$$

$$\sigma_{\dot{\varphi} n} = 4443 \text{ rad/sec}.$$

Viterbi (1966) suggested that the performance of other second-order loops, in the presence of large amounts of noise, be estimated by using the equations for a first-order loop [e.g., Eq. (18.11)] with $\sigma_{\varphi 0}$ obtained using B_{n0}

for the pertinent loop parameters. This procedure produces the correct results for the two cases for which exact solutions have been obtained, as shown above, and is supported by experimental evidence, which we will now illustrate.

18.1.4 Experimental Evidence

Charles and Lindsey (1966) have provided experimental results[2] for a high-gain second-order loop ($K \gg \omega_z$) with $\zeta = 0.707$ and six values of ρ_{L0} from 0.41 to 6.41 dB and have compared these results to three different theoretical models. Figure 18.5 shows their data for $\rho_{L0} = 1.41$ dB along with the first-order-loop results of Eq. (18.11). Lindsey has also given an approximate solution[3] that tends to form a lower bound in the center of Fig. 18.5 and an upper bound further out.

Figure 18.6 shows the experimentally determined probability distribution function of phase error (integral of the density) for several values of ρ_{L0} from the same source. Note how the distribution approaches Gaussian at high S/N, corresponding to the integral of Eq. (18.13), and uniform at low S/N, corresponding to the integral of Eq. (18.14).

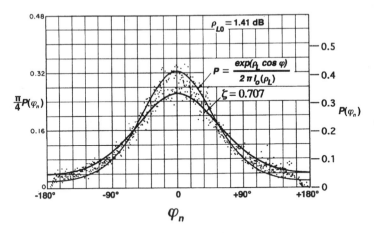

Fig. 18.5 Theoretical and experimental density (per radian) of phase errors. Data points, a first-order loop curve P, and Lindsey's approximation for a high-gain second-order loop are shown. [From Lindsey and Simon, 1973, p. 33[4].]

---

[2] The data is also in Lindsey and Simon (1973, pp. 33–35) and Viterbi (1966, pp. 112–115) (4 of 6 data sets).
[3] Lindsey and Simon (1973, pp. 31–36). They reference Lindsey (1972).
[4] See also Lindsey and Simon, 1978, p. 285; Charles and Lindsey, 1966, p. 1161; Viterbi, 1966 p. 115.

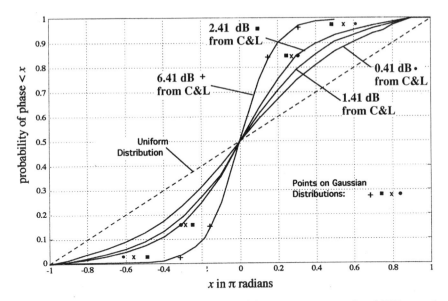

Fig. 18.6   Probability distribution of phase with parameter ρ_{L0}, $\zeta = 0.707$, $\alpha = 1$. [From Charles and Lindsey (C & L), 1966, p. 1161 (© 1966 IEEE).]

Example 18.2   Phase Error in Second-Order Loop, Large Noise   Under the conditions of Example 18.1, estimate the probability of a phase error within $\pm 35°$ if the loop is type 2 but has the same (linear) noise bandwidth as in Example 18.1.

Although the noise bandwidth is the same, the noise-free parameters in Eq. (14.10) will be different for this case ($\alpha = 1$) than they were in Example 18.1 ($\alpha = 0$). However, $\rho_{L0} = 1/\sigma_{\varphi 0}^2$ will be the same. We do not have an exact solution available for high-gain second-order loops. We do have the quasi-linearized approximation, as given by Eq. (17.17) and Fig. 17.2, and, alternatively, Viterbi's suggestion that we use the answer for a first-order-loop, which we obtained in Example 18.1. Let us compare that to the quasi-linear solution.

For a Gaussian distribution, as assumed in the quasi-linear method, what value of $\sigma_{\varphi n}$ would give the 0.57 probability from Ex. 18.1? From a table of the normal probability distribution function, we find that a probability of 0.57 can be obtained by integrating the normal function over $x = \pm 0.79$, where x is the offset from the center of the density function having a variance of one. Since x is also the offset normalized to the variance, we can determine the variance for which the probability is 0.57 when the phase is within $\pm 35°$ (0.61 rad).

$$0.79 = x = \varphi/\sigma = 0.61 \text{ rad}/\sigma.$$

Therefore, $\sigma_{\varphi n} = 0.61 \text{ rad}/0.79 = 0.77 \text{ rad}$ or $\sigma_{\varphi n}^2 = 0.59 \text{ rad}^2$.

Looking now at the quasi-linear relationship in Fig. 17.2 and Eq. (17.17), when $\sigma_{\varphi 0}^2 = 0.5$ rad^2, the corresponding $\sigma_{\varphi n}^2$ is at least 0.71 rad^2 so, in this case, the results of the quasi-linear approximation are more pessimistic than the approximation that uses the exact results for a first-order loop. This is true even for $\zeta \Rightarrow \infty$, which represents one of the two cases ($\omega_n \gg \omega_z$) for which the first-order solution Eq. (18.12) is exact.

18.2 CYCLE SKIPPING

When noise causes a loop to skip a cycle, the loop will go out of lock and the VCO may return toward its center frequency unless $\Omega < \Omega_{PI}$. For a given ω_n, a low pole frequency ω_p will slow the VCO's tendency to go toward its center frequency and thus reduce the number of multiple skips [Blanchard, 1976, p. 320] whereas low ζ tends to cause cycles to be skipped in bursts [Ascheid and Meyr, 1982]. Phenomenon like mixer imbalance can make the results poorer than theory. Obviously simulation and experiment are important in this area.

18.2.1 First-Order Loop

Using the Fokker–Plank equation, Viterbi (1966) found, by a method outlined in Section 18.2.5, the mean time between cycle skips. For a first-order loop with zero mean phase and frequency error, he found that the mean time between cycle skips is

$$T_m = \frac{\pi^2 \rho_{L0} I_0^2(\rho_{L0})}{2 B_{n0}} \text{ cycle.} \tag{18.17}$$

From this equation, the mean frequency of cycle skipping, $\bar{F} \triangleq 1/T_m$, is plotted in Fig. 18.7 (long curve).[5]

For high ρ_{L0}, Eq. (16.16) for $v = 0$, becomes[6]

$$I_0(\rho_{L0}) \approx \frac{\exp(\rho_{L0})}{\sqrt{2\pi\rho_{L0}}} \left(1 + \frac{1}{8\rho_{L0}}\right). \tag{18.18}$$

So Eq. (18.17) then becomes

$$T_m \approx \frac{\pi}{4} \frac{\exp(2\rho_{L0})}{B_{n0}} \text{ cycle.} \tag{18.19}$$

---

[5]F is not an average of frequencies. (How could we get frequencies for a collection of random evens?) It is the number of events divided by time, which is also the reciprocal of mean time, $n/\Sigma T_i = 1/\Sigma(T_i/n) = 1/T$.

[6]The second term is shown here, even though it will be dropped, to make it easier to estimate the validity of the approximation for a given value of ρ.

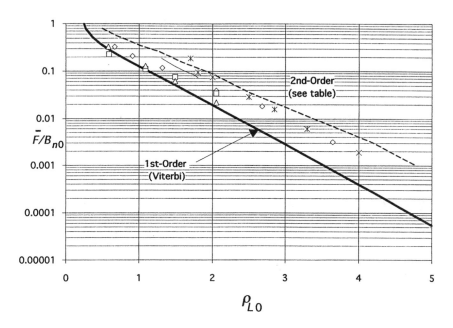

Symbol	α	ζ	Source
Δ		1.4	Rowbotham and Sanneman (1967)
□	1	0.35	
◇			
- - - - -	0.9788	0.707	Ascheid and Meyr (1982)
*	1		Smith (1966)
——	0.9756		

Fig. 18.7 Rate of cycle skipping for a first-order loop and data points for high-gain second-order loops. $\overline{F}$ is in skips/sec and B_{n0} is in Hz.

18.2.2 High-Gain Second-Order Loops

Data points from other sources are also plotted in Fig. 18.7. The dashed line is the approximate locus of experimental data points given by Ascheid and Meyr.[7] Their study also shows the effect of mistuning on the rate of cycle skipping. Rowbotham and Sanneman's (1967) data are from an approximate simulation of second-order, type 2 (i.e., with integrator-and-lead filter) loops. These points do not exactly give $\overline{F}$ but something related to it, the reciprocal of the mean time required, starting at $(0, 0)$ in the phase plane, to cross $\pm \pi$ and not return for $t = 4/\omega_{n0}$. This does not count multiple skips.

[7]Ascheid and Meyr (1982) (data is also given at $\zeta = 1$).

Smith's (1966) data is also from a simulation. He counted a cycle as being skipped whenever $\pm(2n + 1)\pi$ rad was crossed, except when the crossing was a reversal of the previous crossing. Smith found that, for low ρ_L, many cycles tended to be skipped at one time but that the number skipped together was reduced when the loop filter pole occurred at a nonzero frequency. Most of his points are for a type 2 loop. However, the short solid curve represents five data points with a nonzero pole frequency, and there is a noticeable reduction in $\bar{F}$ even though the pole is at only one twentieth of the zero frequency. This improvement as the pole frequency is increased portends the results that we will consider next.

18.2.3 Low-Gain Second-Order Loops

Gupta et al. (1968) obtained values for $\bar{F}$ in a general second-order PLL ($0 < \alpha < 1$) by describing these loops as being intermediate between a first-order loop and a loop with a low-pass filter. These two extremes are characterized by $r = 0$ for the low-pass and $r = 1$ for the first-order loop, where $r = \omega_p/\omega_z$. It has been shown, for $K \gg \omega_z$, that these are limiting cases for the mean time between cycle skipping as r is varied. In other words, the mean time is always between the times for loops characterized by these two extreme values of r. The same is assumed to be true when that inequality does not hold. Since results are available for these two extreme values of r, we can use these results as limits for the "intermediate" cases. We desire to find the mean frequency of clicks because we can use this in an expression such as Eq. (17.53).

Analyses such as in Section 18.2.1 assume zero initial phase error so the loop must return to that approximate condition after each skipped cycle if they are to give good estimates of T_m. If we can prevent bursting, the occurrence of several clicks at the same time, $\bar{F}$ will be the reciprocal of this theoretical T_m. Gupta et al. (1968) found that bursting can be prevented if $\Omega_{PI} \approx \Omega_H$, which they found to occur if

$$B_{n0} < 0.5\omega_p. \tag{18.20}$$

(Note, $B_n = K/4$ for the $r = 0, 1$.) Their experimental results for Ω_{PI}/Ω_H as a function of B_{n0}/ω_p are sketched in Fig. 18.8.

For a given value of ω_n, ζ, and r, there are two values of K that can be chosen, as illustrated in Fig. 18.9. This figure shows two different values of ω_n for the same ζ and r, but we can change ω_n by any factor k without changing ζ or r by changing K, ω_p, and ω_z each by a factor k.

If we choose the lower of the two values of K_0, Eq. (18.20) will be satisfied. Then we can write, for $r = 1$ (first order), the reciprocal of Eq. (18.19),

$$\bar{F}_1 = \frac{1}{\bar{T}_1} = \frac{4}{\pi} \frac{B_{n0}}{\exp(2\rho_{L0})}, \tag{18.21}$$

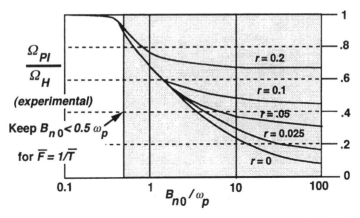

Fig. 18.8 Ratio of pull-in to hold-in ranges for various values of r. [Sketch after Gupta et al., 1968 (© 1968 IEEE).]

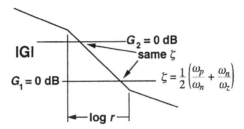

Fig. 18.9 Same damping factor at two natural frequencies.

while, for $r = 0$ ($\alpha = 0$), the mean frequency of cycle skipping was found to be

$$\bar{F}_0 = \frac{\text{cycle}}{\bar{T}_0} = \frac{\text{cycle}}{\pi}\left[\sqrt{\left(\frac{\omega_p}{2}\right)^2 + \omega_{n0}^2} - \frac{\omega_p}{2}\right]\exp(-2\rho_{L0})$$

$$= \frac{B_{n0}8\zeta_0}{\pi}\left[\sqrt{\zeta_0^2 + 1} - \zeta_0\right]\exp(-2\rho_{L0}). \tag{18.22}$$

Here ρ_L is the signal-to-noise ratio in the loop (Section 14.2), using B_n based on linear assumptions,

$$\rho_{L0} = 1/\sigma_{\varphi0}^2. \tag{18.23}$$

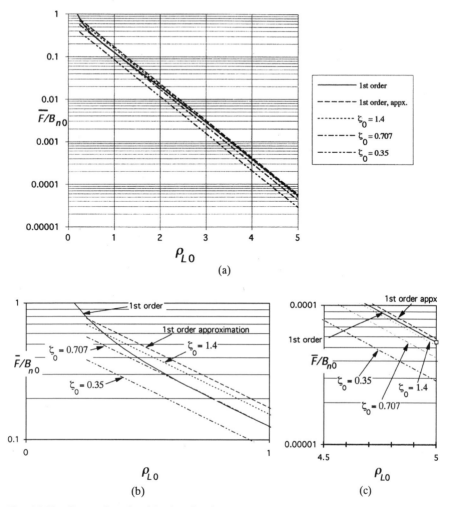

Fig. 18.10 Rate of cycle skipping for low-gain second-order loops. Left and right ends of (a) are expanded at (b) and (c). The first-order curve is from Eq. (18.17) and the first-order approximation is from Eq. (18.19).

Equations (18.21) and (18.22) for several values of ζ are plotted in Fig. 18.10. The more accurate expression for a first-order loop, based on Eq. (18.17), is also plotted again. Note that Eq. (18.21) gives a higher $\overline{F}$ than does Eq. (18.22) for any ζ. Gupta et al. (1968) verified these results by simulation and experiment.

Note that the low-gain second-order loop (Fig. 18.10c) has a lower rate of cycle skipping than the first-order loop whereas the high-gain second-order loop (Fig. 18.7) has a higher rate, at least for large ρ_L.

18.2.4 Probability of Cycle Skipping in a Given Time

The probability that a cycle will be skipped within a time T (timed either from the previous burst of skipped cycles or from the occurrence of initial zero phase and frequency error) has been given as [Rowbotham and Sanneman, 1967; Smith, 1966]

$$P(T) = 1 - e^{-T/T_m}. \tag{18.24}$$

18.2.5 Outline of Derivation of T_m for First-Order Loop

We start again with Eqs. (18.4) and (18.5), but this time the boundary condition (18.9) is not applied. This time $p(\varphi_n, t)$ is meant to represent the probability distribution of the phase of something that is "eliminated" if it reaches the phase limits of $\pm\varphi_L$. Viterbi (1966) imagines the phase being that of a pendulum that is removed (cut off by a knife edge) if it reaches these limits. The value of such a density function is that it can be integrated over $\pm\varphi_L$ to determine the probability that the phase has never reached those limits. It is not a true probability density because its integral over all values is less than one. We (and Viterbi) designate it $q(\varphi_n, t)$ to differentiate it from the function discussed in Section 18.1.2. The Fokker–Plank equation still describes the density and its dynamics while the phase stays within the $\pm\varphi_L$ limits. Since the phase q cannot exist at the limits (being eliminated upon reaching there), the boundary conditions are $q(\pm\varphi_L, t) = 0$ for all time. Viterbi (1966) indirectly solves for $q(\varphi_n, t)$, using the function to determine T_m, the mean time between cycle skips (with $\varphi_L = 2\pi$). He also shows that the frequency of reaching $\pm\pi(\varphi_L = \pi)$ is twice the frequency of skipping cycles. This is consistent with the equal likelihood, in the first-order loop, that the phase, having reached $\pm\pi$, will return to its original region or skip a cycle.

18.M APPENDIX: NONLINEAR SIMULATION WITH NOISE

The MATLAB script NLPhN simulates the nonlinear acquisition behavior of a PLL in the presence of additive noise. It is similar to NLPhP, but random noise has been added to the offset voltage u_{off} in Fig. 8.M.1, as shown by u_n in Fig. 18.M.1.

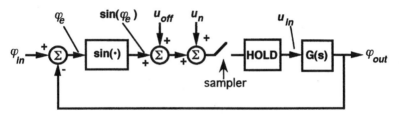

Fig. 18.M.1 Noise added to model.

18.M.1 Simulating Noise

Here u_n is a normally distributed random number with a mean of 0 and a variance of σ_N^2. This produces an equivalent noise power density, as seen in the loop (with noise bandwidth $B_n \ll f_s$) of $S = 2\sigma_N/f_s$, where f_s is the sampling frequency. That is, after sampling, the noise power σ_N^2 is distributed over a band equal to half of the sampling frequency, $f_s/2$. To see this, consider a band of noise of density S_x that extends over a frequency band from 0 to kf_s, where k is an integer. This would be related to total noise power by

$$\sigma_N^2 = kf_s S_x. \tag{18.M.1}$$

We are using one-sided noise density on which we have standardized, so the two-sided density is $S_x/2$ and extends from $-kf_s$ to $+kf_s$. Noise at frequency $f_x < f_s/2$ will be seen at f_x after sampling. Due to aliasing, so will noise at $f_x + jf_s$, where $-k < j < k - 1$, giving a total two-sided power at f_x in a bandwidth df of $2k(S_x/2)\,df$ or a two-sided density of

$$S/2 = 2k(S_x/2). \tag{18.M.2}$$

Thus the one-sided density at f_x is

$$S = 2kS_x. \tag{18.M.3}$$

But this is $2k$ times the original density. It is what would have been obtained had the noise power been distributed over only $f_x/2$ rather than kf_x. More directly, using (18.M.1) in (18.M.3),

$$S = \sigma_N^2/(f_s/2). \tag{18.M.4}$$

This is the power spectral density at u_{in} in Fig. 18.M.1 for $f_x \ll f_s$ where the Hold transfer function is essentially unity. It will be multiplied by $(K_p')^2$ to give the power spectral density after the phase detector (PD), $\langle u_{1n}^2 \rangle/df$ from Eq. (13.10). Thus the density before the PD is

$$\sigma_N^2/(f_s/2) = \langle u_{1n}^2 \rangle / \left[(K_p')^2\, df \right] = N_0/P_c, \tag{18.M.5}$$

where the last equality is due to (13.10), and the variance for the noise generator must be set to

$$\sigma_N^2 = (N_0/P_c)(f_s/2). \tag{18.M.6}$$

In terms of ρ_{L0}, the S/N in the (linear) loop noise bandwidth, this is

$$\sigma_N^2 = \frac{N_0 B_{N0}}{P_c B_{N0}} \frac{f_s}{2} = \frac{1}{\rho_{L0}} \frac{f_s}{2 B_{N0}}. \tag{18.M.7}$$

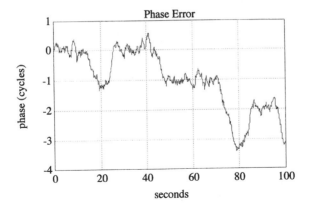

Fig. 18.M.2 Phase in the presence of noise, $\omega_n = 1$, $\zeta_0 = 0.707$, $\alpha_0 = 1$, S/N in $B_{n0} = 1.41$ dB.

18.M.2 Observing Cycle Skipping with NLPhN

Figure 18.M.2 shows the phase output from a run of NLPhN. (Others are shown in Fig. 18.1.) Truncation of the phase display has been turned off so we can observe cycle skipping. Note how the phase tends to stay at a multiple of 1 cycle error for a while and then move on to another multiple, except at 73 sec where it passes through -2 cycles without pausing.

18.M.3 Gathering Phase Statistics

The script erdis can be used to plot the probability distribution function for 1001 points generated by a run of NLPhN. If NLPhN is set to produce 1001 or more values of the phase error er with truncation, erdis will produce the distribution function when run subsequently. (It could be easily modified to operate on a different number of points of er.) Figure 18.M.3 shows the distribution function generated by such a run and compares the results to some of the experimental curves from Fig. 18.6; erdis is given at the end of this section. We could also use the MATLAB function hist(er) to build a histogram of the phase error er after running NLPhN. That would be an approximation (using a finite set of data points) to the probability density function. The added phase due to sampling in these simulations was $-2.796°$.

18.M.4 Changes to NLPhP to create NLPhN

Figure 18.M.4 shows the changes to NLPhP that add noise to the simulation. Changes or additions are in **bold** between portions retained from NLPhP. The user sets either the noise-to-signal ratio in the loop bandwidth, $1/\rho_{L0} = $ NS, or the ratio of noise power spectral density to signal, No. Noise bandwidth, Bn, is computed from Eq. (14.10). After new information is output to the Command window, a multiplier, Kn, is computed for the noise voltage. The command rand('normal') is then given to set MATLAB's random number

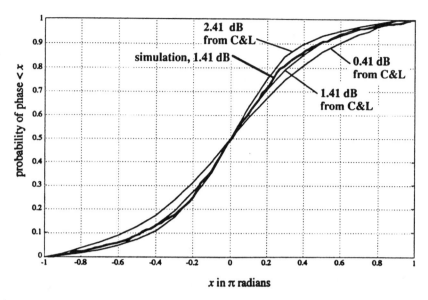

Fig. 18.M.3   Probability distribution of phase with parameter ρ_{L0}, $\zeta_0 = 0.707$, $\alpha_0 = 1$. Simulation results are compared to experimental curves from Charles and Lindsey (1966, Fig. 9). (© 1966 IEEE). This is the same data that appears in Fig. 18.6.

generator to create a normal distribution with zero mean and unity variance. During the computation loop, PDn, the product of the random number so generated and the multiplier Kn is added to the offset. The plot of phase detector output has been replaced by a plot of sampler output.

erdis

```
clear z y
subplot(111);
for n=1:101
    j=0;
    z(n)=(n-51)/50;
    k=z(n)*pi;
    for i=1:1000,
    if (er(i)<k)
     j = j+1;
      end
    end
    y(n) = j/1000;
end
plot(z,y)
grid
title('distribution function, zeta = .707, Int&Ld')
xlabel('x in π radians')
ylabel('probability of phase < x')
```

```
% MODIFY PARAMETERS BELOW & GO
%******************************
% Set either of the following parameters and set the
                         other to zero.
NS = .7227; % N/S in Loop Noise Bandwidth
No = 0; % Noise Power Density, per Hz, to Signal
                         Ratio
SR = 0; % 1 for step response (if also Phinit=Winit=0 and Ap
                         or Af not 0).,
. . . . . . . . . . . . .
% op=3: Frequency + Sampler Output vs. Time
% op=4: All of above = 4 plots
op = 4;
%******************************
fprintf('\nLoop with SINUSOIDAL Phase Detector
                         Characteristic')
fprintf('\nWn = %g rad/sec; zeta = %g; alpha = %g',Wn,z,alpha)
Bn = Wn*(1/(2*z) + 2*z*alpha^2)/4; % Noise BW in Hz
fprintf('\nNoise Bandwidth = %g Hz\n',Bn)
% Compute multiplier of noise distribution having
                         unity power.
if NS == 0,
   Dn = No;
   fprintf('Noise Density to Signal Ratio = %g
                         /Hz\n',No)
else
   Dn = NS/Bn;
   fprintf('Signal/Noise in Loop Bandwidth =
                         %g\n',1/NS)
end
Kn = sqrt(Dn*SmpPerOut/(OutInc*2));
rand('normal'); %Set noise stats for normal
                         distribution {v.3.5 only}
% Set step response or not and output that choice
. . . . . . . . . . . . .
   NL = sin(PhIn - y); % <<<<<<<<<<<< THE NON-LINEARITY
   PDn = Kn*randa; %{randa -> rand in v.3.5}
   PDnoise(i+1) = PDn;
   u1(1) = NL + Offset + PDn; % initial input to
                         sampler
      u1(2) = u1(1); % constant Hold output -> same at
                         beginning and end of dt
. . . . . . . . . . . . .
   if op == 3, % selecting Sampler Output vs. time
      subplot(212);
   else,
      subplot(224),
   end % if op==3
   plot(t,sin(er)+PDnoise,'+')
   title('Sampler Output')
```

Fig. 18.M.4   Changes to NLPhP to create NLPhN.

PROBLEMS

18.1 Repeat Example 18.1 except use a signal strength of -6 dBm.

18.2 Given: First-order loop; signal power 1 mW; one-sided additive noise density 10^{-8} W/Hz; and loop one-sided noise bandwidth 50 kHz. Find the mean frequency of cycle skipping ($\bar{F}$).

18.3 Given: Additive input noise producing 0.2 rad^2/Hz equivalent S_φ (one-sided); integrator-and-lead filter, $\zeta_0 = 0.5$; and initial phase and frequency errors are zero.

(a) Find ω_{n0} to give a mean time until the first cycle skip of 1 ± 0.25 sec.

(b) What then is the probability that a cycle will be skipped in the first 0.5 sec?

(c) What is the true output phase variance when the one-sided noise bandwidth B_{n0} is reduced to 2 Hz?

CHAPTER 19

ACQUISITION AIDS IN THE PRESENCE OF NOISE

In this chapter we revisit some of the acquisition aids of Chapter 9 and consider how their performance is affected by noise. In the two cases we will consider, the loop is closed during the sweep through $\Omega = 0$. In the first case the sweep voltage continues after lock whereas, in the second case, a coherent detector stops the sweep voltage when it detects lock.

19.1 MAXIMUM SWEEP RATE, PLAIN CLOSED LOOP

The following formula can be used to estimate the sweep rate that will permit lock with 90% probability for a loop with an integrator-and-lead filter and no auxiliary circuitry to stop the sweep:

$$\dot{\omega}_{90\%} \approx \frac{(0.7 - \sigma_{\varphi 0})\omega_n^2}{1 + \delta}, \qquad (19.1)$$

where

$$\delta = 0 \quad \text{for} \quad \zeta \geq 1, \qquad \delta = \exp\left(\frac{-\pi\zeta}{\sqrt{1 - \zeta^2}}\right) \qquad \text{for } 1 \geq \zeta \geq 0.5. \quad (19.2)$$

Frazier and Page (1962) gave twice the value in Eq. (19.1) as the results of their experimental work, but this has been disputed by Gardner (1979, p. 81), who suggests an even more conservative value than given here. Damping factors of 0.7 to 0.85 are within the ranges that are recommended, by these two authors, for acquisition while sweeping in noise. See Appendix 19.A for additional discussion of how Eq. (19.1) was obtained. From the discussion there it is apparent that Eq. (19.1) should be considered only an approximation.

19.2 REDUCTION OF COHERENT DETECTOR OUTPUT (CLOSED-LOOP SWEEPING)

We have seen that u_1 is reduced in the presence of noise, and we can express the reduction by the factor

$$\eta_n = \exp\left(-\tfrac{1}{2}\sigma_{\varphi n}^2\right). \tag{17.13}$$

We also have seen that this results in an increased phase error given by

$$\varphi_e = \sin^{-1}\left\{[\dot{\omega}_{\text{in}} - \dot{\omega}_c]\left[\frac{1}{\omega_n^2} - \frac{t}{K}\right]\exp\left(0.5\sigma_{\varphi n}^2\right)\right\}. \tag{17.20}$$

If we apply these two modifications to the output of the coherent detector as expressed by Eq. (9.6) we obtain

$$v_{\text{CD}} = \eta_n K'_{pd}\cos\varphi_e = \eta_n K'_{pd}\sqrt{1 - \sin^2\varphi_e} \tag{19.3}$$

$$= K'_{pd}\exp\left(-0.5\sigma_{\varphi n}^2\right)\sqrt{1 - \exp\left(\sigma_{\varphi n}^2\right)\left\{[\dot{\omega}_{\text{in}} - \dot{\omega}_c]\left[\frac{1}{\omega_n^2} + \frac{t}{K}\right]\right\}^2} \tag{19.4}$$

$$= K'_{pd}\sqrt{\exp\left(-\sigma_{\varphi n}^2\right) - \left\{[\dot{\omega}_{\text{in}} - \dot{\omega}_c]\left[\frac{1}{\omega_n^2} + \frac{t}{K}\right]\right\}^2}. \tag{19.5}$$

19.3 CLOSED-LOOP SWEEPING IN NOISE WITH COHERENT DETECTOR

19.3.1 Successful Acquisition

Blanchard (1976, pp. 293–296) has obtained experimental data showing the probability of successful acquisition in a circuit such as that shown in Fig. 9.1. For his test circuit, $\zeta = 1$ and $\alpha \approx 1$ ($\omega_p \ll K$). The "lock indication" is used to stop the sweep. The threshold is set at half of the value given by Eq. (19.5) [with $K \Rightarrow \infty$],

$$V_T = v_{\text{CD}}/2. \tag{19.6}$$

This is the point that gives the least dispersion in the frequency offset when the sweep stops. The bandwidth ω_d of the low-pass decision filter at the output of the coherent detector was varied to change the probability of a false stop. The wider is that filter the more likely are false stops, but the more readily will the sweep be stopped on the signal. The probability p of a false stop is a parameter. Figure 19.1 illustrates Blanchard's (1976) results for a 99% probability of acquisition. (Blanchard plots probability of acquisition against sweep speed for four noise levels and four probabilities of a false stop.) Here $\dot{\omega} = |\dot{\omega}_c|$.

One interpretation of Fig. 19.1 for some particular loop (i.e., once its parameters have been chosen) is as follows. The noise in the loop bandwidth identifies a particular vertical plane, for example, $\sigma_{\varphi0}^2 = 0.32$. As we increase the bandwidth of the decision filter, more noise passes through the filter and false stops become more likely. Some filter bandwidth will produce a false stop probability of $p = 0.0001$, for example. The intersection of the $\sigma_{\varphi0}^2 = 0.32$

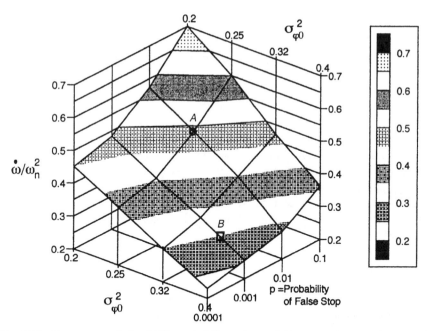

Fig. 19.1 Sweep speed for 99% probability of acquisition vs. phase variance and probability of false stop. The loop has $\zeta = 1$ and $\alpha \approx 1$. x and y axes (horizontal plane) are logarithmic. In this three-dimensional plot, read the intersection of planes representing the three variables with the data surface. The intersections of the constant $x(= \sigma_{\varphi0}^2)$ planes and the constant $y\ (= p)$ planes with the data surface are marked by lines. Shading on the data surface indicates vertical ranges of $z = \dot{\omega}/\omega_n^2$, corresponding to the legend on the right. (Points A and B are used in Example 19.1.) [Based on data taken from Blanchard (1976)].

plane and the $p = 0.0001$ plane defines a vertical line, which is easily visible in the figure. (Sweep speed does not affect p because false stop occurs when the signal is not passing through the decision filter.) As sweep speed increases, probability of acquisition decreases, moving the locus along that vertical line. Probability of acquisition drops to 90% when the locus on the vertical line reaches the surface.

Example 19.1   Sweep Acquisition in the Presence of Noise   A high-gain second-order loop has $\omega_n = 1600$ rad/sec and a ratio of additive noise power density to signal of -36 dBc/Hz. The closed loop is swept using a circuit of the type described above.

   a. How fast can it be swept if the chances of acquisition are to be 99% and the probability of false stop is to be 1%?

From Eq. (14.10), the noise bandwidth is

$$B_n = \frac{1600 \text{ rad/sec}}{4} \left[\frac{1}{2} + 2 \right] \frac{\text{cycle}}{\text{rad}} = 1000 \text{ Hz}.$$

Multiplying this by the relative power density [Eq. (14.4)], we obtain

$$\sigma_{\varphi 0}^2 = 1000 \text{ Hz} \times 10^{-3.6} = 0.25 \text{ rad}^2.$$

We locate this value of $\sigma_{\varphi 0}^2$ on the x axis of Fig. 19.1 and note that the plane representing 0.25 rad^2, which is parallel to the left wall, intersects the data plane in a line that contains point **A**. We also note that the plane representing $p = 0.01$ probability, which is parallel to the right wall, intersects the data plane in a line that also contains point **A**. Point **A** occurs where the height is $z \approx 0.48$, as can be seen by noting its position near the top of the shaded stripe and comparing that to the legend on the right. Therefore, the corresponding sweep speed is

$$\dot{\omega} = z\omega_n^2 = 0.48(1600 \text{ rad/sec})^2 = 1.23 \times 10^6 \text{ rad/sec}^2.$$

   b. By how much will the signal power be decreased if we are to have 0.1% probability of false stop when the sweep speed is set at 7.7E5 rad/sec^2?

Dividing $\dot{\omega}$ by ω_n^2, we obtain $z = 0.3$. This value lies along the top of the lowest stripe on the data plane in Fig. 19.1. The intersection of this value of z with the 0.001 probability plane occurs at a point **B**. This point lies in the plane where $\sigma_{\varphi 0}^2 \approx 0.35$ rad^2. Thus the signal power has been multiplied by a factor of 0.25 rad^2/0.35 rad$^2 = 0.7$, or -3 dB, to produce the corresponding increase in phase variance.

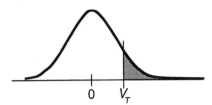

Fig. 19.2 Probability of noise exceeding threshold before lock.

19.3.2   False Stops

We can compute how the probability of false stop, p, depends on the detector threshold and the decision filter bandwidth. While the loop is attempting to lock, the noise in a differential bandwidth at the output of the coherent detector is given by Eq. (13.19), where K_p' is here the gain K_{pd} of the coherent detector.[1] Integrating this noise density over the band of the low-pass decision filter gives the mean square voltage at its output,

$$\sigma_d^2 = K_{pd}^2 B_{nd} N_0/P, \tag{19.7}$$

where B_{nd} is the noise bandwidth of the decision filter and P is the signal power. The noise voltage is characterized by a Gaussian probability density with this variance and the probability that the threshold will be exceeded is obtained by integrating the density that exceeds the threshold V_T. (See Fig. 19.2.) We can do this by integrating the standard normal density function from $x_T = V_T/\sigma_d$ to infinity.

$$p = \frac{1}{\sqrt{2\pi}} \int_{x_T}^{\infty} \exp\left(\frac{-x^2}{2}\right) dx = \frac{1}{\sqrt{2\pi}} \int_{-\infty}^{x_T} \exp\left(\frac{-x^2}{2}\right) dx$$

$$\equiv \text{SNCD}(-x_T) = 1 - \text{SNCD}(x_T), \tag{19.8}$$

where

$$x_T = \frac{V_T}{\sigma_d} = \frac{V_T}{K_{pd}\sqrt{(B_{nd}N_0)/P}}, \tag{19.9}$$

and B_{nd} is the noise bandwidth $(\omega_d/4)$ cycle of the single-pole low-pass decision filter [see Eqs. (14.7) and (14.8)], and SNCD is the standard normal cumulative distribution function, for which tables are available.

The consequence of false stops is to slow up the sweep by a factor $(1 - p)$. The length of each stop would tend to be about $1/\omega_d$. If, to reduce false

---

[1] We have used K_p for the gain of both the coherent detector and the phase detector. The meaning is identical in both cases, but it is not unlikely that, because of differing signal levels, for example, the values will be different for the two detectors in a given application. Therefore, at this point, we use K_{pd} for that particular value of K_p' that applies to the coherent detector.

restarts during lock, the threshold is lowered or the decision filter is narrowed after the sweep stops, then the sweep will be slowed more. If this is not done, then there is the possibility that the loop will begin sweeping again because noise causes the output from the coherent detector to drop below threshold after lock is achieved.

Example 19.2 False Stop What was the decision filter bandwidth in Example 19.1(a).

The probability of false stop in that example was $p = 0.01$. From a table of normal cumulative distribution functions we find that Eq. (19.8) gives 0.01 ($D_{\text{cum}} = 0.99$) when $x_T = 2.33$. We will use this in Eq. (19.9) to solve for B_{nd}.

In Example 19.1, we determined that $\sigma_{\varphi 0}^2 = 0.25$ rad^2. From Fig. 17.2, this implies an actual variance of $\sigma_{\varphi n}^2 = 0.3$ rad^2. Using, in Eq. (19.5), this phase variance with the sweep speed and ω_n from Example 19.1, we find the steady-state output voltage, when the sweeping loop is locked, to be $v_{\text{CD}} = 0.71 K_{pd}$. Setting the threshold at half this voltage, as in Eq. (19.6), we obtain $V_T = 0.355 K_{pd}$ as the value used in the experiment. Using, in Eq. (19.9), these values plus the relative noise power density of -36 dBc/Hz from Example 19.1, we obtain

$$2.33 = \frac{0.355}{\sqrt{B_{nd} \times 10^{-3.6}/\text{Hz}}},$$

or $B_{nd} = 92.4$ Hz.

19.3.3 False Restart

The probability p' that the coherently detected output will drop below threshold after lock is given by the shaded area under the probability density curve in Fig. 19.3.

$$p' = \frac{1}{\sqrt{2\pi}} \int_{-\infty}^{-x'_T} \exp\left(\frac{-x^2}{2}\right) dx \equiv \text{SNCD}(-x'_T) = 1 - \text{SNCD}(x'_T). \quad (19.10)$$

The mean output with the loop locked to a signal in the presence of noise is

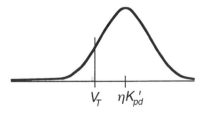

Fig. 19.3 Probability of noise dropping below threshold after lock.

$V_T \quad \eta K'_{pd}$

K'_{pd} multiplied by η from Eq. (17.13). From this, the normalized integration limit is

$$x'_T = \frac{K_{pd}\eta_n - V_T}{\sigma_d} = \frac{K_{pd}\exp\left(-\frac{1}{2}\sigma_{\varphi n}^2\right) - V_T}{K_{pd}\sqrt{(B_{nd}N_0)/P}}. \tag{19.11}$$

Lock will not necessarily be lost when the detector output drops—that depends on how long it stays below threshold—but some transient will be caused by restart of the sweep, even if it is momentary.

19.3.4   False Stop versus False Restart

Since both p and p' depend on the noise power and on the sweep speed and the loop's natural frequency (because V_T depends on them), both will be varied by these parameters. However, the dependence is not the same in the two cases so we may want to select these parameters based on the required values of both p and p'. Blanchard (1976, p. 297) plots p against p' for fixed values of

$$Q_1 \equiv \frac{|\dot{\omega}_c|}{\omega_n^2}\exp\left(0.5\sigma_{\varphi n}^2\right). \tag{19.12}$$

Such a plot helps us trade off p and p' for a given noise level and sweep rate. We can generate any of its curves as follows. Combine Eqs. (19.4), (19.6), and (19.12), assuming a high-gain loop (where $K \gg \omega_n^2 t$), to give

$$V_T = 0.5K'_{pd}\exp\left(-0.5\sigma_{\varphi n}^2\right)\sqrt{1 - Q_1^2}. \tag{19.13}$$

Then write Eq. (19.9) as

$$x_T = Q_2\sqrt{1 - Q_1^2}, \tag{19.14}$$

where

$$Q_2 \equiv \frac{\exp\left(-0.5\sigma_{\varphi n}^2\right)}{2\sqrt{(B_{nd}N_0)/P}}, \tag{19.15}$$

and Eq. (19.11) as

$$x'_T = Q_2\left(2 - \sqrt{1 - Q_1^2}\right). \tag{19.16}$$

If we begin with a probability of false stop p, we can obtain x_T from Eq. (19.8). Then, for a given value of Q_1, if we can obtain x'_T by combining

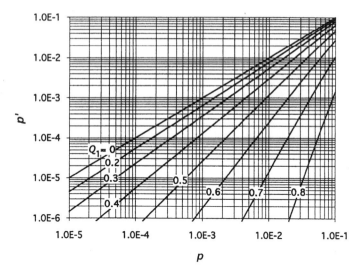

Fig. 19.4 Probability of false restart (p') vs. probability of false stop (p), fixed Q_1. Thresholds set at 50% of maximum output during sweep.

Eqs. (19.14) and (19.16),

$$x'_T = x_T \left(\frac{2}{\sqrt{1 - Q_1^2}} - 1 \right). \tag{19.17}$$

From this we can obtain p' by Eq. (19.10). By these means, p' has been plotted against p in Fig. 19.4.

Equation 19.12 shows that noise can be traded for sweep speed along any of the curves in Fig. 19.4; but, for a given combination of noise and speed (fixed Q_1), the probability of false stop and the probability of false restart depend on the decision bandwidth in a similar fashion. As the bandwidth is widened, both probabilities increase. However, the probability of acquisition also increases, a fact that cannot be seen from Fig. 19.4.

If the operating point in Fig. 19.1 has been determined, for example, by choice of loop parameters and sweep rate in the presence of a given noise environment, then Q_1 and p are no longer independent. For given values of $\sigma_{\varphi 0}^2$ and p, we can read $R = \dot{\omega}/\omega_n^2$ from Fig. 19.1. If we then convert $\sigma_{\varphi 0}^2$ to $\sigma_{\varphi n}^2$, using Fig. 17.2, for example, we can compute Q_1 from Eq. (19.12). This enables us to plot the constant $\sigma_{\varphi 0}^2$ curves, as is done in Fig. 19.5.[2] These give the relationship between p and p' for 99% probability of acquisition. As

---

[2]Theoretically, we can locate the points by knowing p and Q_1. Practically, we compute p' from p and Q_1 and plot p' vs. p. Straight-line approximations are plotted along with the computed points.

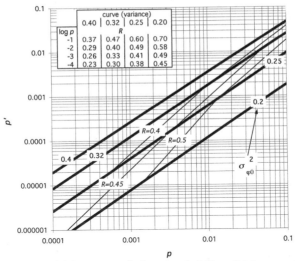

Fig. 19.5   Probability of false restart (p') vs. probability of false stop (p), $\sigma_{\varphi 0}^2$ fixed for 99% acquisition probability. Normalized sweep speed $R = \dot{\omega}/\omega_n^2$ is also shown. Thresholds set at 50% of maximum output during sweep. $\zeta = 1$, $\alpha = 0.9975$. Quasi-linear approximation employed.

the decision bandwidth is widened, the probabilities of false stop and restart increase as do the value of Q_1. The latter indicates that the rate at which the frequency can be swept, while still obtaining a constant 99% probability of acquisition, increases with the wider decision bandwidth. The value of $R = \dot{\omega}/\omega_n^2$ corresponding to each data point is also shown in Fig. 19.5 and lines of fixed R are drawn between points on the constant $\sigma_{\varphi 0}^2$ curves where these values are estimated (by interpolation) to occur.

Consideration of Fig. 19.1 and the two sets of curves in Fig. 19.5 leads us to the conclusion that, under the conditions that have been set,[3] we can choose independently only three of the five parameters, p, p', $\sigma_{\varphi 0}^2$, $\dot{\omega}/\omega_n^2$, and probability of acquisition.

Example 19.3   False Stop and Restart   What are the conditions that correspond to a 1% chance of false stop and a 0.1% chance of false restart after stopping? The loop is a high-gain second-order loop, $\zeta = 1$, and is sweeping with the loop closed.

From Fig. 19.4, $Q_1 \approx 0.51$ at the intersection of $p = 0.01$ and $p' = 0.001$. From Eq. (19.12) then,

$$|\dot{\omega}_c| = 0.51 \, \omega_n^2 \exp(- 0.5 \, \sigma_{\varphi n}^2).$$

---

[3]Coherent detector, 50% threshold setting.

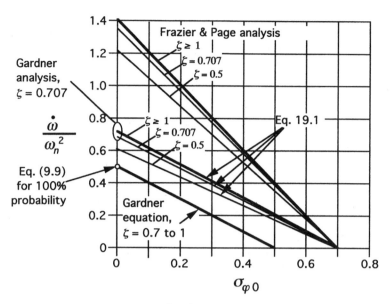

Fig. 19.A.1   Eq. (19.1) and related estimates.

We can also see from Fig. 19.5 that, for 99% probability of acquisition, $\sigma_{\varphi 0}^2 \approx 0.28$. From Fig. 17.2 we estimate $\sigma_{\varphi n}^2 \approx 0.35$. Inserting that value above gives a sweep rate of $0.43\,\omega_n^2$. Compare this to values obtained from Fig. 19.1 at $\sigma_{\varphi 0}^2 = 0.28$, $p = 0.01$ and from Fig. 19.5.

19.A   APPENDIX: DEVELOPMENT OF EQ. (19.1)

As a result of their experimental work, Frazier and Page (1962) gave an equation for the maximum sweep that will allow 90% probability of lock in noise. Equation (19.1) gives only half of their value for maximum $\dot{\omega}$. Gardner[4] states that there is a numerical error of $\sqrt{2}$ running through their work and suggests using $(0.5 - \sigma_{\varphi 0})\omega_n^2$, valid for $0.7 \le \zeta \le 1$, for a preliminary design value. Figure 19.A.1 shows Eq. (19.1), Frazier and Page's (1962) equation, and Gardner's (1979) recommendation.

We have several reference points at zero noise power that should be approached by the estimates of performance in noise as the noise decreases. Gardner's (1979) equation gives the same results as Eq. (9.9) in the absence of noise, but Eq. (9.9) is for 100% probability of lock, suggesting the

---

[4]Gardner, *Phaselock Techniques*, p. 81. The equation that Gardner states is suggested by Frazier's and Page's data is somewhat different than the expression that appears in their paper in the IEEE collection [Lindsey and Simon, 1978, p. 134, Eq. (1)].

conservative nature of Gardner's equation. Gardner also has used phase-plane plots to obtain probability of lock for $\zeta = 0.707$ as a function of S/N. The circle on the axis of ordinates indicates a range of interpolated values for 90% probability between his calculated points at 70 and 100%. Equation (19.1) is basically Frazier and Page's (1962) equation reduced by a constant, so its value at $\zeta = 0.707$ is at the conservative end of the circle.

PROBLEMS

19.1 A loop has $\alpha \approx 1$, $\zeta = 1$, and $\sigma_{\varphi 0}^2 = 0.2 \text{ rad}^2$.

(a) What is the maximum value of $\dot{\omega}/\omega_n^2$ for 90% probability of natural acquisition, without the aid of a lock detector?

What is the value for 99% probability of acquisition when a coherent detector is used to stop sweep and the false alarm probability is

(b) 10%, (c) 1% and (d) 0.01%?

(e) What should be V_T/K'_{pd}, the threshold setting relative to maximum possible output, when $\dot{\omega}/\omega_n^2 = 0.5$?

(f) What is the signal-to-noise ratio in the decision filter for 0.001 probability of false stop? A table of probability functions shows that SCND(3.09) = 0.999. Assume (e).

(g) What value of $\sigma_{\varphi n}^2$ would cause 0.001 probability of the sweep restarting due to noise. Assume the same signal-to-noise ratio in the decision filter as was used in part (f) (i.e., $\sigma_{\varphi n}^2$ changes due to a change in the loop, not in the coherent detector circuitry).

CHAPTER 20

FURTHER STUDIES

20.1 SOURCES FOR ADDITIONAL STUDIES IN PHASE LOCK

Blanchard's (1976) book ends with "Appendix B: Example of a Rough Draft of a Coherent Receiver," a 30-page description of the initial design of a receiver that includes several phase-locked loops (PLL). It illustrates the use of many of the concepts in his and in this book and presents an excellent opportunity for review and for gaining a better feel for how these concepts might be used. Of course, directions in which to pursue further study in particular areas are suggested by the sources referenced throughout this book.

Among the references that cover phase-locked loops in general (i.e., approximately the areas covered by this book), Blanchard's book is perhaps the most thorough in its coverage and in the details offered. Gardner's (1979) book is the second edition of a classic, one of the first on the subject, and a model of succinctness. More recent books include those by Best (1997), Wolaver (1991), Encinas (1993), and Stensby (1997). The latter goes into considerable depth on the nonlinear PLL model in the presence of large amounts of noise, devoting half of the book to something that is covered relatively briefly in Chapter 18 of the present work.

20.2 SOURCES COVERING PHASE-LOCKED FREQUENCY SYNTHESIS

Books that treat phase-locked loops as they apply to frequency synthesizers include those by Crawford (1994), Egan (1981), Kroupa (1973), Manassewitsch

(1976), Rhode (1983), and Stirling (1987). While these have been listed alphabetically, they nevertheless tend to move, left to right, from the more to the less theoretically oriented; that is, from those that explain more of the mathematical basis to those that concentrate more on circuit details. Kroupa (1973) was the first synthesizer book; Manassewitsch (1976) is the most encyclopedic, and Stirling (1987) specializes on microwave synthesizers.

20.A APPENDIX: MODULATIONS AND SPECTRUMS

This Appendix illustrates the relationships between various representations of signals, continuous, amplitude-modulated, and phase-modulated. Specifically, formulas, waveform pictures, phasors, modulation spectrums, signal spectrums, power spectrums for modulation and signals, and single-sidebands spectrums are shown. Table 20.A.1 covers discrete modulations and Table 20.A.2 covers noise modulation.

20.B APPENDIX: GETTING FILES FROM THE WILEY ftp AND INTERNET SITES

To download the files listed in this book and other material associated with it, use an ftp program or a Web browser.

ftp Access

If you are using an ftp program, type the following at your ftp prompt:

```
ftp:// ftp.wiley.com
```

Some programs may provide the first "ftp" for you, in which case you can enter

```
ftp.wiley.com
```

Log in as anonymous (e.g., User ID: anonymous).Leave password blank

After you have connected to the Wiley ftp site, navigate through the directory path of:

```
/public / sci_tech_med / phase_lock
```

Web Access

If you are using a standard Web browser, type URL address of:

```
ftp:// ftp.wiley.com
```

Follow the directions in the ftp section to navigate through the directory path of:

```
/public / sci_tech_med / phase_lock
```

File Organization

Under the phase_lock directory are subdirectories that include MATLAB files for PC, Macintosh, and UNIX systems and Microsoft® Excel files. Important information is included in README files.

If you need further information about downloading the files, you can call Wiley's technical support at 212-850-6753.

TABLE 20.A.1 Discrete Modulation

	1. Formula	2. Waveform	3. Phasor
CW	$g_c(t) = \cos \omega_c t$		
AM	$g_{mA}(t) = 1 + m_A \cos \omega_m t$ $g_A(t) = g_{mA}(t)g_c(t) = (1 + m_A \cos \omega_m t)\cos \omega_c t$ $= \cos \omega_c t + \dfrac{m_A}{2}\cos(\omega_c + \omega_m)t + \dfrac{m_A}{2}\cos(\omega_c - \omega_m)t$		
PM small m	$g_{mp}(t) = m \sin \omega_m(t)$ $g_p(t) = \cos(\omega_c t + g_{mp}(t)) = \cos(\omega_c t + m \sin \omega_m t)$ $= J_0(m)\cos \omega_c t$ $\quad + J_1(m)[\cos(\omega_c + \omega_m)t - \cos(\omega_c - \omega_m)t]$ $\quad + J_2(m)[\cos(\omega_c + 2\omega_m)t + \cos(\omega_c - 2\omega_m)t]$ $\rightarrow \cos \omega_c t + \dfrac{m}{2}[\cos(\omega_c + \omega_m)t - \cos(\omega_c - \omega_m)t]$ as $m \rightarrow 0$		

TABLE 20.A.1 Discrete Modulation (*Continued*)

	4. Mod Spectrum	5. Signal Spectrum	6. Power Spectrum of Modulation (1Ω System)	
			2-sided	1-sided
CW		V $1/2$ — $-f_c$ 0 f_c $$\cos\omega_c t = \frac{e^{j\omega_c t} + e^{-j\omega_c t}}{2}$$		
AM	$1-$ $-f_m$ 0 f_m — $m_A/2$	V $1/2$ — $-f_c$ 0 $f_c - f_m$ f_c $f_c + f_m$ — ratio: $m_A/2$	$1-$ $-f_m$ 0 f_m — $m_A^2/4$	$1-$ 0 f_m — $m_A^2/2$
PM small m	φ(rad) $\mathbf{Re}$ $\mathbf{Im}$ $-f_m$ 0 f_m — $m/2$	V $1/2$ — $-f_c$ 0 $f_c - f_m$ f_c $f_c + f_m$ — ratio: $m/2$	$-f_m$ 0 f_m — $m^2/4$	rad^2 0 f_m — $m^2/2$

462

TABLE 20.A.1 Discrete Modulation (*Continued*)

	7. Power Spectrum of Signal		8. SSB Spectrum
	2-sided	1-sided	
CW	P $1/4$ — $-f_c$ 0 f_c	P $1/2$ — 0 f_c	
AM	P $1/4$ *ratio: $m_A^2/4$* $-f_c$ 0 $f_c-f_m\ f_c\ f_c+f_m$	P $1/2$ *ratio: $m_A^2/4$* 0 $f_c-f_m\ f_c\ f_c+f_m$	**SSB** (dBc) $20\log_{10}(m_A/2)$ f_m Δf
PM small m	P $1/4$ *ratio: $m^2/4$* $-f_c$ 0 $f_c-f_m\ f_c\ f_c+f_m$	P $1/2$ *ratio: $m^2/4$* 0 $f_c-f_m\ f_c\ f_c+f_m$	L (dBc) $20\log_{10}(m/2)$ f_m Δf

463

TABLE 20.A.2 Noise Modulation

	1. Formula	2. Waveform	6. Power Spectral Density of Modulation		8. SSB Spectrum
			2-sided	1-sided	

AM: $g_a(t) = g_{ma}(t)g_c(t)$ $\quad$ ((V, A))

$\underset{\uparrow}{\sigma_{ma}^2}, \overline{g_{ma}} = 0$

PM small m: $g_p(t) = \cos(\omega_c t + \underset{\uparrow}{g_{mp}}(t))$ $\quad$ ((V, A))

$\sigma_{mp}^2, \overline{g_{mp}} = 0$

6. Power Spectral Density of Modulation

2-sided, AM: S_A, $S_{2,x}$, $f_{m,x}$, f_m

1-sided, AM: $S_{1,x} = 2S_{2,x}$, S_A, $f_{m,x}$, f_m

2-sided, PM: S_φ, $S_{2,y}$, $f_{m,y}$, f_m

1-sided, PM: $S_{1,y} = 2S_{2,y}$, S_φ, $f_{m,y}$

7. Power Spectral Density of Signal

	2-sided	1-sided
AM	S_{p2}; $1/4$; $-f_c$; $S_{2,x}/4$; f_c; $f_c+f_{m,x}$; $S_{2,x}$ relative	S_{p1}; $1/2$; $S_{1,x}/4$; $(S_{1,x}=2S_{2,x})$; $S_{2,x}$ relative; f; f_c; $f_c+f_{m,x}$
PM small m higher-order s.b.s not shown	S_{p2}; $1/4$; $-f_c$; $S_{2,y}/4$; f_c; $f_c+f_{m,y}$; $S_{2,y}$ relative	S_{p1}; $1/2$; $S_{1,y}/4$; $(S_{1,y}=2S_{2,y})$; $S_{2,y}$ relative; f; f_c; $f_c+f_{m,y}$

8. SSB Spectrum

SSB (dBc, 1Hz BW): $10\log_{10}(S_{1,x}/2)$, $f_{m,x}$, Δf

L(dBc, 1Hz BW): $10\log(S_{1,y}/2)$, $f_{m,y}$, Δf

464

REFERENCES

Abromowitz, M., and Stegun, I. (1964). *Handbook of Mathematical Functions*, Washington, DC: U.S. Gov't Printing Office.

Allan, D. W. (1966). "Statistics of Atomic Frequency Standards," *Proceedings of IEEE*, Vol. 54, No. 2 (February), pp. 221–230.

Allan, D., Hellwig, H., Kartaschoff, P., Vanier, J., Vig, J., Winkler, G., and Yannoni, N., (1988). "Standard Terminology for Fundamental Frequency and Time Metrology," *Proc. 42nd Annual Freq. Control Symposium, 1988*, pp. 419–425.

Ascheid, G., and Meyr, H. (1982). "Cycle Slips in Phase-Locked Loops: A Tutorial Survey," *IEEE Trans. Commun.*, Oct. 1982, Vol. COM-30, pp. 2228–2241. Also in Lindsey W., and Chie, C., Eds., *Phase-Locked Loops* (New York: IEEE Press, 1986), pp. 29–42.

Best, R. E. (1984). *Phase Locked Loops, Theory, Design & Applications*, New York: McGraw-Hill.

Best, R. E. (1997). *Phase Locked Loops, Design, Simulation, & Applications*, 3rd ed., New York: McGraw-Hill.

Blachman, N. M. (1966). *Noise and Its Effect on Communications*, New York: McGraw-Hill.

Blanchard, A. (1976). *Phase-Locked Loops*, New York: Wiley; also Malabar, FL: Krieger, 1992.

Bois, G. P. (1961). *Table of Indefinite Integrals*, New York: Dover.

Booton, R. C., Jr. (1952). "Nonlinear Control Systems with Statistical Inputs," Mass. Inst. Tech., Cambridge, MA, Report No. 61; March 1 [referenced in Develet (1963)].

Buchanan, J. (1983). "Chip Capacitors and Dielectric Absorption," *Electronic Products*, May 12, pp. 107–109.

Byrne, C. J. (1962). "Properties and Design of the Phase Controlled Oscillator With a Sawtooth Comparator," *The Bell System Technical Journal*, (March, 1962), pp. 559–602.

Cahn, C. R. (1962). "Piecewise Linear Analysis of Phase-Locked Loops," *IRE Trans. Space Electron. Telemetry*, SET-8, No. 1, March, pp. 8–13. Also in Lindsey, W. C., and Simon, M. K., *Phase-Locked Loops & Their Application*, New York: IEEE Press, 1977, pp. 8–13.

Charles, F. J., and Lindsey, W. C. (1966). "Some Analytical and Experimental Phase-Locked Loop Results for Low Signal-to-Noise Ratios," *Proc. IEEE*, Sept. Vol. 54, pp. 1152–1166. Also in Lindsey, W. C., and Simon, M. K., *Phase-Locked Loops & Their Application*, New York: IEEE Press, 1977, pp. 276–290.

Chen, Chi-Tsong (1970). *Introduction to Linear System Theory*, New York: Hold, Rinehart and Winston.

Crawford, J. (1994). *Frequency Synthesizer Design Handbook*, Boston: Artech House.

Davenport, W. B., Jr., and Root, W. L. (1965). *An Introduction to the Theory of Random Signals and Noise*, New York: McGraw-Hill.

Develet, J. A., Jr. (1963). "An Analytic Approximation of Phase-Lock Receiver Threshold," *IEEE Trans. Space Electron. Telemetry*, SET-9, March, pp. 9–12.

Dorf, R. C. (1965). *Time-Domain Analysis and Design of Control Systems*, Reading, MA: Addison-Wesley.

Egan, W. F. (1981). *Frequency Synthesis by Phase Lock*, New York: Wiley: also Malabar, FL: Krieger, 1990.

Egan, W. F. (1991). "Sampling Delay—Is It Real?" *RF Design*, Feb., pp. 114–116.

Encinas, J. B. (1993). *Phase Locked Loops*, London: Chapman & Hall.

Endres, T., and Kirkpatrick, J. (1992). "Sensitivity of Fast Settling PLLs to Differential Loop Filter Component Variations," *Proceedings of the 1992 IEEE Frequency Control Symposium*, New York: IEEE Press, pp. 213–223.

Franco, S. (1989). "Current-Feedback Amplifiers Benefit High-Speed Designs," *EDN*, Jan. 5, pp. 161–172.

Frazier, J. P., and Page, J. (1962). "Phase-Lock Loop Frequency Acquisition Study," *IRE Trans. Space Electron. Telemetry*, Vol. SET-8, Sept., pp. 210–227. Also in Lindsey, W. C., and Simon, M. K., *Phase-Locked Loops & Their Application*, New York: IEEE Press, 1977, pp. 132–149.

Gardner, F. M. (1977). "Hangup in Phase-Lock Loops," *IEEE Trans. Commun.*, Vol. COM-25, Oct., 1977, pp. 1210–1214. Also in Lindsey, W., and Chie, C., Eds., *Phase-Locked Loops*, New York: IEEE Press, 1986, pp. 43–47.

Gardner, F. M. (1979). *Phaselock Techniques*, 2nd ed., New York: Wiley.

Gardner, F. M. (1980). "Charge-Pump Phase-Lock Loops," *IEEE Trans. Commun.*, Vol. COM-28, Nov., pp. 1849–1858. Also in Lindsey, W., and Chie, C., Eds., *Phase-Locked Loops*, New York: IEEE Press, 1986, pp. 321–332 and in Razavi, (1996), pp. 77–86.

Gill, W. (1981). "Easy-to-Use BASIC Program," *EDN*, Feb. 18, pp. 133–136.

Goldstein, A. J. (1962). "Analysis of the Phase-Controlled Loop with a Sawtooth Comparator," *Bell System Tech. J.*, March, p. 607.

Gray, P., and Meyer, R. (1977). *Analysis and Design of Analog Integrated Circuits*, New York: Wiley.

Greenstein, L. J. (1974). "Phase-Locked Loop Pull-In Frequency," *IEEE Trans. Commun.*, Aug., pp. 1005–1013. Also in Lindsey, W. C., and Simon, M. K., *Phase-Locked Loops & Their Application*, New York: IEEE Press, 1977, pp. 150–158.

Gupta, S. C., Bayless, J. W., and Hammels, D. R. (1968). "Threshold Investigation of Phase-Locked Discriminators," *IEEE Trans. Aerospace Electronic Systems*, AES-4, No. 6, November, pp. 855–863.

Gupta, S. C., and Solem, R. J. (1965). "Optimum Filters for Second- and Third-Order Phase-Locked Loops by an Error Function Criterion," *IEEE Trans. Space Electron. Telemetry*, SET-11, June, pp. 54–62.

Hewlett-Packard (1989), *Spectrum Analysis, Application Note 150*, Rohnert Park, CA: Hewlett-Packard Co., pp. 31–33.

Jaffe, R., and Rechtin, E. (1955). "Design and Performance of Phase-Lock Circuits Capable of Near-Optimum Performance over a Wide Range of Input Signal and Noise Levels," *IRE Trans. Inform. Theory*, Vol. IT-1, March, pp. 66–76. Also in Lindsey, W. C., and Simon, M. K., *Phase-Locked Loops & Their Application*, New York: IEEE Press, 1977, pp. 20–30.

Klapper, J., and Frankle, J. (1972). *Phase-Locked and Frequency Feedback Systems*, New York: Wiley.

Kroupa, V. (1973). *Frequency Synthesis*, New York: Wiley.

Kuo, B. C. (1987). *Automatic Control Systems*, 5th ed., Englewood Cliffs, NJ: Prentice-Hall.

Lindsey, W. (1972). *Synchronization Systems in Communication and Control*, Englewood Cliffs, NJ: Prentice-Hall.

Lindsey, W., and Chie, C. (1981). "Survey of Digital Phase-Locked Loops," *Proc. IEEE*, Vol. 69, Apr., pp. 410–431. Also in Lindsey and Chie, Eds., *Phase-Locked Loops*, New York: IEEE Press, 1986, pp. 296–317.

Lindsey, W., and Chie, C., Eds. (1986). *Phase-Locked Loops*, New York: IEEE Press.

Lindsey, W. C., and Simon, M. K. (1973). *Telecommunication Systems Engineering*, Englewood Cliffs, NJ: Prentice-Hall.

Lindsey, W. C., and Simon, M. K., Eds. (1978). *Phase-Locked Loops and Their Applications*, New York: IEEE Press.

Lindsey, W. C., and Tausworthe, R. C. "A Survey of Phase-Locked Loop Theory," *JPL Technical Report*, results reported by Blanchard (1976, p. 319).

Little, A. (1990). "Using Current Feedback Amplifiers," *RF Design*, December, pp. 47–52.

Manassewitsch, V. (1976). *Frequency Synthesizers Theory and Design*, New York: Wiley.

Pease, R. A. (1982). "Understand Capacitor Seakage to Optimize Analog System," *EDN*, October 13, pp. 125–129.

Pouzet, A. (1972). "Characteristics of Phase Detectors in Presence of Noise," *Proceedings of the 1972 International Telemetering Conference*, Los Angeles, CA, pp. 818–828.

Protonotarios, E. N. (1969). "Pull-in Performance of a Piecewise Linear Phase-Locked Loop," *IEEE Trans. Aerospace Electron. Systems*, Vol. AES-5, No. 3, May, pp. 376–386.

Razavi, B., Ed. (1996). *Monolithic Phase-Locked Loops and Clock Recovery Circuits*, New York: IEEE Press.

Rey, T. J. (1960). "Automatic Phase Control, Theory and Design," *Proceedings of the IRE*, October, pp. 1760–1771. Corrections are in *Proceedings of the IRE*, March (1961), p. 590. Also in Lindsey, W. C., and Simon, M. K., *Phase-Locked Loops & Their Application*, IEEE Press, 1978, pp. 309–320.

Rhode, U. (1983). *Digital PLL Frequency Synthesizers*, Englewood Cliffs, NJ: Prentice-Hall.

Rice, S. O. (1963). "Noise in FM Receivers," in *Proceedings Symposium of Time Series Analysis*, New York: Wiley, Chapter 25, pp. 395–424.

Richman, D. (1954a). "Color-Carrier Reference Phase Synch...," *Proceedings of the IRE*, Jan., p. 125. Also in Lindsey, W. C., and Simon, M. K., *Phase-Locked Loops & Their Application*, New York: IEEE Press, 1977, p. 401.

Richman, D. (1954b). "The DC Quadricorrelator: A Two-Mode Synchronization System," *Proceedings of the IRE*, January, pp. 288–299.

Rowbotham, J. R., and Sanneman, R. W. (1967). "Random Characteristics of the Type II Phase-Locked Loop," *IEEE Trans. Aerospace Electron. Systems*, Vol. AES-3, No. 4, July, pp. 604–612.

Sanneman, R. W., and Rowbotham, J. R. (1964). "Unlock Characteristics of Phase-Locked Loops," *IEEE Trans. Aerosp. Navig. Electron.*, Vol. ANE-11, Mar., pp. 15–24. Also in Lindsey, W. C., and Simon, M. K., *Phase-Locked Loops & Their Application*, New York: IEEE Press, 1977, pp. 250–259.

Schwartz, M., Bennett, W., and Stein, S. (1966). *Communication Systems and Techniques*, New York: McGraw-Hill.

Shoaf, J. H., Halford, D., and Risley, A. S. (1973). *Frequency Stability Specifications and Measurement: High Frequency and Microwave Signals*, NBS Technical Note 632, January.

Smith, B. M. (1966). "The Phase-Lock Loop with Filter: Frequency of Skipping Cycles," *Proceedings of the IEEE*, Feb., p. 296.

Spilker, J. J., Jr. (1963). "Delay-Lock Tracking of Binary Signals," *IEEE Trans. Space Electron. Telemetry*, Vol. SET-9, Mar., pp. 1–8. Also in Lindsey, W. C., and Simon, M. K., *Phase-Locked Loops & Their Application*, New York: IEEE Press, 1977, pp. 226–233.

Spilker, J. J., Jr. (1977). *Digital Communications by Satellite*, Englewood Cliffs, NJ: Prentice-Hall, pp. 443–445, 537–596.

Spilker, J. J., Jr., and Magill, D. T. (1961). "The Delay-Lock Discriminator—An Optimum Tracking Device," *Proc. IRE*, Vol. 49, Sept., pp. 1403–1416. Also in Lindsey, W. C., and Simon, M. K., *Phase-Locked Loops & Their Application*, New York: IEEE Press, 1977, pp. 234–247.

Springett, J. C., and Simon, M. K. (1971). "An Analysis of the Phase Coherent-Incoherent Output of the Bandpass Limiter," *IEEE Trans. Commun. Tech.*, COM-19, No. 1, February, pp. 42–49.

Stensby, J. L. (1997). *Phase-Locked Loops, Theory and Applications*, Boca Raton, FL: CRC Press.

Stirling, R. (1987). *Microwave Frequency Synthesizers*, Englewood Cliffs, NJ: Prentice-Hall.

Tausworthe, R. C. (1966). "Theory and Practical Design of Phase-Locked Receivers," J.P.L. Technical Report No. 32-819, Feb., as quoted in Blanchard (1976).

Thomas, G., Jr., and Finney, R. (1984). *Calculus and Analytic Geometry*, 6th ed., Reading, MA: Addison-Wesley, pp. 483–489.

Troha, D. G. (1994). Texas Instruments Application Note sdla005a, *Digital Phase-Locked Loop Design Using SN54/74LS297*.

Truxal, J. G. (1955). *Control System Synthesis*, New York: McGraw-Hill.

Viterbi, A. (1966). *Principles of Coherent Communication*, New York: McGraw-Hill.

Wolaver, D. H. (1991). *Phase-Locked Loop Circuit Design*, Englewood Cliffs, NJ: Prentice-Hall, p. 75.

Ziemer, R. E., and Peterson, R. L. (1985). *Digital Communications and Spread Spectrum Systems*, New York: Macmillan, pp. 278–280.

ANSWERS TO PROBLEMS

1.1 **(b)** −3E-4, 10 kHz; **(c)** 1.94 rad, 204 Hz; **(d)** 0.6 c, 2;
(e) 0.5, 40 MHz/V; **(f)** 1.25 c, 1.5

1.2 16 to 18 MHz; no

1.3 16.682 to 17.318 MHz

2.1 5 kHz, −90°

2.2 **(b)** 3 MHz

2.3 **(b)** 4.6 msec; **(d)** 460 μsec; **(e)** 0.1 rad

3.1 3 V/c = 0.48 V/rad

3.2 **(a)** 3 mA/c = 4.77×10^{-4} A/rad; **(b)** 667 Ω; **(c)** +V and −V
increase by 2 V

3.3 **(a)** $0.0909(10^4$ rad/sec + s)/909 rad/sec + s); **(b)** 0.29 ∠ −56.4°

3.4 **(a)** 5 kΩ; 1.59 μF; **(b)** 1 MΩ; **(c)** $R_{2s} = 20$ kΩ, $R_{2p} = 80$ kΩ,
$C = 9.95 \times 10^{-8}$ F

3.5 **(a)** e, 1 kΩ, 1×10^{-8} F; **(b)** Fig. 3.23e from PD out to ground;
1000 pF; **(c)** 0.126 mA

4.1 **(a)** $\omega_n = 100$ rad/sec, $\zeta = 0.3$; **(b)** $\omega_n = 100$ rad/sec, $\zeta = 0.05$;
(c) $\omega_n = 31.6$ rad/sec, $\zeta = 0.079$; **(d)** $\omega_n = 100$ rad/sec, $\zeta = 0.25$

4.2 **(a)** $[1 + j0.002$ sec/rad $\omega_m]/[1 - (\omega_m$ sec/rad/477$)^2 + j0.0022$
sec/rad $\omega_m]$; **(b)** 0.0132 rad; **(c)** 0.000145 rad; **(d)** 13.2 Hz

5.1 (b) PM is 37.9°, 37.1° for the tangential approximation. Gain margin is infinite.

5.2 Approximation: 18.7 dB at 7000 Hz and 52.7° at 1260 Hz. Accurate values: 23.5 dB at 6980 Hz and 54.2° at 1172 Hz

5.3 10,039 Ω; 3578 sec^{-1}, $\approx$ 3795 sec^{-1}

5.4 20 dB approximate, 20.086 dB accurate

6.1 (a) 5 μF; (b) $\varphi_e = 1$ rad

6.2 (a) 984 kHz; (b) 1002 kHz; (c) 3 V/cycle; (d) 1; (e) 6 kHz/V; (f) 7.14 $\times$ 10^4 rad/sec; (g) 3.59 $\times$ 10^4 rad/sec; (h) $\approx$ 1; (i) 1.748 $\times$ 10^4 rad/sec; (j) 0.889 cycle = 5.59 rad; (k) 111 μsec

6.3 10 kΩ

6.4 0.152 rad

6.5 11.3 rad at 4 msec

6.6 0.063 rad

7.1 (a) 10 MHz; (b) 1 kHz; (c) 200 rad/sec = 31.8 Hz; (d) 31.4 rad; (e) 3.93 msec; (f) 11.78 msec

7.2 (a) 0.01 V; (b) 5.77 kHz/V, $\approx$ 5 kHz/V; (c) 10 Hz; (d) 10^{-3} V

7.3 (a) 0.3; (b) 1.41 kHz

7.4 (a) 10 V/rad; (b) 0, 1/$\sqrt{2}$; (c) 6.28 $\times$ 10^4 rad/sec

7.5 0.4 rad at 7070 Hz

7.6 (a) 10 μF; (b) 14 dB; (c) 0.2 rad

8.1 (a) 30.64 MHz, 1.925 $\times$ 10^8 rad/sec; (b) 30.0637 MHz, $K/\omega_z = \omega_z/\omega_p = 200 \gg 1$; (c) 3180 Hz, $\omega_z/\omega_L = 1 \not\ll 1$; (d) 968 μsec, $\Omega_{PI}/\Omega = 4.5$, $\Omega/\Omega_S = 4.4$, 900 μsec; (e) 178 μsec, $\Omega_{PI}/\Omega = 10.6$, $\Omega/\Omega_S = 1.9$, 100 μsec

8.2 (a) 5.075 $\times$ 10^6 rad/sec = 808 kHz; (b) 8 $\times$ 10^6 rad/sec = 1.28 MHz

8.3 (a) 4.14 $\times$ 10^6 sec^{-1}; (b) 5.06 $\times$ 10^6 sec^{-1}

8.4 (a) 0.1 rad; (b) 200 Hz

9.1 (a) 1.25 $\times$ 10^5 rad/sec^2; (b) $-30°$; (c) 0.173 V; (d) 0.2 V

9.2 698 sec^{-1}

9.3 1207 Hz

9.4 4.86 $\times$ 10^7 rad/sec^2

9.5 (a) 7 MHz $\pm$ 800 Hz; (b) 7 MHz $\pm$ 35.8 kHz; (c) 68 msec;
(d) 7.8 MHz; (e) 0.62 sec

11.1 (a) 5×10^{-7}, 1.6×10^{-5}; (b) 2.5×10^{-13}, 2.53×10^{-10};
(c) 5.37×10^{-3}, 1.84×10^{-3}; (d) 2.89×10^{-4}, 3.39×10^{-5};
(e) 1.6×10^{-6}, 4.93×10^{-7}

11.2 0.044 rad^2

12.1 Some points: $S_{\varphi A}$: -70 dBr/Hz at 10 Hz, 1 kHz, 10 kHz;
-82 dBr/Hz at 200 kHz. $S_{\varphi B}$: -20 dBr/Hz at 100 Hz; -94 dBr/Hz
at 40 kHz; -114 dBr/Hz at 200 kHz

12.2 (a) 0.495 rad^2; (b) 0.00975 rad^2; (c) 0.00101 rad^2; (d) 0.506 rad^2

12.3 8.85×10^{-13} V^2/Hz

12.4 10 kHz

12.5 4.3×10^{-7} rad^2/Hz

13.1 (b) -90; (c) -93; (d) -90; (e) -93

13.2 (a) 0 at DC, rising to 1.33×10^{-11} V^2/Hz peak at 280 kHz, toward
10^{-11} V^2/Hz at high frequencies, falling at 400 kHz;
(b) 10^{-9} rad^2/Hz at DC, falling to 2.5×10^{-10} rad^2/Hz by 200 kHz,
then to zero at 400 kHz; (c) -93 dBc/Hz at 3 MHz, falling to
-99 dBc/Hz at ±200 kHz, dropping to zero at ±400 kHz

13.3 (a) 1.507 V vs. ($\pi/2$) V; (b) 0.7738 V; (c) 1 V, 0.6065 V

14.1 (a) 250 Hz; (b) 7250 Hz; (c) 2500 Hz; (d) 31.25 Hz

14.2 1968 Hz

14.3 (a) type 2 second-order (integrator and lead), $\zeta = 1/\sqrt{2}$;
(b) 1.12×10^5 rad/sec, 0.059 rad^2

14.4 (a) 6.9×10^{-5} rad^2/Hz; (b) 0.71 or -1.5 dB

15.1 (a) 0.01 V; (b) $\pi \times 10^6$ rad/sec; (c) Section 7.6.4; (d) $\leq 1.4 \times 10^{-6}$
V^2; (e) 0.14 rad; (f) 5×10^5 Hz; (g) -13 dBm; (h) 5×10^5 Hz

16.1 (a) 52.28°; (b) 4.95 kHz; (c) 41.9°; (d) 3876 Hz; (e) no change;
(f) $0.5 \Rightarrow -3$ dB; (g) $0.457 \Rightarrow -3.4$ dB

16.2 (a) $2.35 \Rightarrow 3.72$ dB; (b) 3 dB; (c) infinite

16.3 800×10^{-6} rad^2

17.1 (a) 0.284 rad^2; (b) 0.314 rad^2; (c) 16.43°

17.2 (a) 18.4°; (b) 13.3°

17.3   **(a)** 2.5×10^{-4} rad^2; **(b)** 4.27×10^{-4} rad^2; **(c)** 3.637×10^{-4} rad^2; **(d)** 3.2×10^{-4} rad^2

17.4 $$\frac{\sigma_{\varphi nm}^2}{\sigma_{\varphi 0}^2} = \frac{1}{\eta} = \frac{\exp\left(\frac{1}{2}\sigma_{\varphi n}^2\right)}{\eta_m}$$

17.5   **(a)** $1.6 \Rightarrow 2$ dB; **(b)** 4.7 dB; **(c)** 0.8; **(d)** $37°$; **(e)** 0.45 rad^2

18.1   **(a)** 0.29; **(b)** 8890 rad/sec

18.2   1166 Hz/using one term for I_0, 1032 Hz using two; 975 Hz exactly

18.3   **(a)** 11.81 rad/sec (11.58 rad/sec using two terms for I_0; 10.52 rad/sec using one term; **(b)** 0.39; **(c)** 0.67 rad^2

19.1   **(a)** 0.25; **(b)** 0.7; **(c)** 0.57; **(d)** 0.44; **(e)** 0.37; **(f)** $69.7 \Rightarrow 18.4$ dB; **(g)** 0.6 rad^2

INDEX

475